Geophysical Monograph Series

Including
IUGG Volumes
Maurice Ewing Volumes
Mineral Physics Volumes

Geophysical Monograph 161

Circulation in the Gulf of Mexico: Observations and Models

Wilton Sturges
Alexis Lugo-Fernandez
Editors

American Geophysical Union
Washington, DC

Library of Congress Cataloging-in-Publication Data

Circulation in the Gulf of Mexico : observations and models / Wilton Sturges III ... [et al.], editors.
    p. cm. -- (Geophysical monograph ; 161)
  Includes bibliographical references and index.
  ISBN-13: 978-0-87590-426-9 (alk. paper)
  ISBN-10: 0-87590-426-2 (alk. paper)
  1. Ocean circulation--Mexico, Gulf of--Remote sensing. 2. Ocean circulation--Mexico, Gulf of--Mathematical models. I. Sturges, Wilton. II. American Geophysical Union. III. Series.

  GC228.6.M657 2005
  551.46'21364--dc22

                        2005031961

  ISBN-10:0-87590-425-4 (hardcover)
  ISBN-13: 978-0-87590-425-2 (hardcover)

  ISSN 0065-8448

Front cover: Water surface temperature from AVHRR data for a time interval of 3.10 days ending 21 February 2005 02:46 UT. Courtesy of the Ocean Remote Sensing Group, Johns Hopkins University Applied Physics Laboratory.

Back cover: Phytoplankton bloom in the Gulf of Mexico. NASA image courtesy of the SeaWiFS Project <http://seawifs.gsfc.nasa.gov/SEAWIFS.html>, NASA/Goddard Space Flight Center, and ORBIMAGE <http://www.orbimage.com/>.

CONTENTS

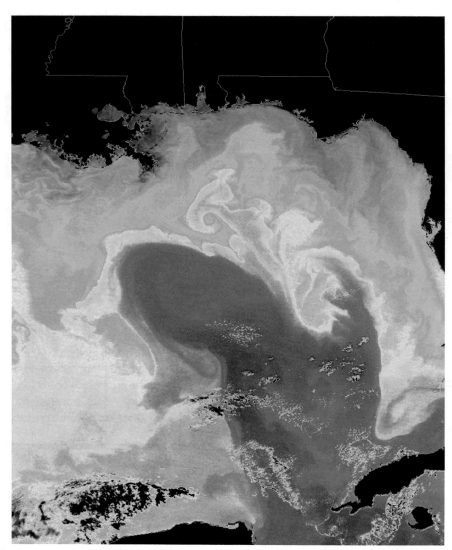

Plate 2. Sea Surface temperature, April 13, 2003. The Mississippi delta is at the top, Cuba at the bottom right, the Yucatan Peninsula at the bottom left. A few clouds are evident. Red is warm, blue is cold; image provided by N. Walker, LSU.

The large warm-core rings that separate from the Loop Current carry a huge volume of water, roughly 300 km across and 800 - 1000 m deep, across the Gulf. The swirl speeds can be as great as 3-4 knots, but the rate at which the ring as a whole moves to the west is about 5 cm/s (~1/10 of a knot). Because the high-speed currents in a ring are a serious threat to drilling rigs, our ability to predict ring shedding and motion has been of great concern to the oil industry. Despite our best efforts, however, we have yet to build numerical models with the capacity to predict with good reliability when a warm-core ring will separate from the Loop Current. The reader interested primarily in numerical modeling may wish to refer to *Oey, Ezer, and Lee's* paper, which presents a comprehensive survey of recent progress in numerical modeling in the Gulf.

Specific Features of Deep-Water Flow

The flow from the Caribbean Sea between Mexico and Cuba is the "input" to the Loop current. The paper by *Badan et al.* describes the results of a major current-meter mooring program in the Yucatan Straits. Longstanding political problems between the U.S. and Cuba have kept oceanographers from the U.S. from working in the Yucatan Channel, lying between Cuba and Mexico. But research resulting from an arrangement with Mexican oceanographers at Ensenada has produced an excellent new set of data. The flow pattern found in the upper layers (above 800 m) is essentially in keeping with traditional ideas. But as less flow is found in Yucatan than in the Straits of Florida, some additional flow (~5 Sv) from east of Cuba is a required. Therefore we have been intrigued by the deep flow, which does not look at all like what most of us had anticipated. The deep flow (below ~ 1000 m) goes in and out like a plunger on each cycle of Loop Current intrusion, agreeing with ideas put forward by G. Maul decades ago. Surprisingly, as the Loop Current intrudes to the north, deep Gulf water flows back to the south, carrying a transport roughly one-fifth that of the upper flow that becomes the Florida Current.

Chassignet et al. describe the overall flow in the Gulf. Using surface color data from the SeaWiFS satellite, they compare five different numerical models, seeking to answer how well each performs and what features of biological interest seem to be clearly connected to circulation patterns. As can be seen on many of the color images, chlorophyll (and other water-borne particles) is taken from the shelf and transported in convoluted paths as a result of the strongly time-dependent placement of eddies found offshore. Simple ideas of how such transport might have taken place are clearly inadequate without a knowledge of how the material is extracted from the shelf region and how it is moved around by and within the eddy field.

Kantha et al. describe their work with a full-Gulf numerical model that has been in operational use for some years. They perform a hind-cast of the ability of their model to describe the large-scale circulation of the Gulf since 1993, asking: Is it good enough for use by the oil industry? Their model, like the others, is primarily wind-forced; they show that it requires the inclusion (or assimilation) of satellite altimeter data to improve the model sufficiently for it to reach a level of usefulness that will benefit the oil industry.

The large-scale flow of the Gulf has been explored with near-surface drifters for many years. *DiMarco et al.* have studied nearly 1400 drifter tracks. They ask whether, despite great variability, they can deduce reliable patterns of mean flow from these ~10 years of data. And they find that they can. (Movies of surface drifters on the CD are described later.)

Weatherly et al. have studied a set of data from drifters at a depth of 900 m. As expected, they find an enormous variability in the flow as rings pass by. Nonetheless, by using several years of data, they successfully squeeze the statistics with the intention of finding features of mean flow where such results are statistically reliable. Although such mean flows are small, the variability is large. In a separate study using deep floats in the Pacific and Indian Oceans, *Davis* [2005] has shown that the mean values will be uncertain to roughly ~1 cm/sec even after a full year of data. Many years of data are required before a reliable mean emerges.

The Bay of Campeche is a region in the southwestern Gulf whose circulation is somewhat shielded from Loop Current rings. The paper by *Vazquez et al.* examines a variety of historical data, asking whether or not the mean flow there can persist in defiance of the rings' influence. The authors are able to affirm that the flow is consistent with what is expected from wind forcing.

The Inner Edge of the Deep Water and the Outer Edge of the Coastal Circulation

The flow near the outer edges of the continental shelf creates an area where nutrients (or pollution) from continental sources enter, or as we more formally say, "exchange with" deep waters. Some studies of coastal flow examine this phenomenon. The paper by *Hamilton and Lee* is an excellent example of a planned experiment at the edge of the continental shelf. Their paper as well as their movies and data in the CD show how eddies can intrude onto the edge of the shelf and dominate the circulation in that area. We usually expect eddies to "drift along," but Hamilton and Lee show that sometimes they are trapped by the topography and sit there for a month or more.

Slightly farther west, the study by *Biggs et al.* describes the area just offshore of the Mississippi River. In an exemplary

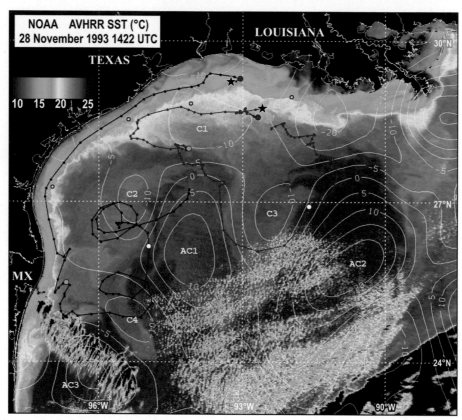

Plate 3. Drifter tracks showing motion of a surface drifter influenced by two large warm-core (anticyclonic–AC1, etc.) rings. Near the top of the figure, both the blue and the black tracks begin in shallow water and move around in counter-clockwise patterns. The SSH contours, from *Leben*, have been added to the SST image. Note that although they are centered on the same day, the two images contain data taken over different time intervals. The color scale for SST is shown; SSH contours are spaced every 5 cm. From *Walker* (this volume).

multipurpose study involving both currents and whales, they show an interesting series of eddy patterns at the edge of the shelf; they use satellite data, currents measured from their ship's Doppler acoustic profiler, and traditional hydro casts. The flow around and between these eddies can bring fluid off the shelf to exchange with deeper waters.

The transfer of particles (or of anything) between the shelf and deep water is an important and difficult topic, as it involves turbulent exchange. Many practical issues arise. How long do particles (larvae, pollutants, sediment) stay on the shelf before they get into deep water? Will particles in river runoff settle out on the shelf before getting into deep water? Can we predict in advance and under what conditions this material will linger on the shelf, or will it be whisked quickly offshore? Conversely, how do fish larvae get from deep offshore spawning regions into the shallow grass beds? Following the track of a surface drifter can give insight into how particles move. In *Walker's* discussion of flow on the shelf off Louisiana and Texas, she shows a surface drifter that not only leaves the shelf to be carried around a large anticyclone, but which also is brought back to the shelf again by another anticyclone, as shown in Plate 3.

The chapter by *Morey et al.*, though primarily a study of the coastal flow in the western Gulf, also explores the cross-shelf transport of coastal waters. The authors emphasize the cross-shelf export of shelf and river water by wind- and buoyancy-driven coastal circulations as well as by the mesoscale eddy field (as seen in Plate 3, for example). Loop Current rings propagate to the west, reaching the coast of Mexico and Texas in a few months. Some of the large-scale flow in the western half of the Gulf is driven by these decaying Loop Current Rings while some is driven by wind. Oceanographers have yet to determine which force is dominant here despite considerable speculation over three decades. The new orbit recently given to the TOPEX/Poseidon satellite in 2003 should shed light on the issue. The shift of 1.42 degrees of longitude, which interleaves the TOPEX/Poseidon and Jason ground tracks, will provide greatly enhanced coverage of the far western Gulf. It will improve our ability to deal with many practical issues of ocean transport as varied as oil spills (remember Ixtoc?), young turtles leaving their hatching grounds, search and rescue, or effects on shrimp fisheries.

Deep and Near-Bottom Flow

The majority of papers presented here are concerned with the circulation in the upper layers of the Gulf. The paper by *Sturges* examines the waters that originate in the Atlantic, flow through the Caribbean, and ultimately refresh the deeps of the Gulf. Sturges questions not only inflow of waters, but

also return flow: what goes out? A remarkable three-layer overturning pattern is found that can be traced in the deep water from the Atlantic to the Gulf and back.

Tides

Kantha has contributed a brief summary paper on tides in the Gulf. His title includes the term "barotropic," which derives from fluid mechanics and refers to that part of the flow that is uniform from top to bottom. In addition to his main paper, he has provided a brief movie of the daily average tidal flow over a 24-hour period. We can ask where the tidal inflow is and where the tides are primarily diurnal and semi-diurnal and actually see the answers.

COASTAL CIRCULATION

At the risk of belaboring the obvious, the shallowest water is found at the coast. *B. Douglas,* a leader in the study of sea level rise, explores the issue. He compares the Gulf and Atlantic coasts, pointing out exceptions to the general rule that they are quite similar. Despite wide variability, the trend in the Gulf is unmistakable; a rise of ~20 cm in 75 years is almost a foot per century. (A plot of data from San Francisco or New York would look similar, the differences being in the details.) Who can argue with the kind of data that Douglas presents? Note, however, his use of the term "relative" when related to sea level rise. A single tide gauge does not tell whether the water is coming up, the land is going down, or some mix of both. The rate here, however, is close to the world average at mid latitudes.

Johnson's paper gives us a description of movies (found on the CD) of satellite-tracked drifters in the waters of the Gulf coast between Florida and Texas. These drifters show

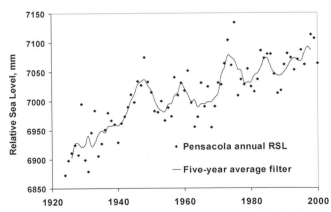

Figure 1. Annual means of Relative Sea Level (RSL), dots, at Pensacola, FL. Note the large low frequency fluctuations around the upward trend. Taken from the chapter by *B. Douglas.*

how the near-surface currents on the shelf consist largely of back-and-forth motions. Monthly means as well as movies of daily motions show how irregular the motions can be. The mean flow, averaged over many weeks, is remarkably small when compared with the transient or daily currents, which respond mostly to local winds. (There are notable exceptions to this general rule; coastal trapped waves propagate for long distances. See for example reviews by *Brink* [1991], *Davis and Bogden* [1989] or *Huyer et al.* [2002]). The question is often asked, What is the mean flow just offshore (of point X)? The questioner may have seen occasional strong flow, and is surprised to hear, "Well, the average over which week? The mean flow this week is to the north; last week it was to the south, and the long-term average may not be reliably different from zero."

Using current-meter moorings and a numerical model, *Weisberg et al.* explore the flow on the very wide west Florida shelf. Although the simplest wind-forced models can explain a major fraction of the wind-forced currents fairly well, they ignore vertical motions. However, as the authors show here, a full 3-D model can depict where upwelling and downwelling regions become important.

On the north coast of the Gulf, the general area to the west of the Mississippi River is commonly called the LATEX (Louisiana-Texas) region (to be distinguished from a major experiment referred to by the same acronym). The year-long mooring array discussed by *Jarosz and Murray* measured the currents close inshore at the eastern end of this region, west of the outflow from the Mississippi River, from the coast out to a depth of ~ 20 m. They found the reversal of the coastal current in the summer, as expected; while the mean speeds were ~ 6 cm/sec or less, they found persistent peak speeds well over a knot.

Farther to the west, *Nowlin et al.* discuss the flow on the northern Gulf coast, using the nearly three-year-long data set from the LATEX experiment. The shelf off Texas is wide, and a persistent gyre-like pattern develops there. Currents over the inner shelf flow eastward, toward the Mississippi River in summer and reverse during the rest of the year. The flow pattern appears to be driven mainly by the annual cycle of winds. Because the fresh river outflow usually tends to be toward the west, the reversing flow pattern off Texas leads to an annual signal in salinity as well as in the currents.

WHAT IS IN THE CD?

The CD on the inside back cover of this book allows us to incorporate a range of extra material of interest, including appendices to some of the articles and a large number of color figures. But above all, it lets us include the animations, our "movies," from which we have learned so much. *Leben's*

movie alone, showing sea surface height since 1993, presents over a decade of continuous data.

An "ocean color" movie by *Muller-Karger and Hu,* using images from the SeaWiFS satellite sensor, records ocean radiation in the visible and near infrared wavelengths. This allows us to detect patterns of chlorophyll (or color interpreted as such) and to spot the occurrence of phytoplankton blooms and trace their motion. Though daily coverage of the entire Gulf is available, weekly and bi-weekly versions are used for these movies in an attempt to suppress cloud cover, a standard averaging technique used with sea surface temperature as well. For comparison, Leben has contributed a brief movie of chlorophyll at the sea surface, computed with a slightly different algorithm, superimposed on his contours of SSH.

Anderson and Walker have made a movie that combines data from many surface drifters collected with SST maps over the whole Gulf. The drifter motions are plotted so that the colors of the track show the speeds. What is particularly interesting is the extent to which the drifters follow (or do not!) the patterns of SST.

The ocean data from the DeSoto Canyon experiment (from the paper by *Hamilton and Lee*) are combined into a movie showing velocity and temperature data at several depths. *Johnson's* movies show results of two separate year-long surface drifter experiments, and *Kantha's* animation shows deep-water tides. A movie produced from *Oey and Ezer's* numerical model shows a time series of velocity, top to bottom, on a N-S section at 90° W; it is remarkable to see how the pattern changes so thoroughly as Loop-Current rings pass through the section.

Hamilton compiled views of the deep mean flow as measured by long-term moorings and has contributed an extremely valuable set of figures. His diagrams show the mean flow as well as the variability at several levels for all available moorings longer than 6 months and deeper than 800 m. These direct measurements of the variability are at least as valuable as the means, as the operative question in deep water usually is "At what level of confidence can the computed mean values be reliably distinguished from zero?"

CONCLUDING THOUGHTS

This is being written in the summer of Hurricane Katrina, another terrifying reminder of the fragility of the Gulf and all other coasts. How will the witch's brew of polluted water affect the sea life along the coast as it drains from New Orleans?

An important and relevant question is whether the presence of a large intrusion of the Loop Current affected the increase in Katrina's strength from a Category 1 to a 4 or 5. We thank N. Walker and R. Leben, who prepared Plate 4, which superimposes the storm track, the sea surface temperature, and the position of the Loop Current. A large eddy-like

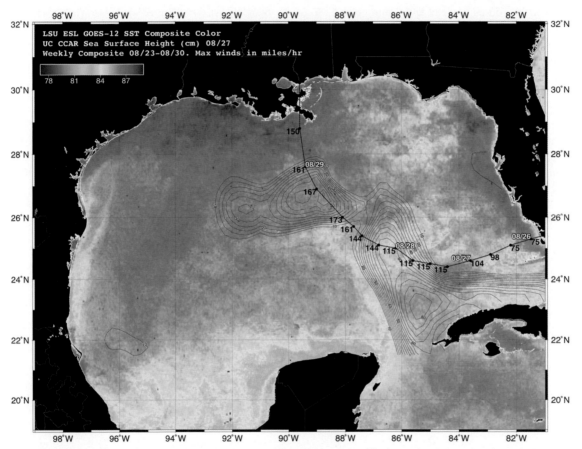

Plate 4. The track of Hurricane Katrina: dates show position of the center; black numbers show maximum wind speed (mph). Contours of SSH show the position of the Loop Current and a large not-yet-detached ring. Color scale shows SST in °F.

feature extends to the west but has not yet separated from the main current.

A hurricane draws energy from warm surface water. See for example the recent study of Hurricane Ivan [*Walker and Leben*, 2005] and the more general study by *Price et al.* [1994]. The warm upper layer in the Gulf in late summer may be only 25-30 m thick. Within the Loop Current, however, the warm upper layer is quite deep and thus has the potential for providing a great deal more stored heat energy. The availability of a large quantity of stored heat from the Loop Current and its eddies is clear and unambiguous. Yet because the Loop Current does not have a significant annual cycle, its position in the summer is nearly random; at the onset of a major storm, the Loop Current may extend only a small distance into the Gulf or have a major intrusion as shown here. The questions involving these interactions are complex: there is vertical mixing as well as a feedback cycle; the cyclonic (counterclockwise) surface winds drive upwelling beneath the hurricane, so stronger hurricanes will drive stronger upwelling. The stored heat in the deep upper layer clearly influences hurricanes, but whether the effect is large or small is an open question. Hurricane prediction models "know about" sea-surface temperature, but because the models do not yet couple the atmosphere and the ocean, the depth of the warm layer is not part of the calculation (T. N. Krishnamurti, *pers. comm.*).

In looking over the papers in this book and the materials on the CD, we are struck by two things. The first is the enormous progress in physical oceanography in general,

and that of the Gulf of Mexico in particular, since the major summary by *Capurro and Reid* in 1972. Technology has advanced much more than we could possibly have anticipated three decades ago. The second is the increasing realization of how much we have not yet learned. Most of us were drawn to oceanography because it seemed exciting and fun. It still is.

REFERENCES

Brink, K. H., Coastal-Trapped Waves And Wind-Driven Currents Over The Continental Shelf. *Ann. Rev. Fluid Mech.*, **23**: 389–412. (doi:10.1146/annurev.fl.23.010191.002133), 1991.

Capurro, L. R. A., and J. L. Reid, Jr., eds, Contributions on the Physical Oceanography of the Gulf of Mexico. Gulf Publishing, Houston, TX, 288 pp, 1972.

Davis, R. E. and P. S. Bogden, Variability on the California shelf forced by local and remote winds during the Coastal Ocean Dynamics Experiment. *J. Geophys. Res.*, **94**, C4, p. 4763–4783, 1989.

Davis, R. E., Intermediate-depth circulation of the Indian and South Pacific Oceans measured by autonomous floats. *J. Phys. Ocean.*, **34**, 683–707, 2005.

Huyer, A., R. L. Smith, and J. Fleischbein, The coastal ocean off Oregon and northern California during the 1997-8 El Nino. *Progr. Ocean.*, **54**, 311–341, 2002.

Price, J. F., Sanford, T. B. and Forristall, G. Z., Forced stage response to a moving hurricane, *J. Phys. Oceanogr.* 24, 233–260, 1994.

Walker, Nan D., Robert R. Leben and Shreekanth Balasubramanian, Hurricane-forced Upwelling and Chlorophyll a Enhancement within Cold-core Cyclones in the Gulf of Mexico. *Geophys. Res. Lett.*, 2005.

W. Sturges, Florida State University, Tallahassee, FL 32303-4320

A. Lugo-Fernandez, U.S. Minerals Management Service, New Orleans, Louisiana 70123-2394

M. D. Shargel, Florida State University, Tallahassee, FL 32303-4320

A Synopsis of the Circulation in the Gulf Of Mexico and on its Continental Margins

W. J. Schmitz Jr.,[1] D. C. Biggs,[2] A. Lugo-Fernandez,[3] L.-Y. Oey,[4] and W. Sturges[5]

This article is a synopsis of the state-of-the-art knowledge and understanding of the circulation in the deep Gulf of Mexico as well as in the coastal flow regimes on its continental margins. The primary purpose is to review the ideas and results in this special new volume on the circulation in the Gulf of Mexico and integrate them with material from selected previous publications. This overview therefore contains more repetition and expository discussion than normally found in journal articles. An extensive reference list is provided for interested readers, a special feature for students.

This volume is also meant to appeal to an audience beyond the general population of physical oceanographers, to include a more diverse community of Marine Scientists and Engineers, along with social interests of a scientific/technical nature. The article in this volume by *Biggs et al.* is focused on a study of some specific biogeochemical implications of on- and off-margin flow (on- and off- continental shelf/slope regions) in the northern Gulf of Mexico. The articles by *Chassignet et al.* [this volume] and *Morey et al.* [this volume], for example, note particular applications. Potential influences by the currents in the Gulf of Mexico on processes of interest in other branches of Marine Science are occasionally considered in this synopsis (for example, there is a special section on the effects of current patterns on coral reefs). The reader may also refer to a comparatively recent review of some of the physical oceanographic influences that regulate the biology of the Gulf of Mexico by *Wiseman and Sturges* [1999]. Finally, this synopsis is meant to complement and expand upon the Introduction to this volume by *Sturges et al.*, which includes consideration of some general societal interests motivating marine research in the Gulf of Mexico.

[1] Woods Hole Oceanographic Institution, Woods Hole, Massachusetts.

[2] Department of Oceanography, Texas A & M University, College Station, Texas.

[3] Minerals Management Service, Gulf of Mexico OCS Region (MS 5433), New Orleans, Louisiana

[4] Program in Atmospheric and Oceanic Sciences, Princeton University, Princeton, New Jersey

[5] Department of Oceanography, Florida State University, Tallahassee, Florida

Circulation in the Gulf of Mexico: Observations and Models
Geophysical Monograph Series 161
Copyright 2005 by the American Geophysical Union.
10.1029/161GM03

INTRODUCTION

The satellite altimeter database that has been accumulated over the past decade or so [*Leben*, this volume] exhibits a new and exciting picture of the Loop Current and its eddy field in the Gulf of Mexico (also called the GOM, or just the Gulf). The basis for this view of the major component of the general circulation of the Gulf may be found in the movie of sea surface height (SSH) maps on the CD at the back of this volume, herein called *Leben's* Movie. In the composite SSH database contained in this movie (or animation), at any given time the GOM is densely populated by both cyclonic and anticyclonic features. This includes of course the well-known Loop Current (LC) in the east, along with ubiquitous anticyclonic and cyclonic eddies of many shapes and dimensions at diverse locations. Regional maps of the area of interest may be found in Plate 1 and Figure 1 (with some nomenclature) by *Oey et al.* [this volume]. To some extent, the general circulation-related articles in this book are a long overdue update and extension of the material in a book [*Capurro and Reid,* Eds., 1972] on the physical oceanography of the Gulf that was published ~ three decades ago. The compilation by *Capurro and Reid* contained articles by a variety of authors, primarily considering the general circulation of the GOM.

Over the last couple of decades or so there has been a dramatic surge in the study of coastal circulation in the GOM. A recent regional review of coastal circulation regimes [*Boicourt et al.,* 1998], which included those in the GOM, is also updated and extended by the relevant articles in this book. A map showing the locations of the coastal flow regimes in the Gulf (with some nomenclature) may be found in Figure 1 by *Morey et al.* [this volume]. Flow patterns over the continental slope are also of considerable contemporary interest.

Considering the basic orientation of most of the articles in this volume, two general topical areas were chosen for the major sections of this synopsis. These are the general circulation of the GOM (deep-water flow characteristics) in SECTION I, and coastal circulations in SECTION II. Many state-of-the-art questions being addressed for the general circulation of the Gulf as this volume goes to press are associated with the influence of upstream conditions. Studies involving interactions between coastal and deep flow regimes are also topics of strong contemporary interest, as discussed in the coastal circulation section of this synopsis, typically involving flows over the continental slope. Observational, dynamical, and numerical model based results, which have begun to converge to some extent in recent years, are examined together within each of the general subject areas (sections) noted earlier in this paragraph.

I. THE GENERAL CIRCULATION OF THE GULF OF MEXICO, ALONG WITH A CONSIDERATION OF UPSTREAM CONDITIONS

This section of our synopsis is concerned with elements of the general circulation of the deep Gulf that are of contemporary interest. The detailed discussion in this section of results from the relevant chapters in this volume, in connection with other pertinent publications, focuses on the following topical/regional areas: The Loop Current and Eddy Shedding; Upstream Conditions; Anticyclonic Circulation in the Upper

Layers of the Central and Northwestern GOM; Cyclonic Flow in the Bay of Campeche; and Deep Circulation in the Gulf. These sub-sectional themes are underlined in the text below, following a few general remarks.

The first and most prominent component of the general circulation of the Gulf to be considered in this synopsis is the Loop Current (LC) and its eddy field. The observational articles by *Leben* [this volume] and *Schmitz* [this volume] are focused on this topic, as is to a large extent the dynamical and numerical modeling review by *Oey et al.* [this volume]. The influence of upstream conditions on the circulation in the GOM is at this time an area of considerable emphasis in the numerical modeling community, using formulations involving geographical regions larger than the Gulf itself (extended domains). Both *Oey et al.* [this volume] and *Schmitz* [this volume], the latter primarily in his appendix on CD, consider upstream effects, see in addition *Badan et al.* [this volume] and *Sturges* [this volume]. *DiMarco et al.* [this volume] contains the most recent contribution to a historically interesting series of investigations concerned with the mean anticyclonic circulation in the upper layers of the central and northwestern Gulf (~ west of 88–89°N, and north of 21–22°N). *Vázquez de la Cerda et al.* [this volume] is focused on an update of our knowledge of the general circulation in the Bay of Campeche. *Weatherly et al.* [this volume], along with *DeHaan and Sturges* [2005], consider the deeper flow patterns in the GOM, which may be partly driven by the LC and its eddy field. Observations of the deeper circulation in the Caribbean Sea and its interaction with the Gulf are also a subject of intense current interest [*Sturges*, this volume], see in addition *Rivas et al.* [2005]. The review by *Oey et al.* [this volume] also considers the issue of deep currents.

According to *Oey et al.* [this volume], a necessary condition for a basic numerical model of the GOM to be relevant is that it represents as accurately as possible the observable features of the LC and its eddy field. One would expect a "realistic" numerical experiment to therefore reveal the observed complexity of the eddy shedding process, notably the separation properties as determined by *Leben* [this volume], and the influence of amplified and deeply penetrating cyclones described by *Schmitz* [this volume]. Realistic model results should include the effects of LC penetration blocking by these cyclones, which at times may be observed to be arranged in a peripheral cyclonic shield around the LC. In general terms, numerical model results should be directed toward accounting for the key features of the SSH database as reflected by *Leben's* Movie. The reanalysis of available in-situ data by *Nowlin et al.* [2001] provides a complementary database for studies involving model-data intercomparison with respect to the general circulation of the Gulf. Westward propagation of eddies shed by the LC is a most basic property of all observations. *Cherubin et al.* [2005a, 2005b] discuss the role of bottom topography along with peripheral cyclonic shields in eddy shedding from the LC in the context of numerical model results, see also *Oey et al.* [this volume].

Model results should also account for the deep flow characteristics as reflected in the results by *Weatherly* [this volume] and *DeHaan and Sturges* [2005]. The mechanisms behind the upper layer circulation in the western GOM need to be more specifically sorted out in appropriate numerical model results in comparison with the database there. Recent high-resolution numerical experiments along with some dynamical studies of the Gulf circulation do seem more capable of simulating and rationalizing some of the major observed structures [*Cherubin et al.*, 2005a, 2005b; *Oey et al.*, this volume], relative to similar kinds of earlier investigations. However, new numerical sensitivity studies of comparatively long duration on a variety of model parameter specifications, including for example resolution and boundary conditions, are still required.

The Loop Current and Eddy Shedding

The Loop Current (LC) in the GOM is part of the Gulf Stream System, the very energetic western boundary current regime in the North Atlantic Ocean. A map depicting the general setting of the GOM in the western North Atlantic Ocean, along with some nomenclature, is contained in Figure 1 by *Oey et al.* [this volume]. The Loop Current in the eastern Gulf sheds anticyclonic eddies (also known as rings or warm-core eddies, or just eddies) at irregular intervals (see, for example; *Vukovich* [1988] and *Sturges and Leben* [2000], along with the update by *Leben* [this volume]). The eddy field associated with the LC affects almost every aspect of the circulation in the entire GOM. After separated warm-core eddies propagate toward and into the western Gulf, cyclonic features may cleave these anticyclonic rings into smaller eddies [*Biggs et al.*, 1996]. *Kennelly et al.* [1985] observed small scale cyclones around the periphery of a warm-core Gulf Stream Ring. So, the dynamics of warm-core Gulf Stream Rings and warm-core Loop Current eddies may have some features in common (association with peripheral cyclones/frontal eddies).

Leben's Movie of modern SSH maps (on the CD at the back of this volume) shows that we must revise our simple ideas of how anticyclonic rings separate from the LC. There exists in this movie evidence that a diverse group of complex processes variably participate in the composite mechanism of anticyclonic eddy shedding. From several to many examples of each of these processes, as described in the following, are contained in *Leben's* Movie. One well-known component

process, the penetration (sometimes referred to as an intrusion) of the LC ~ northwestward into the GOM, is a basic requirement for the future separation of warm-core eddies. The articles by *Maul* [1977, 1978] contained the original suggestions about the possible effects of time-dependent transport processes at the Yucatan Channel on increases in the volume (penetration or extension) of the LC into the GOM prior to eddy shedding.

These articles, [along with a follow on by *Maul et al.*, 1985], here collectively called the *Maul* View, have recently become connected with a near-explosion in both observational [for example, *Abascal et al.*, 2003; *Badan et al.*, this volume; *Bunge et al.*, 2002; *Candela et al.*, 2002; *Scheinbaum et al.*, 2002], and numerical-dynamical studies of the flow in the Yucatan Channel [for example, *Candela et al.*, 2003; *Cherubin et al.*, 2005a; *Ezer et al.*, 2003; *Oey*, 2004; *Oey et al.*, 2004; *Oey et al.*, this volume]. In the *Maul* View, both upper level and deep flows in the Yucatan Channel participate in a co-oscillation between the GOM and the Cayman Sea. This topic is considered in more detail in the context of upstream conditions (see below).

Maps of sea surface height (SSH) exhibit many examples of the ~ half a dozen generic components of the complex process of eddy shedding. These involve, in addition to ~ northwestward penetration into the eastern Gulf, pulling away from the LC by an anticyclonic eddy (visually) due to westward propagation, and the associated detachment of eddies from the LC by cyclones, both small and large (amplified). Detachment may occur after an extended LC is pinched-down or necked-down by cyclones. However, not every necking-down is followed by detachment. SSH maps also exhibit examples of strong reattachments, involving the recapture of a previously detached eddy by the LC. In addition one observes reattachment by neck filling after near detachments, along with weak reattachments by a touching-like contact between the LC and other anticyclonic features. New observational results on these topics are presented by *Leben* [this volume] and *Schmitz* [this volume]. Previous databased studies of interest with respect to elements of the composite eddy shedding process are due to, for example [*Brooks et al.*, 1984a, 1984b; *Cochrane*, 1972; *Elliot*, 1979, 1982; *Fratantoni et al.*, 1998; *Molinari and Festa*, 1978; *Sturges and Leben*, 2000; *Sturges et al.*, 1993; *Vukovich*, 1988; *Vukovich and Maul*, 1985; *Zavala-Hidalgo et al.*, 2002; and *Zavala-Hidalgo et al.*, 2003a].

For a generalized mechanism of first detachment, either partial or complete, cyclone pairs, one on each side of the LC, play the key role (one type of cyclone may dominate at any give time). Westward propagation is also involved. The initial observational description of first detachment was published by *Cochrane* [1972]. Amplified peripheral cyclones on the boundary of the LC are deeply penetrating in the vertical, as also first observed by *Cochrane* [1972]. Another prototypical view of detachment is observed to occur [*Schmitz*, this volume] when the LC has ~ an east-west orientation at its tip. This tends to occur after a first detachment or near-detachment is followed by subsequent reattachment(s). Pulling apart by westward propagation is the principal initial source for this second mode of detachment, with cyclones slipping in behind the embryonic ring to bring about final visual detachment. In these cases, peripheral cyclones on the northern side of the LC often participate in the detachment process, normally in conjunction with a cyclone on the western side of the LC. Mixtures of these two modes of detachment may also occur, involving both cyclonic intrusions and westward propagation in roughly equal proportion. *Schmitz* [this volume] notes a couple of examples of mixed mode. The mode initiated by westward propagation is the dominant view of final detachment in *Leben's* Movie, and also similar in appearance to the configurations of the LC in the suite of eddy separation times as shown in Plate 5 by *Leben* [this volume].

The overall impact of the results by *Cochrane* [1972] on eddy-shedding by the LC as noted above were nearly ignored for ~ 30 years, except for *Elliot* [1979, 1982], and most recently *Schmitz* [2003] and *Zavalla Hidalgo et al.* [2003a]. For another early clear-cut example of the type of detachment configuration similar to that initially described by *Cochrane* [1972], please see the figure on p. A6-2 by *Molinari and Festa* [1978], in their appendix 6. This figure contains temperature contours at 150 m depth that demonstrate the existence of (amplified) cyclones on each side of the LC at the throat of a detached ring. The figure published by Molinari and Festa is based on the same general dataset (taken in 1967) as acquired and used by *Nowlin et al.* [1968] to identify the first clearly observed example of a large, clearly detached anticyclonic eddy in the eastern Gulf. Portions of the map published by *Molinari and Festa* as noted above are exhibited in modified form in the appendix to the article by *Schmitz* [this volume], which contains other maps of historical interest.

In SSH maps, cyclones appear to block the further northward penetration of the tip of the LC at diverse times for variable time intervals (typically a couple of months, occasionally much longer). One example of penetration blocking by cyclones, based on SSH data from 1998, was first described by *Zavala-Hidalgo et al.* [2002]. During this particular example, the LC was held for several months in mostly a southerly (~ 24–25.5°N) location, where these blocking cyclones were typically present as individual features. On the other hand, when the tip of LC is extended northwestward into the Gulf (~ 26–27°N), cyclones often

appear to block further penetration while arranged in a peripheral shield of varying packing density around the boundary of the LC. Roughly 15 separate examples of these types of temporary blocking of further penetration by the LC into the Gulf by cyclones are described by *Schmitz* [this volume]. A peripheral cyclonic shield and its role in the eddy shedding process has most recently been discussed theoretically and numerically by *Cherubin et al.*, [2005a, 2005b]. For example, *Cherubin et al.* found in their model results that the behavior of a Loop Current Ring resembled that of a shielded vortex (having an anticyclonic core surrounded by a cyclonic belt).

Leben's Movie contains many examples of various configurations of this cyclonic shield [*Schmitz*, this volume]. The locations of specific cyclones or patches of cyclonic activity that form this shield are time-dependent (perhaps largely due to their movement around the boundary of the LC, and influenced as well by episodes of necking-down of the LC by these cyclones). Descriptively, there are typically gaps of various sizes between the individual cyclones or cyclonic patches that combine to make up the shield. Small anticyclones are detached from the LC much more frequently [*Schmitz*, this volume] than previously publicized and many of these are shed between gaps in the band of peripheral cyclones around the LC (see also *Zavala-Hidalgo et al.* [2002]). The tip of the LC may take also advantage of gaps in its cyclonic shield to continue its intrusion into the GOM.

Leben [this volume] is an outstanding new view of the most modern SSH database, including a description of the methodology and its limitations, along with an update of eddy separation dates and timscales. His results (Table 3, Figures 1 and 7) quantitatively define the time of an early stage of the separation of large and medium-sized warm-core eddies from the LC. The new quantitative definition of separation by *Leben* [this volume] involves the tracking and determination of the onset of pulling away by an embryonic eddy from a specific SSH contour (17 cm) associated with the LC. This operational definition of separation also requires that the incipient detached eddy is not later recaptured by the LC, or significantly reattached to the 17 cm SSH contour associated with it. *Leben's* Movie exhibits many examples of "reattachment" between the LC and the anticyclonic eddies that have become detached or nearly-detached.

Figure 7 by *Leben* [this volume] contains a distribution of ring separation intervals that are visually different from the results obtained in the 1973–1992 time frame (involving 22 eddy separations, see *Sturges and Leben* [2000]), in comparison with the recent results from the 1993–2003 time frame (involving ~ 17 eddy separations). *Leben* [this volume] refers to this difference as striking, and suggests that the histograms in his Plates 2 and 3 are non-Gaussian, while noting that it is doubtful that the overall distribution is stationary. This result confirms the long-held view that our records of eddy shedding tend to be too short to provide truly adequate statistics. *Sturges* [1993] examined the question of whether the monthly distribution of ring separations could reliably be distinguished from random, and concluded that it could not, on the basis of the database in hand at that time.

According to the update in Table 3 by *Leben* [this volume], also see his Plate 5, the separation intervals in his modern SSH data base vary from a few weeks up to ~ 18 months. Separation intervals tend to cluster near 4.5–7, 11.5, and 17–18.5 months in the histogram shown in the third frame of his Plate 5, perhaps suggesting the possibility of ~ a 6 month duration between these clusters. The question of why the LC and the shedding process behave in such a semi-erratic manner is a bit of a mystery. *Hurlburt and Thompson* [1980] anticipated this question in their classic modeling paper. *Hurlburt* [1986] found erratic eddy shedding intervals in the lowest eddy viscosity run in a sequence of numerical experiments. *Oey's* [1996] model results yield irregular separations due to local chaotic behavior of the LC, while *Oey et al.* [2003] also propose in addition an upstream influence. According to *Schmitz* [this volume] blocking cyclones can also impact separation intervals. *Leben* [this volume] contains interesting new results with respect to the relationship between separation interval and the latitude of the LC at the time of the previous separation (see his Figures 9 and 10).

Eddies that form along the LC margin are important in biological recruitment and retention. For example, a cold cyclonic recirculation, approximately 200 km in size, develops off the Dry Tortugas when the LC flow enters the southern Straits of Florida. *Fratantoni et al.* (1998) showed how this cyclone develops from eddies along the eastern edge of the LC and then moves into the southern Straits of Florida. *Lee et al.* (1994) correlated the persistence of this "Tortugas Gyre" with enhanced food supply, retention, and shoreward transports for the successful recruitment of locally spawned snapper and grouper larvae in the western and lower Florida Keys.

Upstream Conditions

The interconnectivity of the flow processes in the western tropical Atlantic Ocean, which includes Atlantic Ocean Basin and even global linkages, is a significant aspect of the Gulf Stream System [*Schmitz and McCartney*, 1993; *Schmitz*, 1995; *Schmitz*, 1996a, 1996b], of which the general circulation in the GOM is one component. These remote flow regimes, as well as regional and local forcing, can influence the general circulation in the Caribbean Sea and thence the GOM. Recent views of the time-averaged circulation and of the eddy field near the sea surface in this general area may be found in *Wilson and Leaman* [2000] and *Fratantoni* [2001].

The net inflow into the Caribbean Passages and then into the GOM is approximately 40 % of South Atlantic origin [*Schmitz and Richardson*, 1991]. This water of South Atlantic Origin is in the Florida Current concentrated in two comparatively fresh water masses. High current regimes with temperatures greater than ~ 24°C (salinities 35.8 to 36.2 psu) are found on the left hand and central segments of the Florida Straits (and to some extent the Yucatan Channel) in the upper 100 m or so depth range. There is an additional significant contribution of deeper, mostly upper Antarctic Intermediate Water, see *Schmitz and Richardson* [1991] and *Schmitz* [2003]. A mechanism for the observed modification of upper intermediate water in the tropical Atlantic Ocean (and Caribbean Sea) has recently been suggested by *Schmitt et al.* [2005].

The rest of the upper ocean water entering the Caribbean Sea and the LC and Florida Current is of North Atlantic origin. These water masses have been studied in the GOM, by for example, *Nowlin* [1972]. Windward Passage looks like a particularly interesting conduit [*Johns et al.*, 2002] for interactions between the GOM and the open North Atlantic Ocean. The flow in Windward Passage may be very energetic at very low frequencies [*W. Johns*, pers. comm., 2002; *T. Townsend*, pers. com., 2002; *Oey et al.*, 2003]. The variability of the flow in Windward Passage might at least partially be related to the complexity of the transport distribution in the coupled Florida Current and GOM System in conjunction with Old Bahama Channel [*W. Johns*, pers. comm., 2003, 2004], see *Hamilton et al.* [2005].

The deep flow between the Caribbean and the Gulf has been studied by *Sturges* [this volume]. He finds a 3-layer flow in the waters below the sill depth (~ 700–800 m) of the Straits of Florida. There is inflow to the Gulf from the Caribbean, through Yucatan Channel, at depths of ~800 m down to ~1200 m. There is also the now well-known inflow from the Caribbean in the layers just above the sill at 2000 m. What is surprising is the finding of a return flow of roughly 1 Sv, at depths of ~1300–1900 m, from the Gulf into the Caribbean and back into the Atlantic, a topic also discussed to some extent by *Rivas et al.* [2005].

The time-dependent exchange at the Yucatan Straits between the GOM and the Caribbean Sea, within both the upper and deep ocean, has become a key contemporary issue. As noted earlier, some features of this problem class were originally identified observationally by *Maul* [1977, 1978], also see *Maul et al.* [1985]. For recent studies, see for examples [*Badan et al.*, this volume; *Bunge et al.*, 2002; *Candela et al.*, 2002, 2003; *Chérubin et al.*, 2005a; *Ezer et al.*, 2003; *Oey et al.*, 2003; *Oey*, 2004; *Oey et al.*, this volume; *Rivas et al.*, 2005; *Scheinbaum et al.*, 2002; *Sturges*, this volume]. The recent study by *Bunge et al.* [2002] supports by one example the existence (originally suggested by *Maul* [1977, 1978]) of a co-oscillation between the Cayman Sea and the GOM in connection with time-dependent growth of the volume of the LC. That is, the growth of the Loop Current as it expands to the north displaces an ~ equivalent volume of Gulf water, which is observed to flow back into the Caribbean Sea at depths below ~ 800m.

Oey [1996], see his Figure 6, was the first published account based on numerical model results involving the *Maul* View (in general terms). This study used an idealized domain and an arbitrary (probably unrealistic, especially with respect to horizontal and vertical resolution) region of parameter space, a common problem for most of the existing numerical studies of the GOM circulation. His results contained a simplified co-oscillation in the vertical between the GOM and Cayman Seas (upper and lower layer transports in the Yucatan Straits nearly anti-correlated). Transport fluctuations remotely forced by winds in the North Atlantic Ocean [*Sturges and Hong*, 2001], as well as potential vorticity fluxes at Yucatan Channel, may also be associated with variations in the penetration of the LC into the eastern GOM [*Candela et al.* 2002, 2003; *Oey et al.*, 2003; *Oey*, 2004; *Oey et al.*, this volume]. *Badan et al.* [this volume] contains a summary of the upstream conditions of the Yucatan Current and its contribution to the variability of the Loop Current.

There has been a recent increase in interest in studies of the currents and associated eddy field in the Caribbean Sea and possibly beyond, with respect to their influence on the configuration of the LC [please see for example, *Carton and Chao*, 1999; *Murphy et al.*, 1999; *Andrade and Barton*, 2000; *Wilson and Leaman*, 2000; *Fratantoni*, 2001; *Goni and Johns*, 2003; *Oey et al.*, 2003; *Richardson*, 2005; *Oey et al.*, this volume]. The numerical model results by *Murphy et al.* [1999] opened up a strong interest in the effects on the dynamics of the LC by eddies that originate in the North Brazil Current region, presumably penetrating into the eastern Caribbean, and then into the GOM through the Yucatan Channel. *Goni and Johns* [2003] used SSH data to suggest that water associated North Brazil Current Rings can pass through the Antilles Passages into the eastern Caribbean Sea.

However, observational evidence by *Carton and Chao* [1999], *Andrade and Barton* [2000] and *Richardson* [2005] casts doubt on the ubiquity of the passage of eddies from the eastern Caribbean Sea into the Cayman Sea and thence into the GOM. *Carton and Chao* [1999] found that many of the eddies that they observed in the eastern Caribbean (based on 3 years of altimetric data), dissipated in the coastal waters of Nicaragua. *Andrade and Barton* [2000], based on 15 months of SSH data, found evidence for the passage of only a few eddies from the eastern Caribbean into the Cayman Sea. *Richardson* [2005] observed only one eddy to penetrate from the eastern Caribbean into the Cayman Sea, based on 212 surface drifter trajectories acquired in the time interval 1986–2003. *Andrade and Barton* [2000] also found that some eddies entered the Caribbean through Windward Passage. Their Figure 8 in particular contains a clear-cut example of a sequence of positions of an eddy penetrating into the GOM after having propagated through Windward Passage into the Caribbean.

The case has now definitively been made for using an extended (larger, perhaps basin scale) domain in order to use boundary conditions that minimally restrict the results, when studying the GOM with numerical circulation models [*Sturges et al.* 1993; *Oey*, 1996; *Welsh*, 1996; *Mooers and Maul*, 1998; *Murphy et al.*, 1999; *Welsh and Inoue*, 2000; *Oey and Lee*, 2002; *Ezer et al.*, 2003; *Oey et al.*, 2003; *Candela et al.*, 2003; *Chérubin et al.*, 2005a; *Oey et al.*, this volume]. Although numerical models applied on a basin scale domain are a contender in this regard, it is important to recognize that the overall characteristics of extended domain numerical experiments need to be compared with the database, along with the results explicitly occurring in the GOM.

In addition to the use of numerical model results in the study of key processes, prediction mode is a most practical activity, and two articles in this book are oriented toward this application [*Chassignet et al.*, this volume; *Kantha et al.*, this volume]. *Chassignet et al.* point out a potentially important usage of SeaWifs data (maps) for skill-assessing ocean prediction systems in tracking fronts in the Gulf. They also feel that their study clearly illustrates that biological responses of the surface water in the vicinity of the LC and associated warm and cold eddies are linked to physical events and processes. *Kantha et al.* demonstrate how an operational forecast system for the Gulf can be set up. According to *Kantha et al.* [this volume], roughly half of the oil produced in the United States comes from the Gulf, and production is subject to disruptions by the Loop Current and its eddy field. The prediction of the fate of oil spills is also of serious practical interest.

However, accurate practical prediction in the Gulf is still in the early stages of development. A recent exercise to forecast the time-dependent positions of fronts associated with the Loop Current and its eddies (for ~ 4 weeks), sponsored by an oil industry consortium using a variety (~ 5) of numerical prediction systems, yielded mixed results, with only one prediction system beating persistence. The interested reader may consult *Oey et al.* [2005] for a discussion of some of these new results. The major source of error identified by *Oey et al.* in the performance of the prediction system that beat persistence was in the specification of initial hindcast fields. According to *Oey* [2005, pers. comm.], what is needed is a comprehensive program that can systematically skill-assess model outputs against *in situ* observations such as the currents and hydrographic fields in four-dimensional space-time. The fundamentally baroclinic nature of the circulation induced by the Loop Current and eddies, along with associated dynamical instability and vortex interaction, necessitate such a detailed skill-assessment process. Improved assimilation schemes (for forecast initialization) along with model physics and parameter choices (especially including the higher resolution now known to be required) can then be more readily tested. Recent MMS field programs in the central and western Gulf may offer such an opportunity.

Anticyclonic Flow in the Upper Layers of the Central and Northwestern Gulf

DiMarco et al. [this volume] present a new description of the near surface velocity field in the GOM based on drifter data acquired during the years 1989–1999. Their article contains the most recent major observational contribution to our view of the mean upper ocean circulation (specifically near-surface) in the GOM. *Sturges and Blaha* [1976] had originated the idea that the interior of the central and northwestern GOM might be analogous to the subtropical gyre-type of regime as found in the open North Atlantic say, with the circulation driven by an anti-cyclonic wind stress curl, involving a boundary current in the west. This idea was extended further by *Blaha and Sturges* [1981], *Kassler and Sturges* [1981], and *Sturges* [1993]. A "yearly averaged" section of 14°C isotherm depths (range ~ 200–360 m) based on data acquired by *Kassler and Sturges* [1981], along with historical data, both plotted on their Figure 1, implied the probable existence of an upper layer anticyclonic gyre in the western GOM along 95°W between ~ 21–26°N.

Beringher et al. [1977], based on a substantial temperature database at 200 m depth, suggested that a large scale upper layer anticyclonic circulation was a persistent or quasi-stationary characteristic of the central western GOM. *Elliot* [1979, 1982] concluded that the mean or persistent upper level circulation in the western GOM would likely be both wind-forced and eddy-rectified. *Hoffman and Worley* [1986], based on a three-layer scheme using a hydrographic dataset for a single Gulf-wide cruise in the early 1960's, see for example *Nowlin* [1972], found an anticyclonic gyre like circulation in the upper layers of the central western Gulf. The overall perspective adopted by *Sturges* [1993] involved a combined wind and eddy forced mean upper layer anticyclonic circula-

tion, along with an anticyclonic seasonal circulation that is wind forced, also see *Oey* [1995] and *Welsh* [1996].

DeHaan and Sturges [2005] present a map of the ensemble mean temperature distribution at 400 m depth (in their Figure 1) that also implies the likely existence of an anticyclonic circulation pattern in the central western GOM in the vicinity of this depth. Recently, *Lee and Mellor* [2003], see their Figure 14, find in their numerical model results a time-averaged anticyclonic circulation at the sea surface in the central western Gulf, with a (comparatively wide) western boundary layer. *Lee and Mellor* [2003] note that, in addition to wind forcing, their results show that this model-determined anticyclonic upper level circulation in the western Gulf is largely affected by the average influence of LCR's propagating to the west while dispersing anticyclonic vorticity.

DiMarco et al. [this volume] find an upper layer (based on drifter observations with drogues at 50 m depth) anticyclonic gyre-like circulation (see their Figure 4) in the deep-water regions of the central western GOM (from ~ 22–26°N). They also find a cyclonic gyre in the Bay of Campeche (located south of 22°N). The anticyclonic flow pattern from ~ 22–25.5 °N in their Figure 4 has a reasonably well defined ~ northward-directed boundary current structure in the westernmost Gulf, attributed by *DiMarco et al.* to combined wind and eddy forcing. However, the northerly-located (~ north of 25–26°N) mean currents in their Figure 4 are comparatively weak, please see *DiMarco et al.* for a discussion of this point. Figures 6 through 9 by *DiMarco et al.* [this volume] exhibit maps of 3 month averages of current vectors that depict gross seasonal flow patterns. The summertime flow patterns in their Figure 7 contains stronger flow in the westernmost Gulf than in the wintertime map in their Figure 9, consistent with *Sturges* [1993]. The seasonal maps in Figures 6 through 9 by *DiMarco et al.* [this volume] also exhibit comparatively weak currents to the north of ~ 25–26°N, as is the case for the map for July based on ship-drift observations by *Sturges* [1993], his Figure 1. See however, Figure 2 by *DiMarco et al.* [this volume] for examples of upper layer flow with a well-defined eastward current component in the northern Gulf north of ~ 26°N, based on individual drifter trajectories.

The results mentioned in this sub-section of the synopsis are in general agreement with the overall qualitative conclusions about both the mean and seasonal flows in the central and northwestern western GOM by *Elliot* [1979, 1982], *Sturges* [1993] and *DiMarco et al.* [this volume]. The seasonal flow is wind forced (by both wind stress and curl). But a detailed quantitative determination of the partitioning between the wind and eddy contributions to the mean flow remains at least somewhat elusive. Perhaps a suite of specifically formulated numerical experiments could help resolve this issue.

its middle regions, becoming narrow in the south near the Dry Tortugas. The LC was found to interact with the outer shelf off the Dry Tortugas (where the shelf narrows) at times of anomalously favorable upwelling in the spring.

Coastal Circulations in the Northern Gulf

The LC and its eddy field (including the cyclones that move clockwise around the periphery of the LC), can in principle interact with the WFS, and/or with the shelf system in the northeastern-most GOM, either directly or indirectly. Near surface shelf water can be drawn offshore at the shelf break in the northeastern and north central Gulf by amplified frontal eddies (cyclones) on the boundary of the LC. Studies of the circulation in the northeastern GOM have been published by, for example, *Huh et al.* [1981], *Hamilton et al.* [2000], *Sturges et al.* [2001], *Wang et al.* [2003]; *Fan et al.* [2004]; *Ohlmann and Niiler* [2005], also see *Hamilton and Lee* [this volume]. Some possible relationships between bio-optical data and mesoscale flow patterns in the northeastern GOM were examined by *Hu et al.* [2003].

The studies by *Weisberg and He* [2003] clearly indicated the importance of open-ocean forcing on shelf dynamics in the northeastern Gulf. However, the shelf forcing in their case is primarily through the large-scale sea-level gradient along the shelf, set up by the Loop Current's proximity to the shelfbreak [*Hetland et al.,* 1999]. One expects such a forcing to be resolved by satellite SSH data, and explains why *Weisberg and He* [2003] were able to emulate the effects of the Loop Current System with essentially a depth independent forcing.

However, in DeSoto Canyon and on the Mississippi-Alabama outer shelf, *Hamilton and Lee* [this volume] observed small-scale depth dependent eddies (of both signs) induced by northward intrusions of Loop Current frontal eddies and warm filaments over the upper slope and outer shelf. *Wang et al.* [2003] analyzed this region in conjunction with satellite SSH and data-assimilated model results, and showed that the model was able to depict the large-scale structure, but failed to reproduce smaller-scale eddies. The subject of eddy (or Loop Current) shelf interaction involving these smaller-scale frontal meanders or eddies is yet to be fully explored. It was clear that satellite-SSH assimilation alone was insufficient, and one should use higher-resolution localized data. By assimilating both satellite SSH and local drifter data in July 1998, *Fan et al.* [2004] were able to reproduce a small eddy in DeSoto Canyon that was not resolved by the satellite methodology available.

Hamilton and Lee [this volume] noted that the upper layer jet-like flow which they observed could occasionally export low-salinity water from the shelf/slope regime in the north-eastern Gulf onto the WFS. They suggested that the comparatively fresh water observed on the southwest Florida shelf by *Lee et al.* [2002] had a Mississippi River source. Also see *Ortner et al.* [1995] and *Morey et al.* [this volume] relative to the fate of this river's discharge. *Chassignet et al.* [this volume] note that cyclones on the northern edge of the LC appear to be entraining shelf water from the narrow shelf regime in the northeastern Gulf. This might also be likely to occur offshore of the comparatively narrow shelf at the southern end of the WFS west of the Dry Tortugas where large (amplifed) cyclones are typically observed. According to the early-on article by *Cochrane* [1972], cold fresh water from both the WFS and the Campeche Bank was present in the cyclonic features that he observed to be penetrating into the LC in the southeastern-most GOM (~ west of the Dry Tortugas).

The eddy field associated with the continental slope and rise systems in the northern GOM has been extensively studied in the past several years [*Frolov et al.*, 2004; *Hamilton*, 2000; *Hamilton et al.*, 2000; *Hamilton and Lugo-Fernandez*, 2001; *Hamilton and Lee*, this volume; *Oey and Zhang, 2004*; *Ohlmann et al.*, 2001; *Stuyrin et al.*, 2003; *Wang et al.*, 2003]. *Hamilton and Lugo-Fernandez* [2001] observed very energetic deep currents near the Sigsbee Escarpment south of the Mississippi Delta. *Hamilton and Lee* [this volume] show that currents in the De Soto Canyon are dominated by warm filaments, LC eddies and sometimes the outer fringes of the LC itself. An along-slope upper-layer eastward jet in the upper layers is a prevalent feature of the data analyzed by *Hamilton and Lee*, while in lower layers the flow is westward.

The presence and influence of cyclones and anticyclones as well as strong near-bottom currents on the continental slope of the northeastern GOM appears to be involved with interactions between topography with the LC and its eddy field [*Hamilton*, 2000; *Hamilton and Lugo-Fernandez*, 2001; *Hamilton and Lee*, this volume, *Oey and Lee*, 2002; *Oey and Zhang*, 2004; *Oey et al.*, this volume]. The continental slope and rise regimes in the northern Gulf are of course unique and interesting in their own right. *Oey and Zhang* [2004] show how small-scale cyclones and jets may be produced as a result of the Loop Current or a Loop Current eddy interacting with the slope. The role of a bottom boundary layer was thought to be important in this process. According to *Hamilton and Lee* [this volume], little was previously known about the origin and characteristics of the smaller eddies now typically observed (beginning with *Hamilton* [1992]) over the continental slope in the northern Gulf. On the other hand, the LC is the source of both small and large anticyclones, as well as cyclones of diverse sizes. When formed in the northeastern Gulf they could perhaps be trapped topographically to the slope and rise and propagate to the west there.

Specifically designed numerical experiments might also be used in an attempt to sort out the sources of cyclones and anticyclones on the northern slope in the Gulf.

Biggs et al. [this volume] emphasize the role of the spatial and temporal variability associated with the smaller scale eddies located over the continental slope in the northern GOM in on-margin, along-margin, and off-margin flow of surface waters. Using a combination of ship and satellite data, they demonstrate that offshore/onshore flow occurs when/where the surface circulation along the shelf-slope break flows is associated with the presence of anticyclonic as well as cyclonic slope eddies. *Biggs et al.* [this volume] relate these on- and off-margin flow patterns to the distribution of sperm whales, please also see *Biggs et al.* [2000].

Off-margin transport is enhanced when the LC penetrates to a location that is comparatively far to the north. From the ~ 13 year climatology of altimeter data from Topex/Poseidon and follow-on missions that is presented by *Leben* [this volume], also see *Leben's* Movie, northward extension of the LC past 27°N has occurred ~ a half dozen times from 1992 to 2004. On these occasions, substantial amounts of low salinity, high chlorophyll Mississippi River water can be entrained and transported clockwise and off-margin into deep water, as *Walker et al.* [1994] and *Belabbassiet al.* [2001] have shown from field data, also see *Morey et al.* [this volume].

The overlay of SSH data on a 7-day composite of SeaWiFS ocean color in Plate 1 indicates that high-chlorophyll shelf water has been transported off-margin and south along the eastern side of the LC for at least a few hundred kilometers during the summer of 2004. The SSH data used in Plate 1 were provided by *Leben* [pers. comm., 2005], and their superposition on the SeaWiFS data is due to *Hu* [pers. comm., 2005]. From a comparison of weekly and biweekly SeaWiFS imagery for July–August 2004 (which can be found on the CD at the back of this volume, see the animation there by *Hu and Müeller-Karger*), it is evident that this process happened rapidly (days-to-weeks). A couple of plumes of this general type (involving offshore transport of high- chlorophyll water from the shelf regime in the northern Gulf into the LC system and moving with a southward component down its eastern side) are also presented by *Chassignet et al.* [this volume], in their combined SSH and SeaWiFS plates for July 26 and August 8, 2003. An example of high chlorophyll water that is concentrated within a peripheral cyclone is also exhibited by *Chassignet et al.,* in their plate for July 9, 2003.

The high-chlorophyll surface water that was caught up in the LC system in Plate 1 actually may remain as a spatially-coherent "plume" during southerly transport over ~ 10^3 km. Please also see in this regard the trajectory for drifter 06935 in Figure 2 by *DiMarco et al.* [this volume]. *Gilbert*

et al. [1996] identified the presence of a mixed low-salinity Mississippi River outflow plume that was located within the Straits of Florida, said to be linked initially to a flood stage in 1993 that was presumably picked up by a "fully-extended" LC. Their measurements occurred from 10–13 September 1993 when a plume was found to be embedded within the Florida Current and adjacent coastal waters between Key West and Miami, also see Ortner et al. [1995].

Coastal Circulations in the Western Gulf

Morey et al. [this volume] examined the seasonal variability of flow on the continental shelf systems in the western (and northern) GOM using both observations and a relatively high horizontal and vertical resolution numerical model applied on an extended domain. A wind-driven transport of comparatively fresh water involving river discharge takes place in a thin, near surface layer along seasonally varying pathways on the shelf in their model results (possibly several hundred kilometers from its origin according to *Morey et al.* [this volume]). At some point along these pathways, buoyant low salinity water passes over the shelfbreak and becomes available for entrainment into the deep GOM by currents associated with mesoscale eddies. See the article by *Oey et al.* [this volume] for a review of earlier applications [*Morey et al.,* 2002, 2003; *Zavala-Hidalgo et al.,* 2003b] using the same numerical model configuration that was used in the study by *Morey et al.* [this volume]. As noted by *Morey et al.* [this volume], the seasonal coastal circulation patterns throughout the southwestern GOM were clearly determined in detail for the first time by *Zavala-Hidalgo et al.* [2003b], a much needed new study. The previous picture of this region by *Boicourt et al.* [1998] was based on studies by *Biggs and Müller-Karger* [1994], *Gutiérrez de Velasco et al.* [1992, 1993] and *Vidal et al.* [1992, 1994].

The shelf systems in the western GOM were divided by *Morey et al.* into three regions (the LATEX shelf, the East Mexico Shelf, and the western Campeche Bank), as defined in their Figure 1. The Louisiana-Texas shelf circulation is comparatively well known, but its interactions with the East Mexico Shelf flow regime were recently determined in detail, see *Zavalla-Hidalgo et al.* [2003b]. *Morey et al.* examine the seasonal circulation on the continental shelf regime from roughly Veracruz up to about Brownsville (~ 19 to 26°N), see *Sturges* [1993] with respect to a larger scale view of the seasonal circulation in the western Gulf. *Morey et al.* find that over the East Mexico Shelf (where the alongshore component of the wind stress is dominant), the currents from September through March tend to run down the coast (with a southward component). The currents on the East Mexico Shelf tend to run northward (upcoast) from

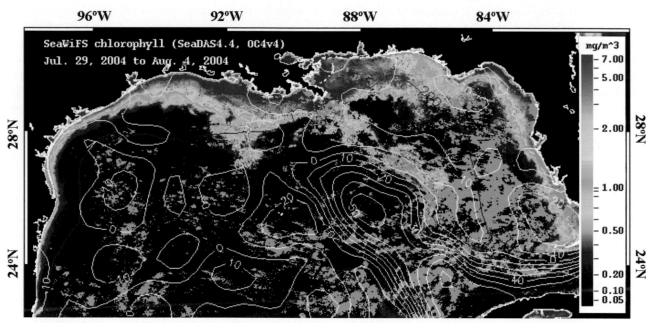

Plate 1. An illustration of the transport of high-chlorophyll water from the shelf system in the northeastern Gulf southward along the eastern side of the Loop Current. Isolines of sea surface height are superimposed on a SeaWiFS map of 7-day composite color, centered on August 1, 2004.

May through August, during which time coastal upwelling is experienced. In the fall and winter the generally cyclonic flow on the LATEX shelf advects low salinity Mississippi water downcoast to the East Mexico Shelf. At other times water from the East Mexico Shelf is transported onto the LATEX shelf. This confluence in coastal flow regimes near the delta of the Rio Grande River (~26°N) is emphasized by *Morey et al.* [this volume], as well as by *Zavalla-Hidalgo et al.* [2003b].

The articles in this volume by *Jarosz and Murray, Nowlin et al.,* and *Walker* focus on aspects of the flow on the LATEX (Louisiana-Texas) Shelf System in the northwest Gulf. This area is also often called the Texas-Louisiana Shelf. These studies confirm and extend the basic seasonal inner shelf circulation pattern proposed by *Cochrane and Kelly* [1986]; downcoast (from Louisiana to Texas) currents in fall, winter and spring and upcoast (from Texas to Louisiana) in the "summer months". *Cochrane and Kelly* also contains a very helpful summary of previous work and a valuable bibliography. See for example the articles that they cite by *Kimsey, Smith, Temple, and Watson. Smith* [1978] had demonstrated earlier that the alongshore currents over the inner shelf off Port Aransas, Texas were driven by the alongshore wind-stress. Although *Cochrane and Kelly* [1986] published the initial comprehensive description and rationalization of the circulation patterns and their most basic dynamics on the inner and middle shelves for the Texas-Louisiana flow regime, the important interaction of the currents on the outer shelf with eddies over the slope there was first considered by *Oey* [1995].

The article by *Walker* [this volume] presents a recent evaluation of the wind-forced behavior of the Louisiana-Texas coastal current, the position and strength of the coastal jet, and the combined effects of winds and eddies on offshore entrainment of shelf waters into the deeper GOM. *Walker* [this volume] integrates observations from satellite sensors with in-situ current measurements in order to define a few distinct circulation regimes on the Louisiana-Texas shelf and slope. These observations were acquired (see her Figure 1 for positions) during the 1992–1994 time frame also considered by *Nowlin et al.* [this volume]. *Walker* initially emphasizes processes that interrupt the prevalent "non-summer down-coast" flow within the coastal current from Louisiana to Mexico. *Walker* [this volume] also identified in her dataset a cross-shelf jet moving offshore (eastward flow) near 28°N in the fall of 1993, similar at least qualitatively to previous results in this range of locations by *Brooks and Legeckis* [1982] and in Figure I–22 by *Nowlin* [1972], see also *Zavalla-Hidalgo et al.* [2003b]. Moreover, the off-shelf jet event in the northwestern Gulf, which was observed during the fall of 1993 by *Walker* [this volume] to be forced by a cyclone over the shelf and an anticyclonic

eddy over the slope, further reinforces the importance of eddies in driving outer shelf currents.

Walker [this volume] found that during the "summer months", the coastal current reversed for 4 to 6 weeks, with an extensive coastal upwelling system developing from ~ 23°N (along the Mexican coast) to ~ 28°N (near Port Aransas, Texas). This upwelling was typically revealed in satellite imagery as a ~20 km wide zone where surface temperatures drop to 23–26° C, in contrast to temperatures of 30–32°C further offshore. Until recently not much had been published concerning the wind-driven coastal upwelling that may occur from ~ April through September along the south Texas and north Mexican coasts in the westernmost Gulf. This is surprising in light of the linkage between upwelling and biological productivity (Wiseman and Sturges, 1999).

The potential for upwelling in the western GOM was actually recognized rather early-on by *Franceschini* [1953]. *Angelovic* (1975) suggested that "summer" upwelling could occur in the vicinity of Brownsville, Texas (~ 26°N), albeit tenuously present. Their results could be an early indication of episodic or interannual variability in this upwelling regime, also see Appendix B.3 by *Nowlin et al.* [1998]. The report sections by *Kelly* [1988a, 1988b] contain clear cut observations of episodic wind-driven coastal upwelling over the LATEX Shelf (offshore of Freeport, Texas near Bryan Mound) in June 1980, also see Chapter 3 by *Schmitz* [2003].

Nowlin et al. [1998] and *Walker* [2001] describe seasonal upwelling along the Texas coast during the summers of 1992 through 1994 and note its spatial scale and temporal extent in broad descriptive terms. Based on climatological winds, they find upwelling-favorable wind stress along much of the western boundary of the GOM between 20°N–28°N from April through August. However, during April and May the effect is episodic, weak, and infrequent. June appears to be a transition month more favorable to upwelling conditions. July and August show marked and persistent upwelling during 1992–1994 [*Nowlin et al.,* 1998] in all types of data available (meteorological, satellite, in situ hydrography, and moored current meter). Note the offshore jet starting near the delta of the Rio Grande River (~26°N) in Figure B.3-4 by *Nowlin et al.* [1998], and also see the Coastal Zone Color Scanner Map by *Biggs and Müelller-Karger* [1994], in their Plate 1. *Glenn et al.* [1996] discuss the effects of varying alongshore topography on upwelling, which they found to occur in conjunction with bathymetric features that are somewhat similar to the above noted river delta located around 26°N in the westernmost GOM.

Zavala-Hidalgo et al. [2003b] have recently presented an updated view of the upwelling regime along the Mexican coast (~ south of Brownsville) in the western Gulf. The geographical components of the shelf regime along the Mexican coast in the southwestern Gulf were called (their Figure 1) by *Zavala-Hidalgo et al.* [2003b] the Tamulipas-Veracruz (TAVE) Shelf, and the Campeche Bank. They emphasize the "summer upwelling" on the TAVE Shelf, which is found to be observed for about twice as long (May through August) as upwelling over the LATEX Shelf (~ 1–2 months). Their results are also consistent with a "mean westward flow" on the Campeche Bank, as found by *Merino* [1997]. According to *Merino* [1997], the upwelling on the Yucatan Shelf is probably caused by bottom friction or other topographical mechanisms. Similar suggestions of upwelling on the eastern Yucatan shelf due to the interaction of the Loop Current with topography were made by *Cochrane*, [1969], *Ruiz-Renteria* [1979], and *Welsh* [1996]. It is also probable that topographically induced upwelling may result from Loop Current eddy interaction with the sloping bottom in the western GOM, see *Sahl* et al. [1993, 1997].

Nowlin et al. [this volume] point out that the Texas-Louisiana coastal circulation pattern can be divided into an inner shelf (depth < 50-m) where currents are predominantly driven by winds in the weather band (5–10 days) while the outer shelf currents are driven primarily by mesoscale activity. According to *Nowlin et al.* [this volume], for the low-frequency circulation over the Texas-Louisiana continental shelf, currents over the inner shelf are upcoast (Rio Grande to Mississippi River) in summer and downcoast in non-summer and are driven by the annual cycle of winds, in the vein of *Cochrane and Kelly* [1986]. *Nowlin et al.* point out that this flow pattern results in an annual signal for salinity, with lowest salinity waters occurring (a) in late spring along the inner portion of the western shelf when downcoast (westward and southward) flows carry the high discharges from the Mississippi-Atchafalaya rivers to the Mexican border, and (b) in summer over the inner and outer eastern shelf when the upcoast (northward and eastward) flow causes a pooling of the discharges from the Mississippi-Atchafalya rivers over that shelf. According to *Nowlin et al.,* upcoast winds during summer also result (along with upwelling) in high salinities over the western shelf due to advection from the direction of Mexico. Currents over the outer shelf are variable, but predominantly upcoast throughout the year.

The circulation based on other components of the data from the LATEX field program is discussed by *Jarosz and Murray* [this volume]. A special set of moored current measurements were used to examine the variability of the Louisiana-Texas Coastal Current on the Louisiana inner shelf. According to *Jarosz and Murray*, this current system is responsible for the distribution of fresh water, sediment, nutrients and pollutants on the shelf system in the northwestern Gulf. The study by *Jarosz and Murray* examines the results of a dataset from a line of moorings on the inner shelf off Cameron, LA. During an eight-month period between June 1996 and January 1997, seasonal reversals of the coastal current were clearly observed in June (upcoast flow regime) and September (downcoast flow regime) of 1996, similar to early-on results by *Cochrane and Kelly* [1986]. Large amplitude current and transport fluctuations were related to both along-shore wind stress and along-shore sea-surface slope.

According to *Jarosz and Murray*, the force balances change seasonally in the area of their array. *Jarosz and Murray* find that winds are the predominant force for downcoast (toward Texas from their moored array) currents on the inner shelf, but that both winds and surface slopes are probably important during the upcoast current regime. A simple model that includes the effects of friction and forcing by the along-shelf wind reproduces fairly well the fall-winter down-coast jet scenario, which confirms that the jet is wind-forced. However, the agreements in summer are fair at best, which suggests that other neglected terms are important. The most likely candidate is the along-shore pressure gradients induced by the large Atchafalaya-Mississippi discharge in spring and summer.

Notes on the Connectivity of Circulation and Larval Dispersion

Larval dispersion is important to coral reefs in the Gulf because it controls the reefs' connections with each other, which in turn affects genetic exchanges, diseases, recovery after hurricanes, fish populations, and other perturbations (see, for example, *Jordán-Dahlgren* [2002]; *Lipp et al.* [2002]). According to a recent review article by *Jordán-Dahlgren and Rodríguez-Martínez* [2003], coral reefs are biologically connected by fluxes associated with the applicable dominant current systems (they refer to *Roberts* [1997]). *Jordán-Dahlgren and Rodríguez-Martínez* also point out that the coral reefs in the Mexican region of the Gulf of Mexico (their Figures 1 and 2), are present only where appropriate substrate morphology is conducive to reef growth. *Lugo-Fernandez et al.* [2001a, 2001b] examined the role of currents and warm-core eddies in the dispersion of larvae in the Gulf of Mexico using satellite-tracked as well as simulated drifters at the sea surface. *Biggs* [1992] showed that Loop Current rings transport biological materials from tropical areas into the western Gulf. According to *Jarrett et al.* [2005], the Loop Current is at this time a source for the warm and clear waters that support an anomalously deep coral reef that is located on a drowned barrier island along the western margin of the south Florida Platform, roughly 50 km due west of the Dry Tortugas. The role of currents and frontal eddies in the dispersion of larvae along the Florida Keys has been analyzed by *Lee et al.* [1992] and *Lee and Williams* [1999].

Lugo-Fernandez et al. [2001a] suggested a potential sporadic link between the Flower Garden Banks and the Florida Keys, along with Mexican reefs as far south as Veracruz. In 1984, a massive die-off of the sea urchin *Diadema antillarum* spread across the western Atlantic region [*Lessios et al.*, 1984]. Current patterns in the Gulf (see for example, *Dimarco et al.* [this volume]) dispersed the pathogen as far north as the reefs associated with the Flower Garden Banks as evidenced by the decline of this sea urchin reported by *Dokken et al.* [2003]. Another dramatic event revealing reef connectivity is the recent discovery of *Acropora palmata* at the Flower Garden Banks [*K. Deslarzes*, pers. comm., 2005]. This coral is a shallow water species adapted to high energy environments. Similarly, the absence of gorgonians at the Flower Garden Banks may be connected to prevailing currents and the larvae's biology. In the deep Gulf, the *Lophelia* coral, which can produce large beds, is known to occur. Because this coral is a non reef-forming, solitary genus that is adapted to cold, deep water (> 400 m), its dispersion in relation to the deep Gulf circulation is of contemporary interest and study.

The links between coral reefs and terrestrial events have received more attention recently [for example, *Andréfouët et al.*, 2002]. The Flower Garden Banks reefs are influenced by terrestrial river discharge in spite of their distance from shore (~200 km). *Deslarzes and Lugo-Fernandez* [in press] attributed annual variations in salinity of 30 to 36 psu at the Flower Garden Banks as due to regional river discharge brought to these reefs by the coastal circulation in the northwestern Gulf [*Jarosz and Murray*, this volume; *Morey et al.*, this volume; *Nowlin et al.*, this volume]. Similar salinity fluctuations can occur along the Campeche-Veracruz shelves because of local rivers [*Zavala-Hidalgo et al.*, 2003b; *Jordan-Dahlgren*, 2002].

Sediment inputs associated with river discharges are another accepted limiting factor for reef development in the Gulf [*Spalding et al.*, 2001]. On the inner shelf (depth < 50 m) of the northern Gulf the large sediment input from the local rivers and consequent high light absorption prevent coral reef development (along with cold winter temperatures). The discovery of coral settling on oil and gas platforms of the northern Gulf supports the notion of substrate limitation by sediment cover when temperature, salinity, and light conditions are suitable, see *Sammarco et al.* [2004]. The coral reefs in the Veracruz region of the southwestern Gulf [*Emery*, 1963; *Morelock and Koenig*, 1967] exist in an area of high terrigeneous sedimentation. According to *Tunnel* [1988, 1992], these reefs are under severe stress due to both natural conditions and anthropogenic pressures. Along the northern Gulf shelf, hypoxia during summer also limits reef development. Hypoxia has been attributed to the combination of high nutrients from the Mississippi River and the summer stratification and circulation of regional shelf water (see, for example, *Rabalais et al.* [2002]).

REFERENCES

Abascal, A. J., J. Scheinbaum, J. Candela, J. Ochoa., and A. Badan, (2003). Analysis of flow variability in the Yucatan Channel. *J. Geophys. Res., 108(C12)*, 3381, doi: 10.1029/2003JC001992.

Andrade, C. A., and E. D. Barton, (2000). Eddy development and motion in the Caribbean Sea. *J. Geophys. Res., 105*, 26,191–26,201.

Andréfouët, S., P. J. Mumby, M. McField, C. Hu, and F. E. Müller-Karger, (2002). Revisiting coral reef connectivity. *Coral Reefs, 21*, 43–48.

Angelovic, J. W., (1975). Physical Oceanography, In: Environmental studies of the South Texas Continental Shelf. NOAA Final Report, U. S. Dept. Comm., Gulf Fisheries Center, Galveston, TX. 290 pp.

Badan-Dangon, A., K. H. Brink, and R. L. Smith, (1986). On the dynamical structure of the mid-shelf water column offf northwest Africa. *Cont. Shelf. Res., 5*, 629-644.

Badan, A., J. Ochoa, J. Scheinbaum, and J. Candela, (this volume). Circulation in the approaches to Yucatan chanel.

Behringer, D. W., R. L. Molinari and J. F. Festa, (1977). The variability of anticyclonic flow patterns in the Gulf of Mexico. *J. Geophys. Res., 82*, 5469–5476.

Belabbassi, L., P. Chapman, W. D. Nowlin, A. E. Jochens, and D. C. Biggs, 2005, (in press). Summertime nutrient supply to near-surface waters of the northeastern Gulf of Mexico, 1998, 1999, and 2000, *Gulf. Mex. Sci.*

Biggs, D. C., (1992). Nutrients, plankton, and productivity in a warm-core ring in the western Gulf of Mexico. *J. of Geophys. Res. 97*, 2143–2154.

Biggs, D. C. and F. E. Müller-Karger, (1994). Ship and satellite observations of chlorophyll stocks in interacting cyclone-anticyclone pairs in the western Gulf of Mexico. *J. Geophys. Res., 99*, 7371–7384.

Biggs, D. C., and P. H. Ressler, (2001). Distribution and abundance of phytoplankton, zooplankton, ichtyoplankton, and micronekton in the deepwater Gulf of Mexico. *Gulf Mex. Sci. 19*, 7–29

Biggs, D. C., G. S. Fargion, P. Hamilton and R. R. Leben, (1996). Cleavage of a Gulf of Mexico loop current eddy by a deep water cyclone. *J. Geophys. Res., 101*, 20,629–20,641.

Biggs, D. C., R. R. Leben, and J. G. Ortega-Ortiz, (2000). Ship and satellite studies of mesoscale circulation and sperm whale habitats in the northeast Gulf of Mexico during GulfCet II. *Gulf Mex. Sci, 18*, 15–22.

Biggs, D. C., A. E. Jochens, M. K. Howard, S. F. DiMarco, K. D. Mullin, R. R. Leben, F. E. Muller-Karger and C. Hu, (this volume). Eddy-forced variations in on-margin and off-margin summertime circulation along the 1000-m isobath of the northern Gulf of Mexico, 2000–2003, and links with sperm whale distributions along the middle slope.

Blaha, J. P. and W. Sturges, (1981). Evidence for wind-forced circulation in the Gulf of Mexico. *J. Mar. Res., 39*, 711–734.

Boicourt, W. C., W. J. Wiseman, Jr., A. Valle Levinson and L. P. Atkinson, (1998). Continental shelf of the southeastern United States and the Gulf of Mexico: In the shadow of the Western Boundary Current. In: *The Sea*, Vol. 11, A. R. Robinson and K. H. Brink (eds), John Wiley, NY, pp.135–182.

Brooks, D. A. and R. V. Legeckis, (1982). A ship and satellite view of hydrographic features in the western Gulf of Mexico. *J. Geophys. Res., 87*, 4195 – 4206.

Brooks, D. A., S. Nakamoto and F. J. Kelly, (1984a). *Investigation of low frequency currents in the Gulf of Mexico, Volume I*. Texas A&M Research Foundation, Contract 110483-HHH-1AH.

Brooks, D. A., S. Nakamoto and F. J. Kelly, (1984b). *Investigation of low frequency currents in the Gulf of Mexico, Volume 2*. Texas A&M Research Foundation, Contract 110483-HHH-1AH.

Bunge, L., J. Ochoa, A. Badan, J. Candela, and J. Scheinbaum, (2002). Deep flows in the Yucatan Channel and their relation to the Loop Current extension. *J. Geophys. Res., 107*, 3233, doi:10.10292001JC001256.

Candela., J., J. Scheinbaum, J. Ochoa, and A. Badan, (2002). The potential vorticity flux through the Yucatan Channel and the Loop Current in the Gulf of Mexico. *Geophys. Res. Lett.*, *29 (22)*, 2059,doi:10.1029/2002GL015587.

Candela., J., S. Tabahara, M. Crepon, B. Barnier, and J. Scheinbaum, (2003). Yucatan Channel flow: Observations versus CLIPPERATL6 and MERCATORPAM models. *J. Geophys Res.*, *108(C12)*, 3385, doi:10.1029/2003JC001961.

Carton, J. A., and Y. Chao, (1999). Caribbean Sea eddies inferred from TOPEX/ POSEIDON altimetry and a 1/6° Atlantic Ocean model simulation. *J. Geophys. Res.*, *104*, 7743 – 7752.

Capurro, L. R. A., and J. L. Reid (eds.), (1972). *Contributions on the Physical Oceanogaphy of the Gulf of Mexico.* Texas A & A Oceanographic Studies, 288 pp.

Chassignet, E. P., H. E. Hurlburt, O. M. Smedstad, C. N. Barron, D. S. Ko, R. C. Rhodes, J. F. Shriver, A. J. Wallcraft, and R. S. Arnone, (this volume). Assessment of ocean prediction systems in the Gulf of Mexico using ocean color.

Cherubin, L. M., W. Sturges, and E. P. Chassignet, (2005a). Deep flow variability in the vicinity of the Yucatan Straits from a high-resolution numerical simulation. *Journal Geophys. Res.*, *110*, CO4009. doi:1029/2004JC002280.

Cherubin, L. M., Y. Morel, and E. P. Chassignet, (2005b). The role of cyclones and topography in the Loop Current ring shedding. *J. Phys. Oceanogr*, in press.

Clarke, A. J., and K. H. Brink, (1985). The response of stratified, frictional shelf and slope water to fluctuating large-scale low-frequency wind forcing. *J. Phys. Oceanogr.*, *15*, 439–453.

Clarke, A. J., and S. Van Gorder, (1986). A method for estimating wind-driven frictional, time-dependent, stratified shelf and slope water flow. *J. Phys. Oceanogr.*, *16*, 1013–1028.

Cochrane, J. D., (1969). Water and circulation on Campeche Bank in May. Papers in Dedication to Professor Michitaka Uda, November 1969, Tokyo, Japan. *In: Special Bulletin of the Japanese Society of Fisheries Oceanography,* 123–129.

Cochrane, J. D., (1972). Separation of an anticyclone and subsequent developments in the Loop Current (1969). In: *Contributions on the Physical Oceanography of the Gulf of Mexico,* Eds. L. R. A. Capurro and J. L. Reid, *Texas A & M University Oceanographic Studies, 2*, 91–106. Gulf Publishing Co., Houston.

Cochrane, J. D. and F. J. Kelly, (1986). Low-frequency circulation on the Texas-Louisiana continental shelf. *J. Geophys. Res.*, *91*, 10645–10659.

DeHann, C. J., and W. Sturges, (2005). Deep cyclonic circulation in the Gulf of Mexico. *J. Phys. Oceanogr.*, *33*, in press.

Deslarzes, K. and A. Lugo-Fernandez, (in press). Influence of terrigenous runoff on offshore coral reefs: An example from the Flower Garden Banks, Gulf of Mexico. In: *Geological Approaches to Coral Reef Ecology: Placing the Current Crises in Historical Context.* Ed. Richard B. Aronson.. Springer Verlag.

Dimarco, S. F., W. D. Nowlin, Jr., and R. O. Reid, (this volume). A statistical description of the near-surface velocity field from drifters in the Gulf of Mexico.

Dokken, Q. R., I. R. MacDonald, J. W. Tunnell Jr., T. Wade, K. Withers, S. J. Dilworth, T. W. Bates, C. R. Beaver, and C. M. Rigaud, (2003). Long-term monitoring at the East and West Flower Garden Banks National Marine Sanctuary, 1998–2001. OCS Study MMS 2003-031. New Orleans: US Department of the Interior, Minerals Management Service.

Elliot, B. A., (1979). *Anticyclonic rings and the energetics of the circulation of the Gulf of Mexico.* PhD Dissertation, Texas A & M Univ., College Station, Tex.,188 pp.

Elliot, B. A., (1982). Anticyclonic rings in the Gulf of Mexico. *J. Phys. Oceanogr.*, *12*, 1291–1309.

Emery, K. O., (1963). Coral Reefs off Veracruz, Mexico. *Geofisica Intl.*, *3*, 11–17.

Ezer, T., L-Y. Oey, W. Sturges, and H-C. Lee, (2003). The variability of currents in the Yucatan Channel: Analysis of results from a numerical ocean model. *J. Geophys. Res.*, *108 (C1)*, 3012, doi:10.1029/2002JC001509.

Fan, S., L-Y. Oey, and P. Hamilton, (2004). Assimilation of drifter and satellite data in a model of the northeastern Gulf of Mexico. *Cont. Shelf Res.*, *24*, 1001–1013.

Franceschini, G. A., (1953). The distribution of the mean monthly wind stress over the Gulf of Mexico. Scientific Report # 1, A & M College of Texas, 18 pp.

Fratantoni, D. M., (2001). North Atlantic surface circulation during the 1990's observed with satellite-tracked drifters. *J. Geophys. Res., 106*, 22,067–22,093.

Fratantoni, P. S., T. N. Lee, G. P. Podesta and F. Müller-Karger, (1998). The influence of loop current perturbations on the formation and evolution of tortugas eddies in the southern Straits of Florida. *J. Geophys. Res.*, *103*, 24,759–24,779.

Frolov, S. A., G. G. Sutyrin, G. D. Rowe and L. M. Rothstein, (2004). Loop Current eddy interaction with the western boundary in the Gulf of Mexico. *J. Phys. Oceanogr.*, *34*, 2223–2237.

Gilbert, P. S., T. N. Lee, and G. P. Podesta, (2004). Transport of anomalous low-salinity waters from the Mississippi River flood of 1993 to the Straits of Florida. *Cont. Shelf Res.*, *16*, 1065–1085.

Glenn, S. M., M. F. Crowley, D. B. Haidvogel, and Y. T. Song, (1996). Underwater Observatory captures coastal upwelling events off New Jersey. *Eos. Trans. Amer. Geophys. Union , 77*, 233–236.

Goni, G. J. and W. E. Johns, (2001). A census of North Brazil Current rings observed from T/P altimetry. *Geophys. Res. Lett.*, *28*, 1–4.

Gutiérrez de Velasco, G., S. Shull, P. J. Harvey and N. A. Bray, (1992). Gulf of Mexico experiment: meteorological, moored instrument and sea level observations. *Data report No. 1: August 1990 to July 1991.* SIO Reference Series 92–26, Scripps Institution of Oceanography, University of California, San Diego, CA.

Gutiérrez de Velasco, G., S. Shull, P. J. Harvey and N. A. Bray, (1993). Gulf of Mexico experiment: meteorological, moored instrument and sea level observations. *Data report no. 2: August 1990 to July 1991.* SIO Reference Series 93–31, Scripps Institution of Oceanography, University of California, San Diego, CA.

Hamilton, P., 1990. Deep currents in the Gulf of Mexico. *J. Phys. Oceanogr.*, *20*, 1087 – 1104.

Hamilton, P., (1992). Lower continental slope cyclonic eddies in the central Gulf of Mexico. *J. Geophys. Res.*, *97*, 2185–2200.

Hamilton, P., (2000). Deep currents over the northern Gulf of Mexico slope: observations and dynamics. Proceedings of a Workshop on the Physical Oceanography of the Slope and Rise of the Gulf of Mexico, September 12 – 14, 2000, sponsored by MMS.

Hamilton, P., and A. Lugo-Fernandez, (2001). Observations of high speed deep currents in the northern Gulf of Mexico. *Geophys. Res. Lett.*, *28*, 2867–2870.

Hamilton, P., and T. N. Lee, (this volume). Eddies and jets over the slope of the northeast Gulf of Mexico.

Hamilton, P., T. J. Berger, J. J. Singer, E. Wadell, J. H. Churchill, R. R. Leben, T. N. Lee, and W. Sturges, (2000). DeSoto Canyon Eddy Intrusion Study, Final Report Volume I: Executive Summary. OCS Study MMS 2000-079. U. S. Dept. of the Interior, Minerals Management Service, Gulf of Mexico OCS Region, New Orleans, La. 37 pp.

Hamilton, P., J. C. Larsen, K. D. Leaman, T. N. Lee, and E. Wadell, (2005). Transports through the Straits of Florida. *J. Phys. Oceanogr.*, *35*, 308–322.

Hetland, R. D., Y. Hsueh, R. R. Leben and P. P. Niiler, (1999): A loop current – induced jet along the edge of the west Florida shelf. *Geophys. Res. Lett.*, *26*, 2239 – 2242.

Hofmann, E. E. and S. J. Worley, (1986). An investigation of the circulation of the Gulf of Mexico. *J. Geophys. Res.*, *91*, 14,221–14,238.

Hu C., F. E. Müller-Karger, D. C. Biggs, K. L. Carder, B. Nababan, D. Nadeau, and J. Vanderbloemen, (2003). Comparison of ship and satellite bio-optical measurements on the continental margin of the northeast Gulf of Mexico. *Intl. J. Remote Sensing, 24*, 2597–2612.

Huh, O. K., W. J. Wiseman, Jr. and L. J. Rouse, Jr., (1981). Intrusion of Loop Current waters onto the west Florida continental shelf. *J. Geophys. Res.*, *86*, 4186–4192.

Hurlburt, H. E., (1986). Dynamic transfer of simulated altimeter data into subsurface information by a numerical ocean model. *J. Geophys. Res. 91*, 2372–2400.

Hurlburt, H. E. and J. D. Thompson, (1980). A numerical study of loop current intrusions and eddy shedding. *J. Phys. Oceanogr., 10*, 1611–1651.

Jarosz, E., and S. P. Murray, (this volume). Velocity and transport characteristics of the Louisiana-Texas Coastal Current.

Jarrett, B. D., A. C. Hine, R. B. Halley, D. F. Naar, S. D. Locker, A. C. Neumann, D. Twichell, C. Hu, B. T. Donahue, W. C. Jaap, D. Palandro, and K. Ciembronowicz, (2005). Strange Bedfellows—a deep-water hermatypic coral reef superimposed on a drowned barrier island; southern Pulley Ridge, SW Florida platform margin. *Mar. Geol., 214*, 295–307.

Johns, W. E., T. L. Townsend, D. M. Fratantoni and W. D. Wilson, (2002). On the Atlantic inflow to the Caribbean Sea. *Deep-Sea Res., 49*, 211–243.

Jordán-Dahlgren, E., (2002). Gorgonian distribution patterns in coral reef environments of the Gulf of Mexico: evidence of sporadic ecological connectivity. *Coral Reefs 21,* 205–215.

Jordán-Dahlgren, E., and R. Rodríguez-Martínez, (2003). *The Atlantic Coral Reefs of Mexico. In: Latin American Coral Reefs.* Ed. L. Cortés. Elsevier Scieence B. V.

Kantha, L., J-K. Choi, K. J. Schaudt, and C. K. Cooper, (this volume). A regional data-assimilative model for operational use in the Gulf of Mexico.

Kassler, R. D. and W. Sturges, (1981). Variability of the anticyclonic circulation in the western Gulf of Mexico. Progress Report, ONR Contract NOOO14-75-C-0201, Florida State Univ., Tallahassee, Fla.

Kelly, F. J., (1988a). Inner shelf circulation: northern shelf. OCS Study Report, MMS 88-0065, U. S. Dept. of the Interior, Minerals Management Service, New Orleans, La., 51–62.

Kelly, F. J., (1988b). Physical oceanography and meteorology. Chapter 4 in: OCS Study Report, MMS 88-0067, U. S. Dept. of the Interior, Minerals Management Service, New Orleans, La., 42–132.

Kennelly, M. A., R. H. Evans and T. M. Joyce, (1985). Small scale cyclones on the periphery of a Gulf Stream warm-core ring. *J. Geophys. Res., 90*, 8845–8857.

Leben, R. R., 2004. Real-Time Altimetry Project:http://www.ccar.colorado.edu/.

Leben, R. R., (this volume). Altimeter-derived Loop Current metrics.

Lee, H.-C., and G. L. Mellor, (2003). Numerical simulation of the Gulf Stream System: The Loop Current and the deep circulation. *J. Geophys. Res., 108(C2)*, 3043,doi: 10.1020/2001CJ001074.

Lee, T. N., and E. Williams, (1999). Mean distribution and seasonal variability of coastal currents and temperature in the Florida Keys with implications for larval recruitment. *Bull. Mar. Sci. 64*, 35–56.

Lee, T.N., C. Rooth, E. Williams, M. McGowan, A. F. Szmant, and M.E. Clarke. (1992). Influence of Florida Current, gyres, and wind-driven circulation on transport of larvae and recruitment in the Florida Keys coral reefs. *Conl. Shelf Res. 12*, 971–1002.

Lee, T. N., E. Williams, E. Johns, D. Wilson, and N. Smith, (2002). Transport processes linking south Florida coastal ecosystems. In: *The Everglades, Florida Bay and the Coral Reefs of the Florida Keys.* Chapter 11, CRC Press, pp. 309–342.

Lee, T. N., M. E. Clarke, E. Williams, A.F. Szmant, and T. Berger, (1994). Evolution of the Tortugas Gyre and its influence on recruitment in the Florida Keys. *Bull. Mar. Sci., 54,* 621–646.

Lessios, H. A., D. R. Robertson, and J. D. Cubit, (1984). Spread of Diadema mass mortality through the Caribbean. *Science, 226,* 335–337.

Lipp, E. K., J. L. Jarrel, D. W. Griffin, J. Lukaisk, J. Jaeukiewiez,, and J. B. Rose, (2002). Preliminary evidence for human fecal contamination in corals of the Florida Keys, USA. *Mar. Poll. Bull., 44,* 666–670.

Lugo-Fernandez, A., K. J. P. Deslarzes, J. M. Price, G. S. Boland, and M. V. Morin, (2001a). Inferring probable dispersal of Flower Garden Banks Coral Larvae (Gulf of Mexico) using observed and simulated drifter trajectories. *Cont. Shelf Res., 21*, 47–67.

Lugo-Fernández, A., M. V. Morin, C. C. Ebesmeyer, C. F. Marshall, (2001b). Gulf of Mexico historic (1955–1987) surface drifter data synthesis. *Journal Coastal Research., 17(1)*, 1–16.

Maul, G. A., (1977). The annual cycle of the Gulf Loop Current, 1: Observations during a one-year time series. *J. Mar. Res., 35*, 29–47.

Maul, G. A., (1978). The 1972–1973 cycle of the Gulf Loop Current, 2: Mass and salt balances of the basin. *CICAR-II Symposium on Progress in Marine Research in the Caribbean and Adjacent Regions. FAO Fisheries Rep., 200 (Suppl.),* 597–619.

Maul, G. A., D. A. Mayer, and S. R. Baig, (1985). Comparisons between a continuous 3-year current meter observation at the sill of the Yucatan Strait, satellite measurements of the Loop Current area, and regional sea-level. *J. Geophys Res., 90*, 9089–9096.

Merino, M, (1997). Upwelling on the Yucatan Shelf: hydrographic evidence. *J. Mar. Systems, 13*, 101–121.

Merrell, W. J., Jr. and A. M. Vázquez, (1983). Observations of changing mesoscale circulation patterns in the western Gulf of Mexico. *J. Geophys. Res., 88*, 7721–7723.

Mitchum, G. T., and A .J. Clarke, (1986a). The frictional nearshore response to forcing by synoptic scale winds. *J. Phys. Oceanogr., 16*, 934–946.

Mitchum, G. T., and A. J. Clarke, (1986b). Evaluation of frictional, wind forced long wave theory on the West Florida Shelf. *J. Phys. Oceanogr.,16*, 1029–1037.

Mizuta, G. and N. G. Hogg, (2004). Structure of the circulation induced by a shoaling topographic wave. *J. Phys. Oceanogr., 34*, 1793–1810.

Molinari, R. L. and J. F. Festa, (1978). Ocean thermal and velocity characteristics of the Gulf of Mexico relative to the placement of a moored OTEC plant, final report to the U. S. Department of Energy, DOE/NOAA Interagency Agreement no. BY-76-A-29-1041, 106 pp., NOAA/AOML, Miami, Fla.

Mooers, C. N. K. and G. A. Maul, (1998). Intra-Americas Sea Circulation. In: *The Sea*, Vol. 11. Eds. K. H. Brink and A. R. Robinson. John Wiley, NY, pp.183–208.

Morelock, J., and K. J. Koenig, (1967). Terrigeneous sedimentation in a shallow water coral reef environment. *J. Sedim. Petrol., 37*, 1001–1005.

Morey, S. L., J. J. O'Brien, W. W. Schroeder, and J. Zavala-Hidalgo, (2002). Seasonal variability of the export of river discharged freshwater in the northern Gulf of Mexico. *Oceans 2002 IEEE/MTS Conference Proceedings*, 1480–1484.

Morey, S. L., P. J. Martin, J. J. O'Brien, A. A. Wallcraft, and J. Zavala-Hidalgo, (2003). Export pathways for river discharged fresh water in the northern Gulf of Mexico. *J. Geophys Res., 108(C10)*, 3303,doi: 10.1029/2002JCOO1764.

Morey, S. L., J. Zavala-Hidalgo, and J. J. O'Brien, (this volume). The seasonal variability of continental shelf circulation in the northern and western Gulf of Mexico from a high-resolution numerical model.

Murphy, S. J., H. E. Hurlburt and J. J. O'Brien, (1999). The connectivity of eddy variability in the Caribbean Sea, The Gulf of Mexico and the Atlantic Ocean. *J. Geophys. Res., 104*, 1431–1453.

Nowlin, W. D. Jr., (1972). Winter circulation patterns and property distributions. In: *Contributions on the Physical Oceanography of the Gulf of Mexico, Texas A & M University Oceanographic Studies.* Eds. L. R. A. Capurro and J. L. Reid, *2*, 3–51.

Nowlin, W. D., Jr., D. F. Paskausky, and R. O. Reid, (1968). A detached eddy in the Gulf of Mexico. *J. Mar. Res., 26*, 185–186.

Nowlin, W. D. Jr., A. E. Jochens, R. O. Reid, and S. F. DiMarco, (1998). Texas-Louisiana Shelf Circulation and Transport Processes Study: Synthesis Report, Volume II: Appendices. OCS Study MMS 98-0035. U.S. Dept. of the Interior, Minerals Management Service, Gulf of Mexico OCS Region, New Orleans, LA. 288 pp.

Nowlin, W. D., Jr., A. E. Jochens, S. F. DiMarco, R. O. Reid, and M. K. Howard, (2001). Deepwater Physical Oceanography Reanalysis and Synthesis of Historical Data: Synthesis Report. OCS Study MMS 2001-064, U.S. Dept. of the Interior, Minerals Management Service, Gulf of Mexico OCS Region, New Orleans, LA. 530 pp.

Nowlin, W. D., Jr., A. E. Jochens, S. F. DiMarco, R. O. Reid, and M. K. Howard, (this volume). Low-frequency circulation over the Texas-Louisiana continental shelf.

Oey, L.-Y., (1995). Eddy and wind-forced shelf circulation. *J. Geophys. Res., 100*, 8621–8637.

Loop Current, Rings and Related Circulation in the Gulf of Mexico: A Review of Numerical Models and Future Challenges

Oey, L.-Y., T. Ezer, and H.-C. Lee[#]

Program in Atmospheric and Oceanic Sciences, Princeton University, Princeton, New Jersey

Progress in numerical models of the Loop Current, rings, and related circulation during the past three decades is critically reviewed with emphasis on physical phenomena and processes.

1. BACKGROUND: OBSERVATIONS

The science and art of modeling is deeply rooted in our desire to better describe and understand the world around us: it is imperative that we have some rudimentary knowledge of the system to be modeled. We begin therefore with a summary of observations of the most energetic components of the circulation in the Gulf of Mexico: the Loop Current and Loop Current rings (or eddies; Plate 1). These powerful oceanic features affect, either directly or indirectly through their smaller-scale subsidiaries, just about every aspect of oceanography of the Gulf. One could state that "a necessary condition for a basic model of the Gulf is that it replicates as accurately as possible the observable features of the Loop Current and rings."

The Gulf of Mexico is a semi-enclosed sea that connects in the east to the Atlantic Ocean through the Straits of Florida, and in the south to the Caribbean Sea through the Yucatan Channel (Figure 1). When discussing circulation features and model resolutions, we often find it useful to have some idea of the first-mode (baroclinic) Rossby radius, R_o. Calculation based on the GDEM climatology [*Teague et al.*, 1990] gives $R_o \approx 30$ km in the Gulf and the Cayman Sea (i.e. northwest Caribbean), $R_o \approx 40{\sim}50$ km in the central and eastern Caribbean Sea, and $R_o \approx 10{\sim}20$ km over the slope (c.f. *Chelton et al.* 1998). Though our focus will be on the Gulf, recent studies have indicated that the Gulf and the Caribbean Sea are dynamically inter-dependent; therefore certain aspects of the circulation in the Caribbean Sea will also be discussed. Currents through the Caribbean Sea, the Gulf of Mexico and the Florida Straits constitute an important component of the subtropical gyre circulation of the North Atlantic Ocean. This fact is vividly presented in *Fratantoni's* [2001] decadal, quasi-Eulerian mean drifter analysis. The author's Plate 6, for example, shows intense speeds in the Caribbean Current (> 0.5 m s^{-1}) in the southern portion of the Caribbean Sea, the Loop Current and Florida Current (both > 1 m s^{-1}).

The Loop Current is the dominant feature of the circulation in the eastern Gulf of Mexico and the formation region of the Florida Current-Gulf Stream system. It originates at the Yucatan Channel (where it is called the Yucatan Current) through which approximately 23~27 Sv (1 Sv = 10^6 m^3 s^{-1}) transport passes with a large min-max range of 14~36 Sv [*Johns et al.* 2002; *Sheinbaum et al.*, 2002; *Candela et al.*, 2003]. The Yucatan/Loop Current is a western boundary current with peak speeds of 1.5 to 1.8 m s^{-1} on the western side of the channel near the surface (*Pillsbury*, [1887] based on direct current measurements, plotted in Figure 5 of *Gordon* [1967]; *Nowlin* [1972] based on GEK; *Schlitz* [1973] and *Carder et al.* [1977] based on hydrography; and *Ochoa et al.* [2001] and *Sheinbaum et al.* [2002] based on ADCPs and hydrography). In the Loop (inside the Gulf) intense speeds reach 1.7 m s^{-1} in in situ measurements [*Forristal et al.*, 1992]. The Loop Current also displays a wide range of vacillations, both in north-south and east-west directions. The Loop episodically sheds warm-core rings [e.g., *Cochrane*, 1972; *Vukovich*, 1995] at intervals of approximately 3 to 17

#GFDL, NOAA, Princeton, NJ 08540

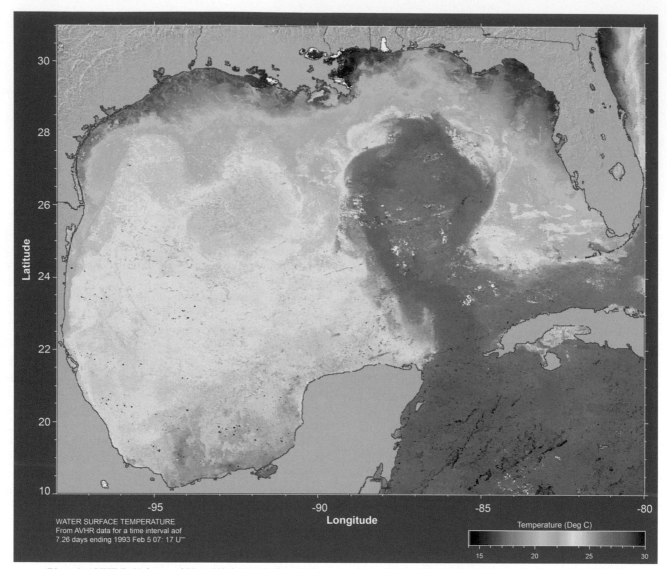

Plate 1. AVHRR (Advanced Very High Resolution Radiometry; http://fermi.jhuapl. edu/avhrr/gm/averages/index.html) seven-day composite sea-surface temperature (SST) for day ending on Feb/05/1998, showing the Loop Current in an extended position into the Gulf, and an old ring further west. Note the appearance of a series of frontal eddies along the outer edges of the Loop Current and the old ring, a cyclone over the east Campeche Bank slope just north of the Yucatan Channel (at ≈ 22.5°N, 87°W this cyclone was analyzed in detail by *Zavala-Hidalgo et al.* 2003a), and a Tortugas eddy (cyclone) at ≈ 24.5°N, 84-85°W. Approximately two months later the Campeche Bank and Tortugas cyclones appeared to cleave the Loop and a ring was shed. Note also cooler shelf waters and even smaller-scale eddies along the shelf-edge.

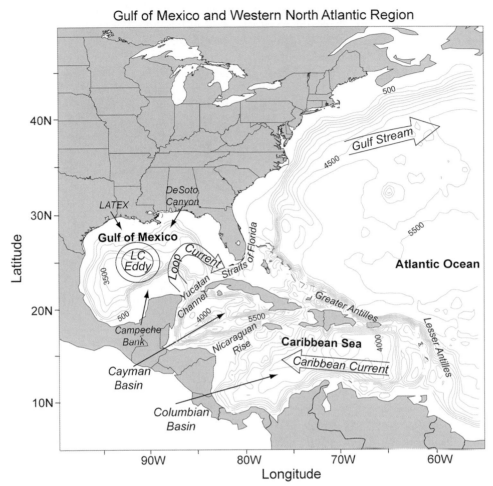

Figure 1. A map for the Gulf of Mexico and the Caribbean Sea with schematic cartoons showing the Caribbean and Loop Currents, Loop Current eddy and the Gulf Stream. The isobaths shown are in meters. This is also the region used by the Princeton models reviewed in the text (e.g. Oey and Lee, 2002).

months [*Sturges and Leben*, 2000; *Leben*, this volume].[1] These rings have diameters ≈ 200~300 km, vertical extent ≈ 1000 m, swirl speeds ≈ 1.8~2 m s^{-1}; they generally translate westward at 2~5 km day^{-1} and have lifetimes of months to approximately a year [*Nowlin*, 1972; *Elliott*, 1982; *Vukovich and Crissman*, 1986; *Cooper et al.* 1990; *Forristal et al.*, 1992]. Websites A.3 and A.4 in Appendix 1 provide historical and real-time satellite-derived sea-surface height (SSH and other information; please see also the accompanying CD of this book); the reader may consult them for an appreciation of the rich variability of the eddy fields, both inside the Gulf (website A.4) and also in the Caribbean Sea and the Atlantic Ocean (website A.3).[2]

Satellite SSH maps generally resolve scales ≈ 100 km and larger (though along-track resolution is good at a particular instant, and a composite of several altimeter maps can generally improve resolution [*Ducet et al.*, 2000]). Smaller-scale frontal eddies (or meanders), filamentary structures, and cyclones are often observed around the edges of the Loop Current and rings [*Leipper*, 1970; *Maul*, 1977; *Vukovich et al.*, 1979; *Huh et al.*, 1981; *Merrell and Morrison*, 1981; *Brooks and Legeckis*, 1982; *Paluszkiewicz et al.*, 1983; *Vukovich and Maul*, 1985; *Vukovich*, 1986; *Hamilton*, 1992; *Fratantoni et al.*, 1998; *Hamilton et al.*, 2003; *Zavala-Hidalgo et al.*, 2003a], as may be seen in satellite AVHRR (Advanced Very High Resolution Radiometry) sea-surface temperature (SST)

---

[1]*Leben* (this volume) documented two more recent eddies—eddy Juggernaut (October 1999) and Millenium (April 2001) that have longer shedding period ≈ 18.5 months.

[2]Appendix 1 contains many useful web-sites, and appendix 2 collects many commonly-used acronyms.

maps (e.g., website A.1; a good one is the seven-day composite on 5 February 1998, shown in Plate 1). These fine-scale features can dominate local circulation, especially in regions removed from direct impacts of the Loop Current and rings (e.g., over the northeastern slope of the Gulf; *Hamilton et al.* [2000]). Frontal eddies sometimes also appear to originate west of the Loop, propagating clockwise around its northern and eastern edges, and to develop into deeply-penetrating cyclones somewhere along the west Florida continental slope [*Vukovich and Maul*, 1985]. SST frontal analysis [e.g., *Vukovich et al.*, 1979] indicates amplification time scales of only a few days, as well as backward-wave breaking features. The scales are 50~150 km in diameters and ~ 1000 m deep. In the southern Straits of Florida, the cyclones are called Tortugas eddies. *Zavala-Hidalgo et al.* [2003a] documented cyclones that developed northeast of the Campeche Bank (23.5°N, 86.5°W). Both the Tortugas eddies and Campeche Bank cyclones appear to be intimately related to the shedding of Loop Current rings [*Cochrane*, 1972; *Fratantoni et al.*, 1998, who also present an excellent review; *Zavala-Hidalgo et al.*, 2003a; *Schmitz*, this volume].

The Gulf has a central deep basin (depth ≈ 3500 m) of relatively limited extent (about 90°-94°W and 23°-25°N, Figure 1) surrounded by continental rise, slope and shelves. The Loop Current and rings therefore readily interact with topography; a ring can split and a new ring can interact with an old one. *Molinari and Morrison* [1988] analyzed data that show effects of the Campeche Bank on Yucatan Current separation and Loop Current penetration. The development of Tortugas eddies is in part a result of the Loop Current being constrained by the west Florida slope [*Fratantoni et al.*, 1998]. *Huh et al.* [1981] give early examples of Loop Current intrusion into the DeSoto Canyon and Alabama/Mississippi shelves; more recent work is *Hamilton et al.* [2000]. *Weisberg and He* [2003] and *Fan et al.* [2004] found evidence of west Florida shelf currents being forced by Loop Current "rubbing" against the slope of the south Florida Straits. In the western Gulf, *Vidal et al.* [1992] describes the collision of a ring with the western slope, during which the ring is split into an anticyclone to the north and a (smaller) cyclone to the south. *Biggs et al.* [1996] observed an example of Loop Current ring cleaved by a cyclone. *Hamilton et al.* [2002] found eddies (of both signs) with diameters 40~150 km over the central Gulf slope. High-resolution AXBT and AXCP also allowed these authors to examine in detail the vertical structures, which showed uplifted (50~100 m) iso-

therms in the deep (~ 1000 m) in the center of the cyclones capped by slight depressions in the upper 200 m, and subsurface maximum velocities with speeds ≈ 0.4 m s⁻¹. These Loop Current/eddy interactions with topography strongly affect how energy of the currents is distributed in the water column, and how deep energy propagates along and across isobaths in the form of topographic Rossby waves [*Hamilton*, 1990; *Hamilton and Lugo-Fernandez*, 2001].

In summary, the broad range of spatial and temporal scales of the circulation in the Gulf of Mexico presents an enormous challenge for modelers. In the past two to three decades, we have witnessed the development of more and more powerful computers and sophisticated ocean models and analysis techniques. Fueled by these advances and by equally impressive progress on observational data acquisition (from satellite, ADCP, sub-surface floats, surface drifters, etc.; for examples, please see the chapters in this book), the amount of modeling work, presented in papers, reports, and websites, is staggering. It is clearly not possible to review all here. We will restrict our review primarily to models of the Loop Current and rings, which means a model domain that encompasses at least the whole Gulf (with one or two exceptions). In terms of the physical processes we will limit ourselves to subinertial time scales of days and longer (thus excluding tides, internal tides, wind and hurricane-induced high-frequency motions, etc., again with a few exceptions). Perhaps a most glaring omission is near-inertial motions for which the Gulf is well known. These anticyclonic circular motions near the ocean's surface are prevalent over the shelf and shelfbreak and may be forced by near-diurnal variation of the wind (*Chen et al.* 1996; *DiMarco et al.* 2000; *Simpson et al.* 2002).³ Discussions of hurricane-induced near-inertial motions (over deep waters) are given by *Shay et al.* (1998), who also observed significant (short-term) storm effects on an underlying Loop Current eddy. At the subinertial time scales, notably absent from our review is work that describes the (sub-tidal) wind-, buoyancy- (river) and eddy-driven circulation of the continental shelves around the Gulf.⁴ Shelf circulation deserves an entire chapter to itself; the following briefly summarizes some of the more recent papers.

Boicourt et al. [1998] gave a comprehensive review of the shelf circulation in the Gulf. *Oey* [1995] modeled the wind- and eddy-forced shelf circulation over the Louisiana-Texas (LATEX) shelf, and found evidence for eddy-induced shelf edge currents. *Hetland et al.* [1999] suggested that the impact of Loop Current could spread along the shelf edge,

---

³Interestingly, buoy wind data also show large inertial range oscillations near 30° (*Stockwell et al.* 2004).
⁴Winds over the Gulf are generally westward, with a predominant anticyclonic curl except for a weak cyclonic curl in a narrow (width ~ 250 km) strip along the western coast and a strong cyclonic curl over the Bay of Campeche in the southwestern Gulf [*Gutierrez de Velasco and Winant*, 1996]. Winds are clearly important over the shelves, but may also affect the large-scale circulation [*Sturges*, 1993].

thereby inducing a southward jet along the edge of the west Florida shelf. Weisberg's group at USF has described both observations and models of the west Florida shelf [*Weisberg and He*, 2003; *Weisberg et al.*, this volume]. *Muller-Karger* [2000] found satellite evidence of northeast Gulf of Mexico (NEGOM) shelf currents driven by both wind and a neighboring Loop Current ring. *Ohlmann et al.* [2001] used drifters, and pointed out the importance of eddies in forcing the mean flow and in effecting cross-shelf exchanges of water masses. The Texas A&M group has recently published a number of important papers on the circulation of the LATEX shelf [*Li et al.*, 1997; *Cho et al.*, 1998; *Nowlin et al.*, this volume]. *Zavala-Hidalgo et al.* [2003b] modeled the western Gulf shelves (west of 89°W: LATEX and Mexican shelves) as a contiguous system, carefully computed the shelf transports, found regions of water confluence, and were able to explain transport interrelationship between the different shelves (also *Morey et al.*, this volume). *Wang et al.* [2003] compared satellite SSH-assimilated model (POM) currents with observations in the DeSoto Canyon (87°W, 29.5°N) and used the Single Value Decomposition analysis [*Bretherton et al.*, 1992] to study influences on outer-shelf and canyon currents by Loop Current frontal eddies that traveled around the Loop Current. *Smith and Jacobs* [2005] combined current meter, ADCP, and drifter data in a weighted least square assimilation algorithm to infer the seasonal circulation on both the LATEX and NEGOM shelves.

There are a number of comprehensive reviews of numerical ocean models [*Greatbatch and Mellor*, 1999; *Griffies et al.*, 2000]. Our focus on a specific region (the Gulf of Mexico) allows an approach that targets physical phenomena and processes. We hope that this review will therefore also be useful to observationalists.

2. LOOP CURRENT, EDDY SHEDDING, EDDIES & RELATED CURRENTS

The first compilation of research work in the Gulf is a book edited by *Capurro and J.L. Reid* [1972]. Amongst many fine papers (e.g., *Nowlin's* descriptions of the Gulf's water masses, *Cochrane's* observations of the Loop Current being cleaved by a cyclone, *R.O. Reid's* elegant formula of the Loop Current's extension, and *Ichiye's* laboratory model of the Gulf and the Caribbean), two papers (by *Paskausky and Reid*, and *Wert and Reid*) described numerical simulations of the Loop Current and eddy-shedding process.[5] The authors utilized barotropic and quasi-geostrophic 2-layer models; they were ahead of their time, considering the scarcity of observations and computing resources. Although

Loop Current-like features and eddies were simulated, the experiments were necessarily limited in scope in terms of the parameter space and length of integration they covered. It was not until nearly a decade later that *Hurlburt and Thompson* [1980] (henceforth *HT*) developed and extensively tested the first prototype model of the Loop Current and eddy shedding. *HT's* work is an important yardstick against which many later model experiments and analyses, even those of today, should be measured.

2.1 Hurlburt and Thompson's Model

HT experimented with 1-layer barotropic, 1.5-layer reduced-gravity (RG), and 2-layer models of an idealized rectangular Gulf of Mexico basin with inflow (Yucatan Channel; = 20~30 Sv in most experiments) and outflow (Florida Straits) ports. The models are based on the nonlinear primitive equations and resolution is 20 km × 18.75 km. The authors emphasized the importance of integrating the models to a statistical equilibrium so that many eddy-shedding cycles are covered. They introduced the idea of designing first as realistic an experiment as possible (in their case the 2-layer model with topography), then working down to look for the simplest model which could reproduce (essentially) the same results. Both the 2-layer and the 1.5 layer model with a steady inflow (and other "reasonable" parameters) produced eddies with diameters 200~400 km and shedding periods 250~360 days, while using the same parameters the barotropic flat-bottom case evolved into a steady state. The following summarizes *HT's* findings:
1. 1.5-layer model is the simplest relevant model of Loop Current and eddy shedding;
2. Shedding is caused by horizontal shear instability of the internal mode;
3. Shedding occurs despite steady inflow specified at Yucatan Channel;
4. Planetary β-effect is essential in the Loop Current's penetration into the Gulf (first pointed out by *Reid* [1972]) and also in eddy-shedding (through westward spreading of the Loop Current and removal of eddies); the penetration time scales dictate shedding time scales;
5. The *f*-plane solution is a steady source-sink flow (no shedding);
6. Nonlinearity is necessary for shedding – the linear solution (when inflow is weak, HT used *0.1 Sv*) is also a steady source-sink flow;
7. In the 2-layer model, loss of energy due to baroclinic instability to the lower layer results in weaker and

---

[5]The book is a "must-read" for any Gulf aficionado, for its historical value and scientific content.

smaller eddies and in shorter shedding periods (e.g. from *12 months* for 1.5-layer model to *8 months* for 2-layer with topography, and *5.4 months* for 2-layer flat-bottom)

8. Eddy-shedding period is dominated by the natural period ($\approx$ 12 months with steady inflow), though there is also some dependency on time-varying inflow;

9. Eddy-shedding period increases with Reynolds number (please see below); irregular shedding (10~14 months) can occur at a sufficiently high Reynolds number and

10. In the 2-layer model with topography, sufficiently strong (10 Sv) bottom inflow from Yucatan Channel traces a clockwise circulation around the Gulf following the f/H contours; the resulting divergence over the west Florida slope can prevent upper-layer deepening, Loop Current's westward spreading and eddy-shedding. Shedding resumes when the bottom deep inflow weakens.

From the large number of experiments they performed with the 1.5-layer and flat-bottom barotropic models, HT proposed a "Regime Diagram" (their Figure 18) that shows:

E: Eddy-shedding regime for Re > Re_c and $R_B < R_{Bc}$;
W: Steady westward-spreading regime for Re < Re_c and $R_B < R_{Bc}$; and,
N: Steady source-sink (i.e.,port-to-port) regime for $R_B > R_{Bc}$, arbitrary Re,

where Re = $v_{in}L_h/A$ is the Reynold's number based on the maximum inflow speed v_{in}, inflow port half-width L_h and the horizontal eddy viscosity (kinematic) A used in the model, and $R_B = v_{in}/(\beta L_p^2)$ is the beta Rossby number based on v_{in}, β, and the half distance, L_p, between the centers of the inflow (Yucatan Channel) and outflow (Florida Straits) ports. The critical Reynold's and beta Rossby numbers, Re_c and R_{Bc}, are approximately *25* and *2* respectively for the 1.5-layer model. The R_{Bc} condition is particularly interesting. That the Loop Current would "short-circuit" from Yucatan-port to Florida-port (i.e., without penetrating into the Gulf and making a loop) when the two ports are close to each other (i.e., L_p is small) is intuitively clear. However, that this short-circuiting should occur when v_{in} is strong is counter-intuitive (c.f. *HT*'s experiments RG40 and RG43, listed in their Table

2). In this case, the tendency for the Loop to spread or bend westward, due to $-\beta v$, is balanced by northward advection of the relative vorticity, $\approx v\partial\varsigma/\partial y$, and streamlines tend to curve eastward.[6] Since in reality (and in all general circulation models or GCMs since *HT*!) β and L_p are fixed so that $R_B \propto v_{in}$, does this mean that (in GCMs and/or the real ocean) a port-to-port mode can occur if surface inflow from the Caribbean increases or if deep inflow increases (*HT*'s finding 10 above), or both? We will see that some GCM runs display this port-to-port mode.

Pichevin and Nof's Analysis. The 1.5-layer model is in general still a very complicated system to solve analytically, but it is much more manageable than the multi-layer or three-dimensional GCMs. *HT*'s finding 1 is therefore significant in that an analytical treatment of the 1.5 layer model may provide valuable insights into the dynamics of Loop Current and eddy-shedding. The paper by *Pichevin and Nof* [1997] (henceforth *PN*; please see also *Nof and Pichevin* [2001] and *Nof et al.* [2004]) is an important contribution in this regard. *PN* analyze the consequences of a northward narrow outflow (i.e., width ~ Rossby radius or less situated next to a southern boundary)[7] debouching into an open ocean. By integrating the *x*-momentum equation over a rectangular domain just outside the outflow, they show that the integrated momentum exerted on the domain by water exiting the rectangle on the right cannot be balanced in a steady state. This "momentum imbalance paradox," as the authors called it, is resolved if either (time-dependent) eddies are allowed to shed to the left ($\beta \neq 0$) or the outflow grows forever (f-plane; *PN*'s Figure 6). The β-effect is again a "must" for eddy-shedding. Thus shedding (or growing bulge) in *PN* is a necessary consequence of the flow's inability to keep the longshore momentum in a steady-state balance. It is important, however, to remember the "narrow outflow" condition implicit in the PN analysis, in which the line integral across the outflow, $\int uvhdx$, is zero, since u = 0 there (h is the upper-layer depth in their 1.5-layer model). Note that this result does not contradict *HT*'s finding 2. On the other hand, *HT*'s outflow does not grow indefinitely when f = constant (finding 5; *HT*'s Figure 6). This apparent (but fundamental) discrepancy is puzzling; it may be due to the additional constraint that exists in *HT*'s model: that the port at Florida fixes the magnitude and location of the outflow, i.e., due to the existence of the additional length scale L_p in *HT*.[8]

---

[6]The assumed steady-state is crucial in *HT*'s argument, which is equivalent to assuming that $(f+\varsigma)$ = constant on geostrophic streamlines. However, see later comparison of *HT*'s and *Pichevin and Nof's* (1997) work.

[7]Note that the 'narrow-outflow' condition is, strictly speaking, not satisfied in the case of the Yucatan Channel, where $R_o \approx 30$ km but the channel's width is $\approx$ 120~150 km (Figure 1).

[8]Viscosity (A, in $m^2 s^{-1}$) may also play a role. However, *HT*'s A = 10^3 $m^2 s^{-1}$ and *PN*'s A = 300 $m^2 s^{-1}$. This difference may be too small to account for the drastic change in the shedding/nonshedding solution.

In summary, *HT*'s 1.5-layer model may be simple, yet it captures remarkably well the gross characteristics of Loop Current variability: its extension (growth), shedding of an eddy, and retraction. With reasonable model parameters, the periods are also about right, approximately 12 months. The model also helped to clarify the longstanding misconception that the Loop Current sheds eddies in response to quasi-annual variation in inflow through the Yucatan Channel. On the other hand, the real ocean works in a curiously complex manner: the observed shedding periods cover a wide range (approximately 3 to 18.5 months); the Loop Current can extend and retract (scales ~ months and O (100 km)) without necessarily shedding an eddy; an eddy can temporarily detach then reattach to the Loop Current (time scales ~ weeks); frontal meanders, eddies and cyclones can develop and can influence eddy shedding; Yucatan shelf/slope, deep-layer and Caribbean influences may be significant, etc. Later models are developed to capture some of these complications.

Since *HT*'s work, 1.5-layer and 2-layer models have been used by a number of authors. *Hurlburt and Thompson* [1982] extended their own work to show that Loop Current and eddies force deep modon (anticyclone-cyclone pair with anticyclone leading). The modon in turn can affect the behaviors of the surface eddies. *Heburn et al.* [1982] studied the instability of the Caribbean Current. *Smith and O'Brien* [1983] examined eddy interactions with topography. *Wallcraft* [1986] (please see brief descriptions in *Lewis and Kirwan* [1987]) extended *HT*'s calculations to high resolutions to simulate small-scale eddies in the Gulf. *Arango and Reid* [1991] used a generalized 1.5-layer model in isopycnal coordinates to study the shedding process and cyclonic eddies. [9]

2.2 General Circulation Models with More Realistic Gulf of Mexico Topography

With the advent of vector-processing machines in the 1980s and early 1990s (CDC Cyber 205, Cray XMP, YMP, C90, etc.), and now of powerful workstations and parallel computers, long-term integrations (to ensure statistical equilibrium) using multi-level and multi-layer primitive equations have become routine.

2.2.1 Loop Current and Eddies in the Early GCMs. The first successful GCM computation of the Loop Current and rings was by *Sturges et al.* [1993], who used MOM [*Bryan*, 1963; *Bryan and Cox*, 1967; *Cox*, 1985; *Semtner and Chervin*,

1992] at a one-fourth degree horizontal resolution and 12 vertical *z*-levels to examine the characteristics of eddy-shedding. A significant departure from HT is that the model domain (8°-36°N, 97°-49°W) includes the Caribbean Sea as well as a portion of the Atlantic Ocean, so that flows in and out of the Gulf through the Yucatan Channel and the Straits of Florida are "free" (i.e., the flow fields there are a solution of the model). The northern boundary (36°N) is a wall, but the eastern boundary (55°-49°W) is a buffer zone ("pumps and baffles") through which climatological data (including the wind curl-driven integrated transport from east of 49°W) are specified. The modeled flow therefore recirculates. Steady winds were also specified over the modeled domain. These specifications result in 19 Sv transport into the Gulf through the Yucatan Channel. The horizontal viscosity is 500 m^2 s^{-1}, diffusivity is 300 m^2 s^{-1}, and vertical mixing is 10^{-4} m^2 s^{-1}. The model was spun up for over 10 years to statistical equilibrium. The model sheds eddies regularly at a period of about 180 days. This is within the observed range, is consistent with *HT*'s finding 7, and suggests a tendency for energy loss to lower layers in a multilevel model. The interesting result (*Sturges et al.*'s [1993] Figure 7) is that while an eddy is separating, the modeled Loop Current shows recirculating flow inside and even return flow (into the Caribbean) on the Cuban side of the Yucatan Channel (albeit much weaker than the northward jet on the Mexican side). *Sturges et al.* noted that observations [*Nowlin*, 1972; *Molinari*, 1977; *Lewis and Kirwan*, 1987] show similar recirculation in the Loop Current. The ring's diameter is about 250 km. The ring translates westward at about 4 km day^{-1} almost on a straight path, and decays at the northwest corner of the Gulf. Peak speeds of about 0.6 m s^{-1} at 130 m occur in the Loop Current; these speeds decay rapidly in the ring as it moves westward, and drop below 0.2 m s^{-1} at about 92°W in the central Gulf. *Sturges et al.* also described deep eddies, anticyclone (leading) and cyclone pair that follows the surface ring, very similar to the modon found by *Hurlburt and Thompson* [1982]. These deep eddies are clearly forced by the ring and show remarkably regular periodicity (generation, propagation and decay) phased-locked with the surface ring. One gets the impression that, at one-fourth degree resolution and 12 z-levels, the solution may be too viscous and/or diffusive.

Dietrich and Lin [1994] used a much reduced eddy viscosity $A \approx 1 \sim 10\ m^2\ s^{-1}$ in their Sandia Ocean Modeling System (SOMS), a rigid-lid z-level primitive-equation model on Arakawa C-grid, but with a fourth-order treatment of the Coriolis term [*Dietrich*, 1993]. The vertical mixing is 10^{-4}

---

[9]Layer models are now routinely run in multi-layer mode, and are applied not only in the Gulf and the Caribbean Sea (*Romanou et al.* 2004), but also in other semi-enclosed seas (e.g. *Hurlburt et al.* 1996; *Hurlburt and Hogan*, 2000). The U.S. Navy now routinely runs a six-layer global ocean model to help with their forecasting needs (NLOM). See Appendix 1 section D for various model acronyms.

m^2 s^{-1}, the same as that used by *Sturges et al.* [1993]. The horizontal resolution is 20 km and there are 16 z-levels. The model includes the northwestern portion of the Caribbean Sea (i.e., the Cayman Sea) and thus flow through the Yucatan Channel is also "free." An inflow transport of *30 Sv* is specified across the southeastern open boundary of the model domain in the Caribbean Sea. An outflow is specified at the Straits of Florida. Integration was carried out for four years and results from years 2-4 were shown. The modeled eddies are about 250 km in diameter, propagate westward at about 3.5 km day^{-1} and have peak swirl speeds at the first model level (z = −22 m) of 0.6~0.7 m s^{-1}. The model appears to be less diffusive than *Sturges et al.'s* model, but this is difficult to judge from the few snapshots that show similar rapid eddy decays west of about 92°W. The plots also show a shedding period of about 180 days. When comparing with *Sturges et al.'s* [1993] results, the period appears to be remarkably insensitive to orders-of-magnitude change in eddy viscosity, contrary to *HT's* prediction (their finding 9). This may imply the dominance of the baroclinic transfer of energy to the deep, though the short spin-up time in *Dietrich and Lin's* model may also be a factor. *Dietrich et al.* [1997] subsequently improved the grid resolution to one-twelfth degree and 20 z-levels using the Arakawa A-grid DieCast model. Although the simulation is still too short (≈ 4 years), the modeled eddies appear to be less dispersive as they traverse westward. The higher resolution also appears to better resolve smaller-scale frontal eddy features at the edges of the Loop Current and rings.

Oey's [1996] terrain-following ocean model (POM) of the Loop Current, rings and their influences on the wind and (river-borne) buoyancy-driven LATEX shelf circulation [*Oey, 1995*] has a horizontal resolution of 20 km×20 km and twenty equally-spaced sigma levels cells in the vertical.[10] The model domain includes a portion of the northwestern Caribbean Sea (the Cayman Sea, as in *Dietrich and Lin,* [1994]) and also the Straits of Florida, so that both inflow (Yucatan Channel) and outflow (Straits of Florida) are "free." An inflow transport of 30 Sv is specified across the southeastern open boundary of the model domain in the Caribbean Sea. Outflow at the northeastern boundary off Florida consists of a combination of transport and radiation conditions [*Oey and Chen,* 1992]. The model uses *Smagorinsky's* [1963] shear-dependent formula

for the horizontal viscosity and diffusivity, with the constant C_{smag}. The *Mellor and Yamada* [1982] level-2.5 turbulence scheme is used for the vertical eddy viscosity and diffusivity. Six runs (each ≈ 10 years or more) were conducted. Four had C_{smag} varied: 0.03, 0.05, 0.075 and 0.10, and two tested the sensitivity of the inflow specifications (not sensitive). These experiments were found to yield irregular eddy shedding, with typical periods 5~15 months (e.g., *Oey's* [1996] Figure 6). These (periods) should be compared with the 10~14 months obtained from HT's experiment RG8 (HT's Table 2; finding 9) at reduced viscosity (A = 300 m^2 s^{-1}), as well as with observed periods 3~17 months [*Elliot,* 1982; *Sturges,* 1993; *Sturges and Leben,* 2000]. The *Smagorinsky's* formulation typically yielded A ≈ 50~400 m^2 s^{-1}. Since the specified inflow was steady, *Oey* [1996] suggested that the shedding irregularity might be due to the generally reduced, and time- and spatially-dependent values of *A*. Other characteristics of *Oey's* [1996] modeled eddies, such as: diameters ≈ 200~400 km, westward-propagation speeds ≈ 3~5 km day^{-1}, and peak speeds ≈ 0.6~0.7 m s^{-1}, are similar to those of *Sturges et al.* [1993] and *Dietrich and Lin* [1994]. *Oey* [1996] tracked his eddies only up to 95°W prior to their interactions with the western Gulf slope, yielding an "eddy-life" ≈ 6 months. *Oey's* [1996] modeled swirl speeds of eddies in the central Gulf (≈ 92°W) typically are ≈ 0.5 m s^{-1} at z = −150 m (his Figure 18), compared with the 0.2 m s^{-1} at z = −130 m of *Sturges et al.* [1993] (their Figure 8) and the 0.2~0.4 m s^{-1} at z = −22 m of *Dietrich and Lin* [1994] (their Figures 7, 8 and 9)[11]. As mentioned earlier, the eddies in *Sturges et al.* and *Dietrich and Lin* tend to weaken considerably west of 92°W. *Oey's* [1996] eddies are more similar to *Dietrich et al.'s* [1997]: they survive past 92°W and interact more strongly with the western Gulf slope. Although *Oey's* eddies appear to be energetic, the modeled eddies and Loop Current are weaker than observed, by as much as 100%!

2.2.2 Yucatan (& Florida Straits) Flow Variability and Topographic Rossby Waves. Four other results in *Oey* [1996] seem relevant in light of more recent findings. First, southward deep flow (z < −750 m; transport ≈ *1~7 Sv*)[12] occurred in the Yucatan Channel each time the model Loop Current extended and shed an eddy; the surface and deep transports are anticorrelated (so that the total inflow is fixed = 30 Sv).

---

[10]*Blumberg and Mellor* [1985] used an earlier version of POM to simulate the Gulf's circulation. The coarse grid (1/2° horizontal resolution and 15 vertical sigma-layers) and excessive viscosity and diffusivity (4000 and 2000 m^2 s^{-1} respectively) probably prevented the modeled Loop Current from shedding rings. The simulation was for one year: the solution was probably dominated by the initial observed climatology.

[11]*Dietrich and Lin's* [1994] did not include vector scales on their plots.

[12]The usual notation is used such that *x* and *y* are west-east and south-north coordinate axes respectively, and *z* is the vertical coordinate with z = 0 at the mean sea-level.

The southward flows lasted weeks ~ months, appeared to precede sheddings, and at times surfaced on the Cuban side of the Yucatan Channel. *Oey* [1996] noted the potentially interesting relevance of his finding to HT's finding 10: that increased (decreased or even reversed) bottom inflow inhibits (promotes) shedding. He also noted that *Maul et al.* [1985] observed similar southward (deep) events and the apparent relation of these events to eddy shedding. More recent data and analyses [*Bunge et al.* 2002; *Ezer et al.*, 2003] support *Maul et al.'s* and *Oey's* [1996] findings. Second, *Oey* noted that these transport fluctuations occurred at shorter periods (weeks ~ months) than eddy-shedding, i.e., not every Loop Current extension (and deep southward flow) resulted in shedding. That the Loop Current has shorter-period fluctuations unrelated to shedding is in contrast to the behaviors of reduced-gravity models, for which a Loop Current extension generally results in shedding (*HT* and *PN*). Third, unlike the Yucatan Channel deep transport, fluctuations in the Straits of Florida show no clear correlation with eddy shedding. Instead, Straits of Florida transports are of even shorter periods (weeks) and energetic near the bottom also. Fourth, flow fluctuations in the western Gulf were found to be correlated with Loop Current variability including eddy-shedding. In particular, for disturbances in the 30~100 days' periods, current fluctuations were bottom-intensified (*Oey's* [1996] Figure 20) and the east-to-west propagation speeds were found to be $\approx$ 12~13 km day^{-1}. *Oey* [1996] cited *Hamilton* [1990] who attributed the (observed) fluctuations to topographic Rossby waves (TRWs). More recent detailed analyses by *Oey and Lee* [2002] and *Hamilton* [2005; submitted to *J. Phys. Oceanogr.*] corroborate *Hamilton's* and *Oey's* results.

2.2.3 Non-Shedding Scenarios: An Interesting Difficulty. The models of *Sturges et al.* [1993], *Dietrich and Lin* [1994], *Dietrich et al.* [1997] and *Oey* [1996] represent early expansions of *HT's* work to multi-level with realistic Gulf topography. These multilevel models have a lot in common (all based on the primitive equations, with inflow/outflow and climatological T/S fields, viscosity, etc.), which may explain why the gross behaviors of the modeled Loop Current and eddy-shedding, in terms of the spatial and temporal scales, eddy sizes and propagation paths and speeds, etc. are very similar. With the exception, perhaps, of *Dietrich et al.'s* [1997] higher-resolution model showing maximum speeds $\approx$ 2 m s^{-1}, all other models' speeds are too weak by as much as 100% or more in comparison to observations. But they all managed

to shed eddies. This may sound trivial until one is reminded of how complex the real-ocean shedding behaviors are, and how incompletely we still actually understand the models' behaviors. It is fitting to close this section on "early GCM's" to mention two recent simulations that do not shed eddies. In a well-designed model of the North Atlantic Ocean (the POP model at 0.1°×0.1° resolution and 40 vertical levels), including the Gulf of Mexico and the Caribbean Sea, *Smith et al.* [2000] found sporadic eddy shedding (roughly once a year) for the first 9 years of integration, as in previous models. For the subsequent 7 years, however, a stationary (nonshedding) Loop Current developed (*Smith et al.* referred to a similar stationary loop occurring in some of the Community Modeling Effort experiments, [*Bryan et al.*, 1995]). To quote the authors: "The northward flow through Yucatan Strait became shallower and weaker (transport ~ 17 Sv) and the southward recirculation on the eastern side of the strait was enhanced…the vertical shear in the central channel was greatly reduced." The authors attributed the nonshedding behavior to the "decreased vertical shear" consistent with *HT's* and *Oey's* [1996] findings.[13] The implication of an active lower layer invalidates the 1.5-layer model, while enhancement of the southward recirculation on the Cuban side of the channel (c.f. the first of four *Oey's* [1996] results, described above) may result in nonnegligible cross-channel flows; either or both of these may be sufficient to resolve *PN's* "momentum imbalance paradox," making it possible for a nonshedding solution to exist. A similar nonshedding scenario occurs also for the OPA model as reported in *Candela et al.* [2003]. The ATL6 version of this model encompasses the Atlantic Ocean at one-sixth degree by one-sixth degree resolution and 43 vertical levels. To quote the authors, "ATL6 developed a problem in the Gulf of Mexico after the sixth year of simulation (in 1984) that consisted in a blocking of the Loop Current (i.e., it stopped shedding eddies)….an anticyclonic eddy stationed itself to the north of the Yucatan Channel and remained there for the rest of the simulation until the end of 1993." *Candela et al.* [2003] gave no further details. The nonshedding solution is intriguing: its resolution may go a long way in our understanding of the complex behaviors of the GCMs.

2.3 More Recent GCM Results

Later papers that discuss or describe (models of) Loop Current and rings are: *Mooers and Maul* [1998], *Murphy et al.* [1999], *Welsh and Inoue* [2000], *Oey and Lee* [2002], *Ezer et*

---

[13]The weaker transport, *17 Sv*, is not likely to be the culprit as HT found shedding with inflow transport as low as *10 Sv* (their experiment RG18), and *Sturges et al.'s* (1993) transport = *19 Sv* is only tiny bit stronger. The weaker transport also suggests not overly-strong inflow speed, which would actually exclude the port-to-port mode (i.e.,HT's regime N) caused by overly large beta Rossby number $R_B > R_{Bc}$. Also, by weak shears the authors probably meant a more barotropic flow with significant bottom inflow.

al. [2003], *Lee and Mellor* [2003], *Morey et al.* [2003], *Zavala-Hidalgo et al.* [2003b], *Oey et al.* [2003], *Candela et al.* [2003], *Romanou et al.* [2004], *Oey et al.* [2004], *Oey* [2004], *Cherubin et al.* [2005] and *Oey et al.* [2005]. These later models generally have better resolutions (with one or two exceptions). Apart from *Morey et al.* [2003] and *Zavala-Hidalgo et al.* [2003b], who used a domain similar to *Dietrich and Lin's* [1994] and hence specify essentially the total Yucatan transport (see below), others' domains include the entire Caribbean Sea. *Mooers and Maul's* POM (at 20 km resolution and 15 vertical sigma levels; C_{smag} = 0.05) domain is similar to that of *Sturges et al.* [1993], including also the Caribbean Sea. Though the integration was short (1500 days during which 3 eddies were shed), the work suggested a need to treat Caribbean and Loop Currents as an integrated, interconnected system.

2.3.1 Gulf of Mexico & Caribbean Sea Connection. The Gulf-Caribbean connectivity is explored in *Murphy et al.* [1999], who experimented with a 5.5-layer reduced gravity as well as a 6-layer global model with realistic bottom topography, both at one-fourth degree resolution; the horizontal viscosity is 300-500 $m^2 s^{-1}$. The authors found that fragments of the North Brazil Current rings leak in through the Lesser Antilles as potential vorticity perturbations that excite mesoscale eddies in the Caribbean Sea. These eddies amplify and traverse westward, and some manage to squeeze through the Yucatan Channel and affect the timing of Loop Current eddy shedding. The amplification of perturbations and eddy-formation in the eastern Caribbean seem to be supported by recent drifter analysis by *Richardson* [2005], but *Simmons and Nof* [2002] caution that the one-fouth degree resolution may be too coarse to portray the relevant dynamics correctly. There is not (yet?) direct observational evidence of Caribbean eddies squeezing through the Yucatan Channel (though the Canek moorings across the channel suggest such a possibility; *Abascal et al.* [2003]), but *Murphy et al.'s* numerical finding is interesting and potentially significant. *Oey et al.* [2003] expanded upon *Murphy et al.'s* ideas, but instead of looking at the progression of individual eddies, they asked if forcing by winds (six-hourly ECMWF was used) and/or by Caribbean eddies (satellite SSH anomaly was used) would affect the statistics of ring-shedding, the periods in particular. Four 16-year experiments and one 32-year experiment were conducted using *Oey and Lee's* [2002] model (POM) in the

domain shown in Figure 1. The horizontal grid spacing, Δ, is variable, $\Delta \approx 25$ km in the eastern Caribbean and decreasing to $\Delta \approx 10$ km in northwest Caribbean, Yucatan Channel and over the Loop Current (≈ 5 km in the northern Gulf), and C_{smag} = 0.1. There are 25 sigma levels in the vertical, and steady transports (according to *Schmitz* [1996]) are specified at the model's open boundary at 55°W where climatological T/S are also specified. By systematically experimenting with different forcing, *Oey et al.* [2003] found a dominant shedding period of 9~10 months when there is no wind.[14] Remote winds over the Atlantic force short-period transport fluctuations in the Yucatan Channel, and the shedding periods then tend to be short (shortest = 3 months). On the other hand, Caribbean eddies (mostly anticyclones in the model) tend to lengthen the shedding periods (longest = 16 months). *Oey et al.* [2003] used the conservation of potential vorticity argument [*Reid,* 1972; *HT*] to explain the period-lengthening when Caribbean anticyclones are present: influx of anticyclones into the Gulf would tend to confine the Loop Current close to Yucatan/Florida (i.e., *HT's* port-to-port mode); shedding is then less likely.

2.3.2 Deep Processes: Modons, TRW's and Mean Cyclonic Gyre(s). *Welsh and Inoue* [2000] improved upon *Sturges et al.'s* [1993] model with better resolution (one-eighth degree horizontal grid spacing and 15 vertical levels), smaller eddy viscosity and diffusivity (both = 75 $m^2 s^{-1}$) and stronger transport (28 Sv). The Loop Current rings' characteristics are similar to those found previously. However, the better resolution appears to increase the maximum speeds: approximately 1.3 $m s^{-1}$ in the Loop Current and 1 $m s^{-1}$ in the rings. The authors reported that speeds close to 1.75 $m s^{-1}$ were obtained at even higher resolution. Thirteen rings were tracked for the 8 year simulation (after a 12-year spin-up), giving an averaged shedding period $\approx$ 7.5 months, slightly longer than *Sturges et al.'s*. The new insight offered by *Welsh and Inoue* is their descriptions of the development and westward progression of deep modon: anticyclone-cyclone pair (each $\approx$ 150 km in diameter) beneath a modeled ring [c.f. *Hurlburt and Thompson,* 1982, and *Sturges et al.,* 1993]. The pair forms while the Loop Current ring is being shed. They explained the process in terms of *Cushman-Roisin et al.'s* [1990] potential-vorticity conservation for a flat-bottom, two-layer ocean. *Welsh and Inoue* found that the cyclonic component of the modon survives longer (in the western Gulf): this was the first suggestion that deep mean

---

[14]These periods are within the range of HT's "natural" periods of *12 month* for the *1.5-layer* model and *8 months* for the 2-layer model with bottom topography (c.f. *HT's* findings *7* and *8*).

[15]As an observational support for their findings, *Welsh and Inoue* [2000] quoted *Hofmann and Worley's* [1986] geostrophic-current estimate at one hydrographic section (section 7 in Hofmann and Worley) to be a cyclone beneath an anticyclone; however, section 7 actually shows the opposite: anticyclone beneath a cyclone. But as pointed out by *Hofmann and Worley*, the small-scale feature maybe an artifact of their inverse technique.

currents in the Gulf might be cyclonic.[15] *Welsh and Inoue's* results also show rich complexities (their Figures 7 and 8), probably because the topography is better resolved than that used in *Sturges et al.* [1993].

Deep modons with scales and westward propagation speeds similar to those of TRW's (i.e. scales ≈ O(100 km) and speeds ≈ 0.1 m s^{-1}) can occur in models with a flat bottom [e.g, *Hurlburt and Thompson,* 1982]. One cannot, therefore, unambiguously interpret the deep motions found by *Sturges et al.* [1993], *Oey* [1996] and *Welsh and Inoue* [2000] as TRWs. On the other hand, observations indicate that TRWs constitute as much as 90% or more of the deep energy [*Hamilton*, 1990; *Hamilton and Lugo-Fernandez*, 2001]. *Oey and Lee* [2002] examined details of deep energy generated by Loop Current variability and rings in the model described above in conjunction with *Oey et al.'s* [2003] work. A 10-year experiment was conducted with steady transports and annual-mean climatological T/S specified at the model's open boundary at 55°W. Through linear-wave and ray-tracing analyses, they show the existence of TRWs at 20~100 day periods. These waves have group velocity (generally westward, ≈ −10 km day^{-1}) and are excited by deep motions caused by meanders (frontal eddies) that swirl around the Loop Current and propagating rings.[16] The TRWs dominate deep motions along an east-west band as a result of TRW refraction ("confinement") by (i) an escarpment across the central Gulf north of the 3000 m isobath, and (ii) deep westward current (≈ −0.03 m s^{-1}; in region approximately over and north of the 3000 m isobath) and its cyclonic shear. The authors noted that both their modeled westward group speeds and *Hamilton's* [1990] from observations were larger in comparison to those derived from the TRW dispersion relation by about 2 to 3 km day^{-1}. *Oey and Lee* suggested that the faster westward group velocities in the observations implied the existence of a deep mean westward current (−0.03 m s^{-1}) across the central Gulf. *Oey and Lee's* modeled mean flow is generally cyclonic around the deep Gulf (please see their Figure B.1, which shows mean currents at the model's sigma-level 20 (≈ 200 m above the bottom in water depths ≈ 2000 to 3000 m). *Oey and Lee's* analyses have since been confirmed by *Lee and Mellor* [2003] using also POM but forced by the NCEP-Eta wind. These latter authors also confirm the cyclonic deep circulation. Of interest is that their Figure 14b appears to show two cyclonic gyres (at z = −1500 m): one in the eastern Gulf (approximately east of 89°W, where there is a constriction in isobaths > 3000 m) and the other one in the central and western Gulf.

The existence of a deep mean cyclonic gyre (or gyres) found in numerical models has recently been confirmed for subsurface levels deeper than about 1000 m by *DeHaan and Sturges* [2005] based on a careful analysis of historical hydrographic and current-meter data. The authors suggest topographic wave rectification and deep dense inflow from the Caribbean as possible mechanisms. *Mizuta and Hogg* [2004] show that bottom friction causes the divergence of the vertically integrated Reynolds stress (produced by on-slope propagating TRWs), which in turn induces a mean along-slope flow in the "cyclonic" sense (i.e., shallower water on the right-looking down-current). The bottom boundary layer plays a crucial role as Ekman pumping serves to link the mean current with the Reynolds stresses of the wave field. If we take the typical TRW characteristics in the Gulf, and the bottom slope [e.g., *Oey and Lee*, 2002], *Mizuta and Hogg's* analysis then gives a mean cyclonic deep current of about 0.05 to 0.1 m s^{-1}. These speeds are comparable with models [*Oey and Lee*, 2002; *Lee and Mellor*, 2003], though somewhat stronger than observations (generally < 0.05 m s^{-1}; *DeHann and Sturges* [2005]; *Hamilton* [2005, submitted to J. Phys. Oceanogr.]). *DeHaan and Sturges* also have in mind the two-gyre system mentioned above in conjunction with *Lee and Mellor's* [2003] work, and support their findings by noting that the mean currents computed from PALACE floats at z = −900 m are also generally cyclonic. The PALACE mean currents show a more erratic picture but one can generally discern westward flow in the north central Gulf, southward mean in the west/northwest, a cyclonic gyre in the southwestern Gulf (i.e., the Campeche Bay), and broad eastward flow off the northern Campeche shelf at 90°W, from 23°N~26°N, and westward flow further north around 27°N; the flows are more clearly defined in the west and southwest, but contain smaller-scale features in the central Gulf. (Note that inadequate data prevented *DeHaan and Sturges* from estimating flows off the northern Campeche shelf). Vazquez de la Cerda et al.'s (this volume) analyses of hydrographic, drifter, floats and satellite data lend further supports that the mean circulation in the Campeche Bay is cyclonic. In light of these observational analyses, future work should extend *Oey and Lee's* [2002] and *Lee and Mellor's* [2003] calculations with runs using different forcing, grid-resolution and eddy viscosities, and analyzing dynamics and comparing the results with observations. It is likely that deep flows in the Gulf are TRW and eddy-driven: reminiscence of currents with banded structures seen in idealized studies of geostrophic eddies and turbulence [e.g. ,*Vallis and Maltrud*, 1993].

2.3.3 Loop Current and Yucatan Channel Flow Variability. The potential importance of flow variability through the Yucatan Channel to Loop Current and shedding dynamics

---

[16]*Oey and Lee* [2002] assume a constant buoyancy frequency (N) in their analysis. *Reid and Wang* [2004] derive a new TRW dispersion relation based on a more realistic exponential (in z) N-profile; the new relation should be tested in future ray-tracing calculations.

was previously mentioned in connection with *Maul et al.* [1985]; *Oey* [1996]; and *Oey et al.* [2003]. There now exist direct current measurements in the Yucatan Channel (the Canek Program; *Ochoa et al.* [2001]; *Bunge et al.* [2002]; *Sheinbaum et al.* [2002]; *Candela et al.* [2003]; *Abascal et al.* [2003]). The observations consist, among other things, of current-meter and Acoustic Doppler Current Profiler (ADCP) measurements across the channel for 23 months (from August 1999 to June 2001, with a service break in June 2000). The center panel of Figure 2 shows the mean (Figure 2a) and standard deviation (Figure 2b) of the along-channel velocity obtained from the Canek observations. The mean is surface intensified near the west (maximum $\approx$ 1.3 m s^{-1}). Significant inflow (into the Gulf) extends to $z \approx$ -750 m and occupies nearly the whole width of the channel except for a narrow outflow (i.e. into the Caribbean Sea; $v \approx -0.1$ m s^{-1}) near the surface on the Cuban side of the channel. The deep flow (below $z \approx -750$ m) is generally weak, and is directed into the Gulf in the middle, sandwiched between cores of outflows on both sides of the channel. Figure 2b shows that the flow displays considerable variability with a standard deviation $\approx$ 0.5 m s^{-1} in the main surface core in the west. The standard deviation is as large as (or even larger than) the mean near the surface on the Cuban side and also in the deep. There are also non-negligible cross-channel flows (not shown). *Abascal et al.* [2003] reported mean cross-channel speeds of O(0.1 m s^{-1}) and standard deviation $\approx$ 0.15 m s^{-1} near the surface.

Various authors have directly compared their modeled flow field across the Yucatan Channel with Canek observations. *Ezer et al.* [2003] analyzed in details *Oey et al.'s* [2003] simulation results. The model set-up is similar to *Oey and Lee* [2002], described previously, but in addition the model was forced with six-hourly ECMWF wind and monthly climatological surface heat/salt fluxes. The modeled mean and standard deviation are compared with Canek observations in Figure 2 (titled POM), and also in Table 1. The mean shows a strong inflow (maximum $v \approx$ 1.5 m s^{-1}) near the surface in the main western core and deep outflow cores on both sides of the channel ($v \approx -0.05$ to -0.1 m s^{-1}) separated by a weak inflow in the center. These modeled features are similar to those observed. There is also a near-surface outflow ($v \approx -0.15$ m s^{-1}) near the Cuban side of the channel though the observed outflow is narrower and weaker ($v \approx -0.1$ m s^{-1}). The model's main inflow core also extends deeper (the 0.1 m s^{-1} contour is at $z \approx -900$ m compared to observed $z \approx -750$ m). The modeled flow variability is weaker than observation near the surface ($z > -300$ m), by more than 50% in some localized regions. The agreement is

good in the deep, where both model and observation show a standard deviation of about 0.05 m s^{-1}.[17] *Ezer et al.* [2003] also confirm *Oey's* [1996] finding of a deep return outflow (into the Caribbean) in connection with eddy-shedding and show, moreover, that the return outflow correlates with changes in the Loop Current extension area in agreement with *Bunge et al.'s* [2002] observational analysis. *Romanou et al.* [2004] and *Cherubin et al.* [2005] found very similar behaviors with the MICOM model.

Candela et al. [2003] use two configurations of the z-level OPA model, one at one-sixteenth degree resolution – the CLIPPER ATL6 model, and the other at one-twelfth degree resolution – the PAM model (both with 43 z-levels) to compare the flow in the Yucatan channel with Canek observations. The CLIPPER model encompasses the north and south Atlantic oceans (75°S to 70°N), and was spun up for eight years forced by climatological ECMWF air-sea fluxes. It was then continued from 1979 through 1993, forced by daily ECMWF fluxes. Only the first five years of this latter run was analyzed because the model Loop Current stopped shedding eddies after the sixth year (1984 and thereafter; see below). The PAM model covers the north Atlantic Ocean (9°N to 70°N), was spun up for 11 years using ECMWF climatology, and continued another 3 years using daily ECMWF fluxes (1999 to 2001). Results from this latter 3-year run were compared with Canek. Figure 2 and Table 1 show that the modeled Yucatan Channel flows from CLIPPER and PAM share similar characteristics as those from POM, discussed above. The surface shows a main western inflow core and a weak outflow near the Cuban side. The deep outflow cores on both sides of the channel are separated by a weak inflow in the center. The maximum values (both mean and standard deviation) are slightly lower than POM; the exception is the PAM's stronger deep outflow on the eastern side of the channel ($v \approx -0.1$ m s^{-1} at $z \approx$ -850 m).

The work of *Romanou et al.* [2004] and *Cherubin et al.* [2005] represent a first application and validation of MICOM to a relatively small-scale basin with complex topography—the Gulf of Mexico. The model has one-twelfth degree horizontal resolution and 15 vertical layers, and includes the north Atlantic (28°S to 65°N). Monthly mean surface fluxes from COADS were used and the model was run for 20 years; results from the final 5 years were then analyzed. The authors documented the model Loop Current and ring variability. Shedding periods (3~15 months), rings' westward propagation speeds (2~3 km day^{-1}) and trajectories, sizes (300~400 km), rings' orbital speeds (1.5~1.9 m s^{-1}),

---

[17]*Ezer et al.* found that the mean and standard deviation of the modeled currents in the channel agree with *Maul et al.*'s (1985) deep mooring measurement (at $z = -1895$ m, near the bottom above the sill).

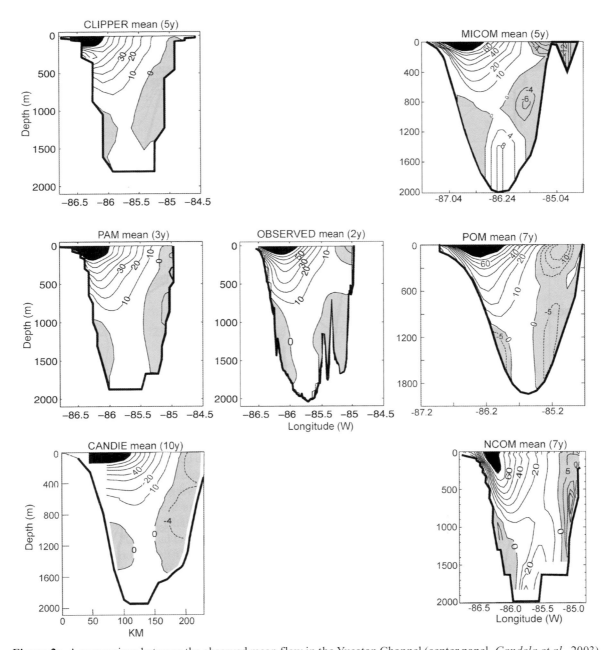

Figure 2a. A comparison between the observed mean flow in the Yucatan Channel (center panel, *Candela et al.*, 2003) and the simulated mean flow in six different models: CLIPPER and PAM are the lower and higher resolution version of OPA (left-upper and left-middle panels, respectively, *Candela et al.*, 2003), CANDIE (left-lower panel, *Sheng and Tang*, 2003), MICOM (right-upper panel, *Romanou, et al.*, 2004), POM (right-middle panel, *Oey et al.*, 2003, *Ezer et al.*, 2003) and NCOM (right-bottom panel, *Morey et al.*, 2003). Light shaded area represents negative velocity (outflow from the Gulf of Mexico to the Caribbean Sea) and dark area represents large core inflow velocity ($v > 80 \, cm \, s^{-1}$). The period averaged in each case is indicated. A common contour interval of 10 cm s^{-1} is used for positive values, but negative contours are kept as in the original figures (i.e. 4 $cm \, s^{-1}$ for MICOM and CANDIE, 5 $cm \, s^{-1}$ for POM and NCOM, 10 $cm \, s^{-1}$ for CLIPPER, PAM and observed). See Table 1 for more quantitative comparison of velocities and variances. Note the gross similarities between the different models. There are, however, some general differences also, e.g. over the western slope where *layer*, *sigma* and *sigma-z* models (right panels) show outcropping velocity contours (i.e. strong shears) while *z-level* models (left panels) tend to be more homogeneous.

Vukovich, F. M. and B. W. Crissman, 1986. Aspects of warm rings in the Gulf of Mexico. J. Geophys. Res., 91, 2645–2660.

Vukovich, F. M., 1995. An updated evaluation of the Loop Currents eddy-shedding frequency. *J. Geophys. Res. 100*, 8655–8659.

Wallcraft, A., 1986. Gulf of Mexico Circulation Modeling Study, Year 2. Progress Report by JAYCOR, submitted to the Minerals Management Service, Metairie, LA. Contract No. 14-12-0001-30073.

Wang, D.-P., L.-Y. Oey, T. Ezer and P. Hamilton, 2003. Nearsurface currents in DeSoto Canyon. *J. Phys. Oceanogr.*, 33: 313–326.

Webster, P. J., G. J. Holland, J. A. Curry and H.-R. Chang, 2005: Changes in tropical cyclone number, duration, and intensity in a warming environment. *Science*, 309, 1844–1846.

Weisberg, R. H., He, R., 2003. Local and deep-ocean forcing contributions to anomalous water properties on the West Florida Shelf. J. Geophys. Res., 108(C6), 10.1029/2002JC001407.

Welsh, S. E. and M. Inoue, 2000. Loop current rings and deep circulation in the Gulf of Mexico. J. Geophys. Res., 105, 16,951–16, 959.

Wert, R. T. and R. O. Reid, 1972. A baroclinic prognostic numerical circulation model. In: Contributions I Physical Oceanography of the Gulf of Mexico, Texas A&M Univ., Oceanographic studies, eds. L. R. A. Capuro and J. L. Reid, Gulf Publishing Co., Houston, 177–209.

Zavala-Hidalgo, J., S. L., Morey, and J. J. O'Brien, 2003a. Cyclonic eddies northeast of the Campeche Bank from altimetry data. J. Phys. Oceanogr., 33, 623–629.

Zavala-Hidalgo, J., S. L., Morey, and J. J. O'Brien, 2003b. Seasonal circulation on the western shelf of the Gulf of Mexico using a high resolution numerical model. J. Geophys. Res., 108 (C12), 3389, doi: 10.1029/2003JCOO1879.

Zavala-Sanson, L., F. Graef, and E. G., Pavia, 1998: Collision of anticyclonic, lens-like eddies with a meridional western boundary. J. Geophys. Res. 103, 24,881–24,890.

L.-Y. Oey and T. Ezer, Program in Atmospheric and Oceanic Sciences, Princeton University, Princeton, New Jersey, H.-C. Lee, GFDL, NOAA, Princeton, NJ 08540.

Upper-layer Circulation in the Approaches to Yucatan Channel

A. Badan, J. Candela, J. Sheinbaum and J. Ochoa

Depto. de Oceanografia Fisica, CICESE, Ensenada, Mexico

The Yucatan Current originates where the Cayman Current turns northwards off the Mayan coast. The latitude at which the current meets the coast fluctuates every few months from the Mexico-Belize border to the island of Cozumel, modulating the currents near the coast. A two-year measurement program recorded the Yucatan Current transporting close to 23 Sv, smaller than the classical value for the Florida Current off Miami, so parts of the Florida Current must flow through Windward Passage or bypass the Caribbean entirely. The velocity fluctuations of the Yucatan Current are larger than the mean everywhere except in its core, and have been attributed mostly to the passage of eddies. About one fourth of the Yucatan Current flows through Cozumel Channel, where significant ageostrophic episodes have been recorded, related to the curvature of the currents entering the channel, and caused possibly by the passage of large mesoscale features. The current also appears to shift laterally, including reversals of the southerly Cuban Countercurrent on the eastern side of the channel. The proper observation of eddies passing offshore might shed better light on the nature of current fluctuations and on the triggering of Loop Current eddies, but they remain elusive even in their bulk integral characteristics, as both the instrumentation deployed and remote sensors appear insufficient to resolve them. A construction of pseudo-streamlines from moored observations is used to signal large eddies passing through the channel.

1. INTRODUCTION

Few areas of the world have been as important to seafaring as the Caribbean and the Gulf of Mexico, especially in the region of the Yucatan Channel, where the two bodies of water meet between the Yucatan Peninsula and the westernmost tip of Cuba (Figure 1). Maritime history in the region was central to the discovery of the Americas, and the passage today remains crucial to global navigation. Yet, the flow through Yucatan Channel remains an enigmatic component of the circulation of the world's ocean, which contains all of the exchange between the Caribbean Sea and the Gulf of Mexico, and most of the inflow into the Gulf of Mexico, as direct backflows from the Atlantic through the Straits of Florida are sporadic and small [e.g. *Brooks and Niiler*, 1975]. The Channel is a major conduit for the northward transport of tropical water, as the throughflow includes a significant portion of the two main components of the southwestern North Atlantic Subtropical Gyre, the northward compensation flow to the Sverdrup transport due to the wind stress curl over the North Atlantic, and the surface branch of the meridional overturning circulation [*Schmitz and Richardson*, 1991].

Project Canek (named after a prominent Mayan figure in the regional literature of the Yucatan) began in December 1996 as an effort to fully resolve the flow between the Gulf of Mexico and the Caribbean Sea. The objectives of the program were to measure the exchange of water and ocean

Circulation in the Gulf of Mexico: Observations and Models
Geophysical Monograph Series 161
10.1029/161GM05

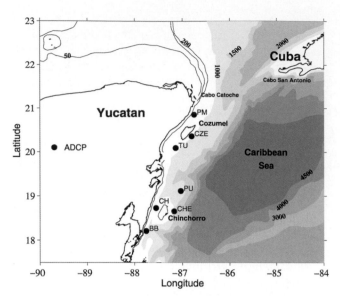

Figure 1. Configuration of Yucatan Channel, with the location of current meter moorings along the coast of Yucatan. Depth contours in meters.

properties between the Caribbean Sea and the Gulf of Mexico and to determine the statistics and causes of the currents through Yucatan Channel, including possible undercurrents and counterflows, the amount of net throughflow, and the water masses involved. In addition we wished to investigate the dynamics of the upwelling over the Campeche Bank, north of the Yucatan Channel, and help elucidate the role of the circulation in sustaining the coral reefs off Yucatan. Since then, a total of ten cruises, all on board R/V *Justo Sierra* of the Universidad Nacional Autónoma de México, have provided an exceptional data set on the physical ocean-ography of the region. The present article is a synthesis of the evidence that lead us to the following conclusions and recommendations for future studies in the region: We believe that our results are robust, and serve as the basis for a better understanding of the dynamics of the exchange between the Caribbean and the Gulf of Mexico. A smaller mean transport through Yucatan Channel might require a revision of the oceanic pathways surrounding the Caribbean Sea; under-standing the nature of the fluctuations should lead to a more comprehensive investigation of large mesoscale features in the northwestern Cayman Sea and of the potential vorticity budgets, which might reveal useful precursors to the shed-ding of eddies from the Loop Current.

Reviews of the earlier knowledge of the general ocean-ography of the combined Caribbean Sea-Gulf of Mexico system can be found in *Mooers and Maul* [1998], *Gallegos and Czitrom* [1997] and, for numerical results, in *Sheinbaum et al.* [1997]. The studies cited there highlight the relation

of the current system in the Yucatan Channel to the world's circulation and the expected role it has on the circulation in the Gulf of Mexico, especially the Loop Current and the large eddies that detach from it; significant efforts are being made in modeling these effects for diagnostic and predictive purposes, and there is keen interest in the pos-sibility of using the detailed knowledge of the circulation through the channel to establish long-term monitoring of global-scale climatic fluctuations. More locally, it remains unclear how this circulation might drive the upwelling onto the highly productive Campeche Bank, whose nutrient sup-ply appears to depend significantly on the dynamics of the Yucatan Current and associated undercurrents. In addition, the currents that hug the coast play an important role in the dispersion of eggs and larvae of commercially important species, promote replenishment and sustain biodiversity of the species in the Middle-America reef. At the times of those reviews, no complete set of measurements of the currents were available, so currents were estimated indirectly or on the basis of the scarce observations available.

2. THE STUDY AREA

Yucatan Channel is located between the northeastern Yucatan Peninsula and the western tip of Cuba, and connects the north-western Caribbean Sea to the southern Gulf of Mexico (Fig. 1); its narrowest is 196 km wide (106 nautical miles) across and stretches between Isla Contoy, off Cabo Catoche, Mexico (21°30'N; 86°49'W), and Cabo San Antonio, Cuba (21°52'N; 84°57'W). The saddle-point sill at its maximum depth, nearly 2040 m, lies at (21°36'N; 85°43'W), 7.5 km to the east of its middle point. The bottom topography there is oriented in a north-south direction, the Yucatan Channel proper, which falls off from the sill rapidly to the south into the deep (>4000 m) Western Cayman Basin, but somewhat more gently to the north of there into the broad Gulf of Mexico Basin. Most of the channel is conformed by the nearly 2000 m deep pedes-tal which extends in a SW to NE direction off the Yucatan Peninsula to Cuba, and conforms the northwestern end of the Cayman Basin. Off Yucatan, a 50 km-wide platform includes a very narrow continental shelf bordered by the coral reefs, and an underwater ridge from which Isla Cozumel delimits the 400 m-deep Cozumel Channel, and Chinchorro Bank the 800 m-deep Chinchorro Channel. The shelf broadens to the north around the tip of Yucatan, where the coast turns due west but the steep Cozumel slope extends northwards several hundred kilometers into the Gulf of Mexico, so the Yucatan shelf opens onto the very broad Campeche Bank.

The flow through Yucatan Channel can be thought of as having two layers, the upper one being that which contains the active throughflow across the Gulf of Mexico and through the Straits of Florida over its 700 m sill [*Maul et al,* 1985]; it includes the Yucatan Current and the Cuban Countercurrent, a predominantly upper-layer, intermittent current that flows from the Gulf of Mexico into the Caribbean near the western end of Cuba [*Emilsson,* 1971]. The lower layer lies below the level of the Florida sill, net exchanges through it are null [*Sheinbaum et al,* 2002], but flows at different levels depend on the water mass climatology of the western Atlantic and on regional processes that involve water mass transformations and the ventilation of the deep Gulf of Mexico [*Maul et al.,* 1985; *Bunge et al.,* 2002; *Rivas et al,* 2005; *Sturges,* 2005]. These include southward counterflows near both slopes of the channel, and a deep northward flow over the sill.

3. A SUMMARY OF CANEK OBSERVATIONS

Project Canek observations included moored current and temperature time series, coastal subsurface pressure recordings, shipboard current measurements, one drifter release, and numerous hydrographic surveys, especially in the section between Cabo Catoche, Mexico, and Cabo San Antonio, Cuba (Figure 1). The first survey covered only portions of the Channel [*Ochoa et al.,* 2001] but showed that currents are predominantly in geostrophic balance across the section, which we later found not to be always true elsewhere along the coast, and demonstrated the presence of heretofore unmeasured significant southerly flows at mid-depth. The first large mooring deployment across the Channel showed that transports were about 25 percent lower than was previously thought [*Sheinbaum et al.,* 2002], and renewed the idea that the origins of the Florida Current need to be better ascertained. This crucial result was confirmed through a second year of measurements [*Ochoa et al,* 2003], and is in agreement with earlier observations in the Florida Straits that indicate that a sizeable contribution to the Florida Current bypasses the Caribbean through the various channels in the Bahamas [*Hamilton et al.,* 2005]. Throughout both years of observations, large mesoscale features, most likely eddies, passing through the region, appeared to modulate both the flow and the water mass composition of the exchange through the Channel.

Figures 2 and 3 illustrate the broad features of the near-surface circulation in the Yucatan Channel and adjacent portions of the Gulf of Mexico, as recorded underway with a shipboard ADCP. The Yucatan Current is a strong jet with speeds between 1 and 2 m.sec^{-1} that occupies the western half of the Yucatan Channel, and proceeds into the Gulf of Mexico as the Loop Current, along the steep slope of the Yucatan shelf. The large grouping of arrows in the principal section of the Channel shows that over the period of measurements, the Yucatan Current is a persistent feature. A sharp

drop of speed on the western edge of the current follows the topography of the Yucatan shelf to the north. Farther east in the Gulf of Mexico, the return flow from the Loop Current approaches Cuba and veers towards the Florida Straits, where at this time the jet occupies the southern half of the Key West-Havana section. A portion of the Loop Current diverges on the Cuban coast, from where a branch returns towards Yucatan Channel as the Cuban Countercurrent. This flow can then, as is the case in this figure, round the western tip of Cuba and extend as a narrow jet back into the Caribbean. The anticyclonic feature off northwestern Cuba whose northern edge is the return flow of the Loop Current and whose southern edge constitutes the Cuban Countercurrent has been recognized as a geographically stable semi-permanent gyre that interacts often with the flow through Yucatan Channel and is related to the existence and extension of the Cuban Countercurrent. In turn, the presence of a southerly jet on the eastern side of the Channel introduces a shear, which might be interpreted as resulting from the passage of eddies through Yucatan Channel. Some variations on this general distribution of currents have been documented during various Canek cruises (Fig. 3); the Yucatan Current is generally most intense on the western side of the Channel, and usually follows the Yucatan slope to the north, but the Cuban Countercurrent and its southerly extension is sometimes absent, at which times the Yucatan Current extends across the Channel and may have a secondary maximum to the east.

Figure 4 summarizes the basic statistics of the flow along the coast of Yucatan and across the channel. Off Yucatan, the mean speed of the current increases from the confluence towards the Yucatan Channel, where it reaches 1.5 m.sec^{-1} as water gathers against the coast and is accelerated northwards. As such, the Yucatan Current is usually found on the western side of the channel and up to one fourth of the flow passes through the 400 m deep Cozumel Channel, between the island and the coast. The variance decreases somewhat along the coast; as the mean is generally larger than the standard deviation, current reversals seldom occur within the Yucatan Current, except at the station south of Chinchorro bank and of the confluence of the Cayman Current, where the flow often reverses to the southward. In the western Yucatan Channel proper, both the mean and the variance of the flow are substantially larger, showing that a large portion of the energy converges there from the central Yucatan Basin. Everywhere else, the standard deviation is larger than the mean, and flow reversals are common, especially in the central channel, and within the eastern Cuban Countercurrent. Spectra of the flow at the various moorings along the coast show that the decrease of variance occurs predominantly in the very low frequency

Using a combination of remotely-sensed altimetry and ocean color was especially important in this study to identify oceanographic features during summer, since during the six month period from May through October when surface waters become uniformly warm in the Gulf of Mexico, sea surface temperature (SST) loses most its contrast and therefore its ability to detect frontal features. Both SSH and SeaWiFS do have some technical limitations that bias their measurement ability in coastal waters and over the continental shelf, but for our study these limitations should have been minimal since the ship surveys were carried out farther offshore over the continental slope and deep basin. For example, it is difficult for altimeters to accurately measure the SSH anomaly at or near the land-ocean boundary, or for SeaWiFS-derived chlorophyll to be unambiguous in coastal waters where colored dissolved organic matter and/or suspended sediment concentration are high. Moreover, the altimetry data generally have a 100-km decorrelation length scale and are temporally averaged over a 10-day period, and the SeaWiFS data although they inherently have 1-km spatial resolution had to be composited for a 7-day period to minimize the incidence of missing data due to clouds. Nevertheless, for both altimetry and ocean color, the capability to remotely detect spatial features located not only along but also to either side of the ship tracks outweighed their inherent limitations.

For this paper, we have examined SeaWiFS images and SSH fields that were concurrent with sperm whale sightings. In addition, as a complement to this paper, a complete time series of weekly and bi-weekly SeaWiFS composites of the entire Gulf of Mexico for the period October 1997 to September 2004 is provided in the companion CD-ROM.

This CD-ROM also contains a detailed description of how to examine the SeaWiFS data and animation tools.

Methods for visual and passive acoustic survey for sperm whales during *Gordon Gunter* fieldwork are summarized by *Mullin and Fulling* [2004] and *Thode et al.* [2002], respectively, and during *Gyre* fieldwork by *Jochens and Biggs* [2003, 2004]. In brief, marine mammal researchers searched during daylight hours with BigEye telescopic binoculars and listened at night as well as during daytime hours with towed hydrophone arrays.

To quantitatively evaluate if whale sightings were more frequent in regions of shelf water entrainment, we did a simple spatial statistical analysis to compare the average remotely-sensed SSH and ocean color at locations where whales were encountered with the ensemble average conditions along the 1000-m isobath. The results are presented in Table 2. For each of the eight ship surveys that are summarized in Table 1, we computed the average SSH and SeaWiFS chlorophyll concentration along the 1000-m isobath, 91°–86° W, for one-week periods at the mid-points in time of each of the field surveys. To compute the mean SSH from daily rawdata that are gridded with 1/4 degree x 1/4 degree resolution, for each of the surveys we selected the 36 SSH grid points closest to the 1000-m isobath 91°–86° W and averaged these (n = 36 x 7 days = 252). The SeaWiFS rawdata have much higher (1 km) spatial resolution but they are one-week composites. In practice, we picked 121 locations that were about 5 km apart along the 1000-m isobath, 91°–86°W, for the calculation of mean along-isobath chlorophyll. For each of the surveys, the number of data points averaged to determine the mean chlorophyll concentration along the 1000-m

Table 1. Synopsis of areas along the middle slope that were surveyed in summers 2000–2003 by NOAA Ship *Gordon Gunter* and by TAMU R/V *Gyre*.

Ship and Cruise	Begin Fieldwork	End Fieldwork	Longitude Range	CTDs + XBTs
Gunter Leg One SWAMP 2000	28 Jun 2000	8 Jul 2000	90.2°W to 87.8°W	4 CTDs
Gunter Leg Two SWAMP 2000	12 Jul 2000	27 Jul 2000	90.0°W to 86.6°W	2 CTDs
Gunter Leg One SWAMP 2001	17 Jul 2001	28 Jul 2001	90.7°W to 87.8°W	2 CTDs + 12 XBTs
Gunter Leg Two SWAMP 2001	29 Jul 2001	9 Aug 2001	88.7°W to 87.7°W	1 CTD + 1 XBT
Gunter Leg Three SWAMP 2001	10 Aug 2001	22 Aug 2001	90.5°W to 87.1°W	6 CTDs + 41 XBTs
Gyre 02G06 SWSS 2002 Leg 1	20 Jun 2002	8 Jul 2002	94.7°W to 86.4°W	5 CTDs + 54 XBTs
Gyre 02G11 SWSS 2002 Leg 2	19 Aug 2002	15 Sep 2002	focus on region 89.7°W to 87.4°W	8 CTDs + 39 XBTs
Gyre 03G06 SWSS 2003 Leg 1	31 May 2003	20 Jun 2003	94.8°W to 86.8°W	8 CTDs + 89 XBTs
Gyre 03G07 SWSS 2003 Leg 2	26 Jun 2003	13 Jul 2003	focus on region 89.5°W to 87.2°W	5 CTDs + 51 XBTs

Table 2. Mean SSH and SeaWiFS chlorophyll concentrations (CHL) at locations where whales were encountered, compared with ensemble mean conditions along the 1000-m isobath, 91°-86° W. See text for explanation of how means were computed. Means highlighted in bold face indicate conditions where whale encounters were significantly different at the 95% confidence level than the ensemble conditions along the 1000-m isobath (i.e., means were different by > 1.96 x standard error).

Year	Julian Days	whales SSH mean	whales SSH std error	1000-m SSH mean	1000-m SSH std error	whales CHL mean	whales CHL std error	1000m CHL mean	1000m CHL std error
2000	183–189	-0.13	0.58	-0.12	0.25	0.56	0.28	0.42	0.04
2000	204–210	**-0.63**	0.21	0.79	0.12	**1.15**	0.09	0.79	0.06
2001	204–210	**-11.74**	0.19	-5.98	0.36	**0.99**	0.07	0.42	0.07
2001	232–238	-6.61	0.44	-7.18	0.32	**0.80**	0.04	0.53	0.05
2002	162–168	**-5.81**	0.31	-2.70	0.19	**2.88**	0.26	0.60	0.15
2002	253–259	**-2.59**	0.32	0.47	0.16	**1.41**	0.07	0.41	0.06
2003	155–161	**-14.35**	0.72	-9.49	0.61	**2.15**	0.20	0.37	0.14
2003	190–196	**-19.27**	0.49	-10.67	0.42	0.17	0.01	0.32	0.05

isobath in Table 2 thus ranged from $n = 121$ when the 1000-m isobath was cloud-free to $n = 92$ when clouds were present in the composites and obscured some of the 1000-m isobath. The number of data points averaged to summarize conditions where whales were encountered ranged from $n = 4$ (SWAMP 2000, Leg 1) to $n = 412$ (SWSS 2002, Leg 2).

3. RESULTS

3.1 LC Dynamics and Slope Eddies, Summer 2000

Since no ADCP was installed for this summer 2000 cruise, we have no direct shipboard measurements to tell us whether there was flow along-margin, on-margin, or off-margin as we followed the survey tracks of this cruise. Instead, and as the next two paragraphs will explain, we used a combination of CTD data and altimetry maps to deduce that surface flow was off-margin along most of the ship track between 89.5°W and 86.6°W.

In the first half of June 2000, a minor LC eddy separated from the LC and was designated as LC eddy-K, also called Kinetic Eddy [Jim Feeney, Horizon Marine, communication]. Altimetry maps for July 2000 show this LCE-K was centered SSW of the Mississippi River delta, with a deepwater cyclone to the east. This anticyclone-cyclone geometry should have enhanced the entrainment of shelf water along the eastern side of LCE-K and its transport off-margin. Figures 1A and 1B present plots of the CCAR Historical Mesoscale Altimetry data to illustrate how the location of LCE-K and cyclone changed between SWAMP 2000 Leg 1 and SWAMP 2000 Leg 2. The ship surveys along the 1000-m isobath did not reach the deepwater cyclone, and although Leg 1 crossed into the interior of deepwater LCE-K, the ship never stopped to make a CTD cast there. Rather, most of the time the ship criss-crossed the off-margin surface flow confluence 89.5°W to 87.5°W.

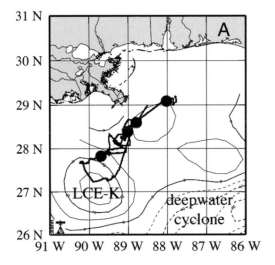

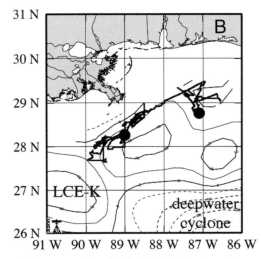

Figure 1. Survey tracks of NOAA Ship *Gordon Gunter* over the middle slope showing the locations of CTD stations, overlaid on SSH anomaly maps for the mid-points in time of SWAMP 2000 Leg 1 (a) and Leg 2 (b).

Figure 2A is a potential temperature versus salinity (TS) plot of the SWAMP 2000 CTD data. The westernmost of the CTD stations was made at/near the NW periphery of LCE-K, while the other five CTDs were made between 89°W and 87°W and so were in the off-margin surface flow confluence. Surface salinity was 36.5 at the westernmost CTD and 31.0–35.1 at the other CTD locations. In areas of off-margin flow, the low salinity surface water was on average just 15 m thick. In contrast, a two-fold deeper mixed layer (32 m, based on the decrease in temperature with depth) was documented at the westernmost CTD done at the edge of LCE-K. However, the 15°C depth, which domes upward to < 170 m in strong cyclones and is depressed to > 350 m in the Loop Current and in the interior of LCEs, was just 216 m at the westernmost CTD and ranged from only 204 – 219 m at the other five CTDs. The combination of surface salinity and 15°C depth data is strong evidence that none of the six CTDs were done far enough inside either feature to get a representative vertical profile of either LCE-K or the cyclone. Additional evidence that the westernmost CTD was done at or near the periphery of (but not inside of) LCE-K is the absence of subsurface salinity > 36.5 at this location. In the GOM, such high salinity Subtropical Underwater is only

found inside the LC or in LCEs recently separated from this Caribbean inflow water.

Panels A and B in Plate 1 show the sightings of sperm whales, superimposed on SeaWiFS composite maps for one week periods that were both relatively cloud-free and close to the midpoints in time for Leg 1 and Leg 2. On Leg 1, the four sightings of sperm whales were made near the NE periphery of LCE-K and in the region of the off-margin flow of green water that extends south from the mouth of the Mississippi River. The average SSH anomaly where the whales were encountered (-0.1 cm) is not significantly different from zero and is representative of the ensemble average along the 1000-m isobath. The average surface ocean color where whales were encountered was green (average SeaWiFS chlorophyll concentration = 0.4 µg/L) but one of the four whales was seen in blue water (0.15 µg/L) and the others were in surface water as green as 1.4 µg/L. By Leg 2, the SeaWiFS imagery shows the region of green water is expanded seaward and that some of this green water is being entrained into the eastern periphery of LCE-K. All sperm whale sightings on Leg 2 were in this now much expanded area of green water, off-margin flow, and the average chlorophyll concentration where whales were seen was signifi-

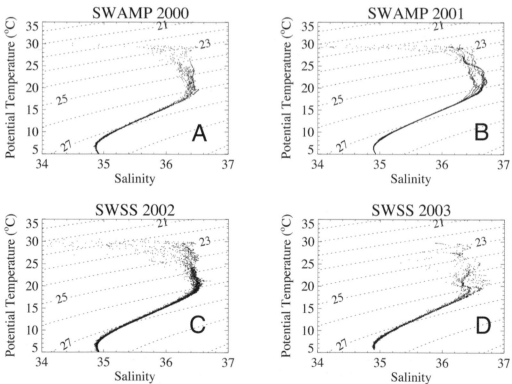

Figure 2. Potential temperature versus salinity plots of the CTD stations done during a) SWAMP 2000, b) SWAMP 2001, c) SWSS 2002, and d) SWSS 2003. Isopycnal lines present density as sigma-theta; x-axis salinity is expanded to feature the range 34–37 to better show on which cruises Subtropical Underwater (>36.5) was present or absent.

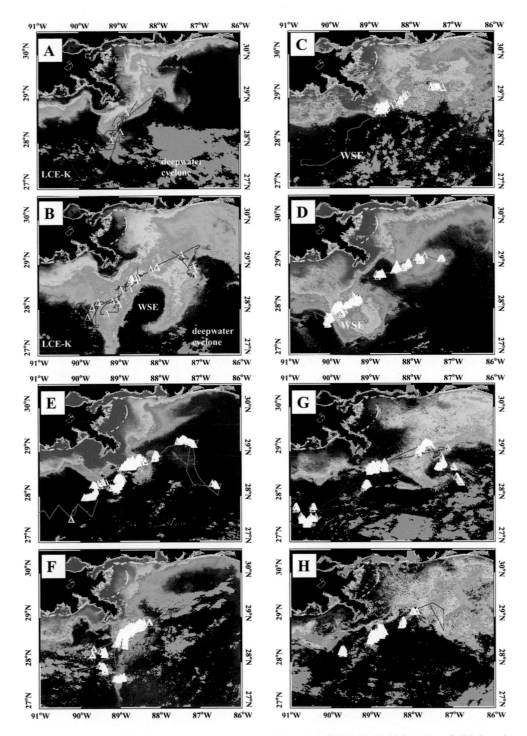

Plate 1. Locations where sperm whales were encountered during (A) SWAMP 2000 Leg 1 and (B) Leg 2, overlaid on SeaWiFS 7-day composites centered on 4 July 2000 and 25 July 2000, respectively. Ship tracks are shown as pink lines, and whale encounters as white triangles. Panels C and D show encounters during SWAMP 2001 Legs 1 and 3, overlaid on SeaWiFS 7-day composites centered on 26 July 2001 and 23 August 2001, respectively. Panels E and F show encounters during SWSS 2002 Legs 1 and 2, overlaid on SeaWiFS 7-day composites centered on 14 June 2002 and 13 September 2002, respectively. Panels G and H show encounters during SWSS 2003 Legs 1 and 2, overlaid on SeaWiFS 7-day composites centered on 7 June 2003 and 12 July 2003, respectively.

cantly greater than the ensemble average along the 1000-m isobath (Table 2). A warm slope eddy (WSE) about 100 km in diameter that shows up as a +10 cm SSH high in the altimetry maps in Figure 1 is better seen in the SeaWiFS imagery in Plate 1B. Acting like a smaller version of LCE-K, this WSE also entrained green water from the 1000-m isobath and transported this green water farther off-margin and out into deeper water. The average SSH anomaly where whales were seen was only about −1 cm, but this was significantly lower than the ensemble average for the rest of the 1000-m isobath (Table 2).

3.2 LC Dynamics and Slope Eddies, Summer 2001

The 15°C depth data from the XBT and CTD stations indicated three broad regions along the middle slope in summer 2001. Spatially, these agreed quite well with the remotely-sensed SSH field (Figure 3). During SWAMP 2001 Leg 1, these regions were: a) WSE located west of 89°W, within which 15°C depths of 295–310 m were encountered west of 89.5°W); b) a cyclone located east of 88°W, within which 15°C depths domed to less than 180 m; c) a confluence region of off-margin surface flow between these two circulations, in which 15°C depths ranged 181–231 m. Both of the CTD stations made on Leg 1 were made near 88.5°W, in the boundary region between confluence and cyclone. Surface salinity at these stations was 33–34, which confirms there was off-margin rather than on-margin flow, and in TS plots there is no indication of Subtropical Underwater > 36.6 at depth from Leg 1 CTDs (Figure 2B). The absence of high salinity in the subsurface maximum, together with shallow 15°C depths at these two stations (134 m; 181 m), is strong evidence that the area around 88.5°W was also influenced by the adjacent cyclone.

NOAA scientists sighted sperm whales at several locations along the 1000-m isobath east of 89°W during SWAMP 2001 Leg 1, although no whales were sighted in the WSE. Most sightings of whales were in the confluence region of off-margin flow between 88.1°W and 88.7°W, but whales were also seen in the cyclone east of 87.6°W. As Table 2 shows, the average SSH where whales were seen was significantly lower than the ensemble average along the 1000-m isobath, and in fact no whales at allwere seen if SSH < -9 cm. Table 2 also shows these locations where whales were seen averaged significantly greener in color (SeaWiFS chlorophyll concentration about 1.0 µg/L) than the SeaWiFS chlorophyll concentration ensemble average of 0.4 µg/L for the 1000-m isobath.

When the data from the two CTD stations done on Leg 1 are combined with data from the seven CTDs done on Legs 2 and 3, there is a positive relationship between 15°C depth

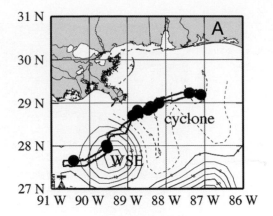

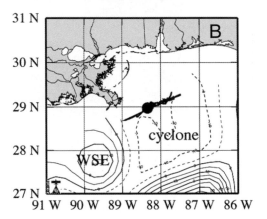

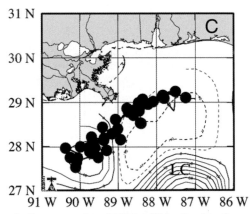

Figure 3. Survey tracks of NOAA Ship Gordon Gunter for the three legs of SWAMP 2001 fieldwork, superimposed on the respective mean SSH fields for the mid-points in time of each leg. The locations of XBTs and CTDs are shown.

and dynamic height. The cyclone-confluence boundary zone along the 1000-m isobath at 88.3–88.5°W had a height anomaly of -4 to -8 dyn cm, while farther to the west along the 1000-m isobath in the region 89.7–90.1°W, the WSE had a height anomaly of +9 to +10 dyn cm.

One XBT was dropped and one CTD station was made during SWAMP 2001 Leg 2, both at the same location along the 1000-m isobath at 88.3°W. This location was a transition from confluence to cyclone, as indicated by a 15°C depth of 170 m, surface salinity of 34.5, and dynamic height anomaly of -7 cm. NOAA scientists sighted sperm whales here, and also in the region a little to the east near 87.8°W.

From 41 XBTs dropped and 6 CTDs done during SWAMP 2001 Leg 3, the continental margin along the 1000-m isobath can again be subdivided into three broad regions. As during Leg 1, these agree quite well with the remotely-sensed SSH field (see bottom panel of Figure 3): a) WSE west of 89.5W, within which 15°C depths averaged 250 m; b) a cyclone located east of 88°W, within which 15°C depths averaged 166 m; c) a confluence region of off-margin surface flow between these two circulations, in which 15°C depths averaged 209 m. Salinity subsurface exceeded 36.7 at three CTD stations done near 90°W within the WSE, which indicates the WSE likely originated as a LC warm filament. That this filament must have been a rather shallow extrusion from the LC (limited to the upper 150 m meters or so) is indicated by the dynamic height anomaly of these CTDs in the interior of the WSE, relative to a reference level of 800 db: dynamic height anomaly is only +9 to +10 cm, which is in good spatial agreement with the remotely-sensed SSH field for this area near 90°W (Figure 3, bottom panel).

The average SSH anomaly along the 1000-m isobath during SWAMP Leg 3 was –7 cm (Table 2), and during Leg 3 the NOAA scientists sighted sperm whales pretty much all along the 1000-m isobath. The mean SSH anomaly where whales were encountered was also –7 cm (Table 2). Five groups of whales were seen east of 89°W, in the confluence region of off-margin flow and in the cyclone, but whales were also seen between 89.3°W and 90.2°W, in the WSE.

Current velocity determined from ADCP data collected during the summer 2001 SWAMP cruises is presented in Figure 4. Near-surface currents measured by the hull-mounted ADCP agree quite well in average speed and direction with our understanding of the meso-scale circulation derived from XBT and CTD data and from SSH altimetry. The velocity vectors that are shown along track are each the average of 10 two-minute periods of raw data, so each represents the average current velocity for 20 minutes of time along track. The upper panel of Figure 4 shows that during SWAMP Leg 1, currents along the 1000-m isobath set northerly between 90°W and 89.5°W, and then changed to generally easterly and then to off-margin from 89.5°W to 88°W (i.e. anticyclonic flow around the WSE). These anticyclonic current speeds averaged about 1 m s⁻¹. There is also evidence for cyclonic (counterclockwise) flow along the 1000-m isobath east of 88°W. The lower panel of Figure 4 summarizes the flow field one month later, during SWAMP

Leg 3. Note that the anticyclonic circulation about the WSE still averages 1 m s⁻¹, but that this is now restricted spatially to where the altimetry shows the WSE had consolidated to become smaller-in-diameter yet higher-in-SSH. Between 88°W and 87°W, the flow is obviously cyclonic, and the off-margin confluence flow in the zone 88-89°W is also apparent. One month following SWAMP Leg 3, additional ADCP data were collected in this region 91°W–86°W by NOAA Ship *Gordon Gunter* during NOAA-SEAMAP fieldwork. By mid-September 2001, the velocity of ADCP-measured currents in most areas along the middle slope 90.5°W to 87.5°W was generally lower (generally < 0.5 m s⁻¹) and generally in the off-margin direction [data not shown here; see *Sindlinger et. al.*, 2005 for details]. Locally higher ADCP-measured velocities around the WSE that by mid-September had now moved to the southwest and offshore are evidence this feature retained an organized anticyclonic circulation.

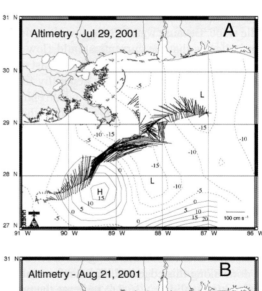

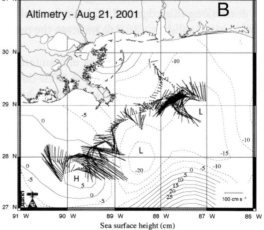

Figure 4. Current velocity determined from ADCP data collected during a) Leg 1 of SWAMP 2001 and b) Leg 2 of SWAMP 2001, overlaid on SSH anomaly maps for the mid-points in time of each leg.

visual evidence that although WSEs over the middle slope can entrain shelf water and move it along-margin or off-margin, the magnitude of the entrainment and the distance that green water is transported off-margin is not as great as when a full-blown LCE is interacting with the middle slope region. A similar comparison which is not shown here but which can be found as Figure 4.6.21 in the SWSS 2003 Annual Report [*Jochens and Biggs*, 2004] contrasts the SeaWiFS monthly composite image for April 2003 (before LCE separation) with June 2003 (LCE close off-margin), to visually make the case that maximum entrainment occurred only when LCE Sargassum was close off-margin.

Sperm whales were not encountered in the Mississippi Canyon area 89.2–90.3°W where/when LCE Sargassum was interacting with the 1000-m isobath (Plate 1, Panel G). Rather, groups of whales were found west of 90.3°W and east of 89.2°W, where 15°C depths (and the SSH maps) indicate these areas were outside the core of Eddy Sargassum. There was a very sharp surface front at the northern periphery of Eddy Sargassum. On the morning of 16 June 2003, we documented a sharp boundary between low-chlorophyll "blue water" and higher-chlorophyll "green water" near 28.9° N, 88.3 °W. From 16 to 18 June 2003, *Gyre* followed numerous groups of sperm whales in the green water northeast of this front. Table 2 shows the average SeaWiFS chlorophyll concentration where whales were encountered was 2.1 µg/L, compared to the ensemble average of 0.4 µg/L for the whole of the 1000-m isobath. Only two sightings of whales were made inside the blue water front marking the northern periphery of the LCE between 88°W and 90°W. This was not unexpected, since the GulfCet 2 program also found sperm whales to be uncommon in the interior of LCEs [*Biggs et al.*, 2000; *Davis et al.*, 2002].

During the two weeks that passed between the time the Mississippi Canyon region was surveyed by *Gyre* during Leg 1 and Leg 2 of SWSS 2003, the altimetry data tell us that LCE Sargassum had rotated clockwise about its major axis and moved away from (rebounded seaward from) the 1000-m isobath. In confirmation of this change in geometry, CTDs taken on Leg 2 show no evidence for Subtropical Underwater (Figure 2D), and 15°C depths from XBTs dropped on Leg 2 were < 255 m. Cyclonic hydrographic conditions were now the norm all along the 1000-m isobath east of 90°W by mid-July 2003, and during Leg 2 several groups of sperm whales were again encountered in the Mississippi Canyon area near 89.8°W, as well as along the middle slope to the northeast (Plate 1, Panel H). The average SSH anomaly 91°–86 °W where whales were encountered on Leg 2 was –19 cm (Table 2), and in fact no whales were seen in environments with SSH > -7 cm even though some areas along the 1000-m isobath 91°–86 ° W had SSH anomalies up to 11 cm.

4. DISCUSSION

LC eddies and slope eddies contribute biological and physical heterogeneity along the continental margin of the northern GOM. Temporal and spatial variations in the geometry of the eddy field along the middle slope determine whether low salinity green water flows off-margin, or if high salinity blue water flows on-margin [*Morey et al.*, 2003 and this volume; *Chassignet et al.*, this volume]. Green water is biologically rich and we hypothesize it supports more food for (more prey for) the squid upon which sperm whales prey. Locally high chlorophyll can also develop at the periphery of eddies, when/where high velocity currents (>1 m s⁻¹) create vertical shear (and thus upwelling of nutrients from midwater). Moreover, locally high chlorophyll can develop when or where nutrient-rich water domes upward in cyclonic eddies. Cyclonic eddies and other nutrient-rich features that persist for 3–4 months in time may be important feeding grounds for sperm whales along the GOM continental slope. LCEs, in contrast, appear to generate the opposite effect.

Our four summers of data show that most sperm whales were encountered in regions with negative SSH and/or higher-than-average surface chlorophyll (Table 2). Maximum and minimum data for each cruise (not shown) indicate the range of SSH along the 1000-m isobath from 91°–86 ° W went from +14 cm to –28 cm, but SSH in locations where whales were encountered never exceeded + 6 cm. Most of the whales were encountered in locations on average 5 cm lower in SSH than the ensemble average for the 1000-m isobath (Table 2), and in summer 2003 some of the whales were found in strongly cyclonic locations for which SSH < -25 cm. Thus, although sperm whales in the western North Atlantic Ocean have been reported to congregate along the periphery of warm core rings [*Griffin*, 1999], in the northern Gulf of Mexico most sightings in summers 2000–2003 were correlated with regions of negative SSH anomaly and to off-margin flows of green water.

In summer 2000, four groups of sperm whales were seen in the region between 90°W and 88°W when this area was first searched 29 June –8 July, but 19 groups of sperm whales were seen between 90°W and 88°W when this region was searched again 12 – 25 July. Remote sensing using altimetry and ocean color showed the combination of clockwise flow around an LC eddy and counterclockwise flow around a deepwater cyclone resulted in off-margin flow. The shape and location of both of these deepwater circulation features changed between the first and second surveys of the 1000-m isobath, but during either survey more sperm whales were seen in the off-margin flow confluence between these two features than were seen outside the confluence zone. Analysis of SeaWiFS imagery shows these two eddies entrained high chlorophyll surface water from the Mississippi River and transported this "green water" seaward to the 1000-m isobath and beyond. SeaWiFS imagery also shows that green water being transported cross-margin by this entrainment of Mississippi River water covered a more extensive area of the 1000-m isobath during the second survey than during the first survey. We speculate that since four times as many sperm whales were seen in the flow confluence along the 1000-m isobath when entrainment and transport were well-developed than when off-margin transport was more limited, it is possible that sperm whales in the vicinity of the mouth of the Mississippi River may detect and then aggregate behaviorally in such deepwater regions of surface confluence.

Biggs et al. [2000] and *Davis et al.* [2002] reported data from the GulfCet II program that suggested that sperm whales may be locally abundant in deepwater cyclones, as well. However, since neither the cruise track of the SWAMP 2000 *Gordon Gunter* sperm whale study nor the ship tracks in follow-on summers extended into deepwater cyclones, it is unknown to what degree the sperm whale distribution may have extended seaward of the 1000-m isobath.

None of the SSH altimetry data for the eight week pre-SWAMP cruise period 25 May through 13 July 2001 have been presented in this paper, but these data are readily accessible on the CCAR website and they are of course part of the 12 year animation of altimetry data 1993–2004 included on CD-ROM as an appendix to this book. Here, we will assume that interested readers will have viewed this movie, for it offers considerable insight into the oceanographic mechanisms that created the WSE that was present along the middle slope during SWAMP 2001. In late May 2001, the altimetry shows there was a strong cyclone along the continental margin of the north central Gulf of Mexico between 90°W and 88°W. The northern edge of LC was far to the south of the middle slope, reaching only to latitude 26°N. However, a warm filament in the final stages of separating from the LC in late May 2001 can be seen as an area of +15 to +20 dyn cm that stretches northward > 275 km, from 26°N into the DeSoto Canyon. A frame-by-frame progression of the movie of SSH fields shows that in June and July there were interactions between this filament and deepwater LC eddy M (LCE-M), which had separated from the LC in late March to early April 2001. During these June–July interactions between LCE-M and the warm filament, the two are sometimes connected by a "bridge" of positive height anomaly. On some dates during June–July the filament, which can now be classed as a WSE because of its location close off the continental margin, appears to "spin up" (SSH > 20 dyn cm). At other times the filament/WSE appears to have relaxed to < 15 dyn cm. By mid-July 2001, just as SWAMP Leg 1 was ready to sail for sea, the filament/WSE

had taken on an elliptical shape and moved west along the slope so that it was centered near 90°W, while the region east of 88°W was cyclonic in nature. The top panels of Figures 3 and 4 and Plate 1C show the mesoscale geometry for Leg 1 of the SWAMP 2001 cruise: the combination of clockwise circulation around the anticyclonic filament/WSE west of 89°W and anti-clockwise circulation about the cyclonic eddy to the east had begun to create off-margin surface flow in the region 89–88°W.

From mid-July until late August 2001, the northward movement of the LC into the eastern GOM apparently caused a change in the relative geometry of this cyclone-anticyclone circulation pair over the continental margin. By early August, the WSE had intensified in SSH anomaly and had become more circular in shape. This change in the geometry of the WSE was coincident with, although probably independent of, the formation of Tropical Storm Barry. By late August 2001, the continued northward movement of the LC apparently squeezed the WSE south and west, so that it moved off-margin into deeper water. By mid-September, when a follow-on NOAA cruise worked stations between 90.5°W to 86.5°W, there was no longer a WSE along the 1000-m isobath. Instead, SSH anomaly along the 1000-m isobath from 91°W to 87°W ranged 0 to -10 cm and the middle slope region had become cyclonic in character.

Similarly, viewing an animation of the pre-cruise SSH altimetry data prior to SWSS 2002 fieldwork permits insight into the oceanographic mechanisms that created the alternation of on-margin and off-margin flow along the middle slope in summer 2002. Readers joining the animation in January 2002 will see that a gradient of increasing SSH from N to S (from shelf to slope) existed over most of the north central GOM for most of the first four months of 2002. This is evident as a temporally persistent although spatially variable region of negative-to-positive sea surface height anomaly. In the negative SSH part of this gradient, which usually includes the 800–1000-m isobaths, doming of nutrient-rich midwater close to the surface likely favors enhanced planktonic new production along this continental margin [*Wiseman and Sturges*, 1999; *Biggs and Ressler*, 2001]. Such conditions, we believe, jump start biological production throughout the food chain, which likely results in more potential prey for sperm whales and other apex predators.

Early in 2002, the LC shed LCE-Q (also called "Quick Eddy"), which then in March–April 2002 cleaved into two pieces: a western QE1 and a central QE2. In the first half of May 2002, QE2 in turn shed a warm filament that extended north into the DeSoto Canyon. During the second half of May and the first half of June 2002, this warm filament consolidated into a WSE, the inshore edge of which reached the Mississippi Canyon region south of the Mississippi River delta. In mid-June 2002, the gradient of increasing SSH over the slope 94°W–88°W indicates there was west to east flow along most of the 1000-m isobath, but off-margin flow west of 94°W and east of 88°W.

To supplement the insights that are available from the altimetry time series, the 15°C depth and ADCP current velocity ship data from summer 2003 indicate that the large, energetic LCE "Sargassum" displaced the upper hundreds of meters or so of usual water in the Mississippi Canyon area with low-nutrient, low-chlorophyll "ocean desert" water of Caribbean origin, when it interacted with the 1000-m isobath in early June 2003. We hypothesize that sperm whales usually seen in this area (i.e., summers 2000–2002) moved west and/or east out of this area during the time this LCE reached farthest north along the margin (early June through late June 2003). Whales were in greater abundance in summer 2003 west of 90°W and east 89°W than in the Mississippi Canyon region 90–89°W where the LCE attained its closest approach to the middle slope. We presume the whales left when their deep-living squid prey was also displaced by this bolus of northward-moving Caribbean water. Return to normal conditions of hydrography (and squid prey?) appears to have occurred after this LCE moved back (rebounded) into deeper water, since by early July 2003 the 15°C depth had returned to normal in the region 90–89°W, and whales were again encountered in Mississippi Canyon.

ADCP volume backscatter intensity (VBI) data can give information about the spatial and temporal variability of plankton stocks in the GOM [*Zimmerman and Biggs*, 1999; *Biggs and Ressler*, 2001], so VBI data were analyzed to look for differences among the various hydrographic regimes that were present in summers 2001–2003 along the middle slope. In summer 2001, the region of the cyclone had the highest average VBI, while the region of the warm filament had the lowest average VBI [*Sindlinger et. al.*, 2005]. In summers 2002 and 2003, areas of anticyclonic circulation again had lower VBI than did areas of low (cyclonic) or near-background SSH anomaly. However, GOM sperm whales dive to feed at depths generally > 600 m, so the VBI in the upper 100–250 m that can be obtained from 153 kHz and 300 kHz ADCPs is of indirect value, at best. Currently, the use of VBI from lower frequency ADCPs such as the 38 kHz phased-array is being investigated to see whether there is more or more variable VBI in 600–800 m water depths along the middle slope where the whales most often feed, than there is in this same depth range over deeper water. Preliminary data [*Kaltenberg*, 2004] indicate that this seems to be the case, although additional work is needed with a calibrated fisheries echosounder or other acoustic system with inherently higher temporal and vertical resolution than the 38 kHz ADCP, which collects bins VBI over five minute and

16 m vertical intervals. Especially if it can pick out the individual squid prey of sperm whales at depths of 600–800 m below the surface, then use of a deep-towed multi-frequency acoustic system such as the "Biomapper" [*Wiebe et al.*, 1996] may be a logical next step to better understand the biological implications of off-margin versus on-margin flow to the sperm whale predators that feed at these middle slope depths in the GOM.

Acknowledgments. TAMU and CCAR support came from MMS-TAMU cooperative agreement 1435-01-02-CA-85186. USF support came from NASA contracts NAS5-97128 and NAG5-10738. Sea-WiFS data are the property of Orbimage Corporation and data use here is in accordance with the SeaWiFS Research Data Use Terms and Conditions Agreement of the SeaWiFS project. The Sperm Whale Acoustic Monitoring Program (SWAMP) was supported by the NMFS and the MMS under Interagency Agreement 15958. The Sperm Whale Seismic Study (SWSS) is sponsored by the MMS and involves researchers from Oregon State University, Woods Hole Oceanographic Institution, Scripps Institution of Oceanography, University of Durham, and University of St. Andrews in addition to those from TAMU, with support and cooperation from the Industry Research Funders Coalition (International Association of Geophysical Contractors and oil and gas companies), National Fish and Wildlife Foundation, National Science Foundation, and Office of Naval Research. The program is a cooperative, multi-disciplinary agreement to study sperm whales and their response to seismic exploration in the Gulf of Mexico.

REFERENCES

Belabbassi, L., P. Chapman, W. D. Nowlin, Jr., A. E. Jochens, and D.C. Biggs, Summertime nutrient supply to near-surface waters of the northeastern Gulf of Mexico: 1998, 1999, and 2000, *Gulf Mex. Sci.*, in press.

Belabbassi, L., Importance of physical processes on near-surface nutrient distributions in summer in the northeastern Gulf of Mexico, M.S. thesis, Department of Oceanography, Texas A&M University, College Station, TX, 85 pp., 2001.

Berger, T. J., P. Hamilton, J. J. Singer, R. R. Leben, G. H. Born, and C. A. Fox, Louisiana/Texas shelf physical oceanography program: Eddy circulation study, final synthesis report. Vol. I, Tech. Report, OCS Study, MMS 96-0051, U.S. Department of the Interior, Minerals Management Service, Gulf of Mexico OCS Region, New Orleans, LA, 324 pp., 1996.

Biggs, D. C., G. S. Fargion, P. Hamilton, and R. R. Leben, Cleavage of a Gulf of Mexico Loop Current eddy by a deep water cyclone, *J. Geophys. Res.*, *101*, 20,629–20,641, 1996.

Biggs, D. C., R. R. Leben, R. R., and J. G. Ortega-Ortiz, Ship and satellite studies of mesoscale circulation and sperm whale habitats in the northeast Gulf of Mexico during GulfCet -II, *Gulf Mex. Sci. 18*, 15–22, 2000.

Biggs, D. C., and P. H. Ressler, Distribution and abundance of phytoplankton, zooplankton, ichthyoplankton, and micronekton in the deepwater Gulf of Mexico, *Gulf Mex. Sci. 19*, 7–35, 2001.

Chassignet, E. P., H. E. Hurlburt, O. M. Smedstad, C. N. Barron, D. S. Ko, R. C. Rhodes, J. F. Shriver, A. J. Wallcraft, and R. A. Arnone, Assessment of ocean prediction systems in the Gulf of Mexico using ocean color, contributed paper to this volume.

Davis, R. W., J. G. Ortega-Ortiz, C. A. Ribic, W. E. Evans, D. C. Biggs, P. H. Ressler, R. B. Cady, R. R. Leben, K. D. Mullin, and B. Wursig, Cetacean habitat in the northern oceanic Gulf of Mexico, *Deep-Sea Res. I, 49*, 121–142, 2002.

Dietrich, D. E., and C. A. Lin, C. A., Numerical studies of eddy shedding in the Gulf of Mexico, *J. Geophys. Res., 99*, 7599–7615, 1994.

Elliott, B. A., Anticyclonic rings in the Gulf of Mexico, *J. Phys. Oceanogr., 12*, 1291–1309, 1982.

Fletcher, W. A., Seasonal and interannual differences in surface chlorophyll and integrated water column chlorophyll stocks in the NE Gulf of Mexico, M.S. thesis, Department of Oceanography, Texas A&M University, College Station, TX, 128 pp., 2004.

Gordon, H. R., and M.Wang, Retrieval of water-leaving radiance and aerosol optical thickness over the oceans with SeaWiFS: a preliminary algorithm. *Appl. Opt., 33,* 443–452, 1994.

Griffin, R. B., Sperm whale distributions and community ecology associated with a warm-core ring off George's Bank, *Mar. Mammal Sci., 15*, 33–51, 1999.

Hamilton, P., G. S. Fargion, and D. C. Biggs, Loop Current eddy paths in the western Gulf of Mexico, *J. Phys. Oceanogr., 29*, 1180–1207, 1999.

Hooker, S. B., and E. R. Firestone (Eds.), *SeaWiFS postlaunch technical report series, Vol. 22: Algorithm updates for the fourth SeaWiFS data processing*, 51–59, 2003.

Hu, C., F. E. Muller-Karger., D. C. Biggs, K. L.Carder, B. Nababan, D. Nadeau, and J. Vanderbloemen, Comparison of ship and satellite bio-optical measurements on the continental margin of the NE Gulf of Mexico, *Int. J. Rem. Sens., 24*, 2597–2612, 2003.

Hu, C., F. E. Muller-Karger, G. A. Vargo, M. B. Neely, and E. Johns, Linkages between coastal runoff and the Florida Keys ecosystem: A study of a dark plume event. *Geophys. Res. Lett. 31*, L15307, doi:10.1029/2004GL020382, 2004a.

Hu, C., E. Montgomery, R. Schmitt, and F. E. Muller-Karger, The Amazon and Orinoco River plumes in the tropical Atlantic: Observation from space and S-Floats, *Deep Sea Res-II*, 51, 1151–1171, 2004b.

Jochens, A. E., S. F. DiMarco, W. D. Nowlin Jr., R. O. Reid, and M. C. Kennicutt II, Northeastern Gulf of Mexico Chemical Oceanography and Hydrography Study: Synthesis Report, U.S. Department of the Interior, Minerals Management Service, Gulf of Mexico OCS Region, New Orleans, LA, OCS Study/MMS 2002-055, 331 pp. plus 9 appendices, 2002.

Jochens, A. E., and D. C. Biggs, editors, Sperm whale seismic study in the Gulf of Mexico, Annual Report Year 1, U.S. Department of the Interior, Minerals Management Service, Gulf of Mexico OCS Region, New Orleans, LA, OCS Study/MMS 2003-069, 128 pp., 2003.

Jochens, A. E., and D. C. Biggs, editors, Sperm whale seismic study in the Gulf of Mexico, Annual Report Year 2, U.S. Department of the Interior, Minerals Management Service, Gulf of Mexico OCS Region, New Orleans, LA, OCS Study/MMS 2004-067, 154 pp., 2004.

Kaltenberg, A. M., 38-kHz ADCP observations of deep scattering layers in the northern Gulf of Mexico sperm whale habitat, M.S. thesis, Department of Oceanography, Texas A&M University, College Station, TX, 124 pp., 2004.

Kirwan, A. D., Jr., W. J. Merrell, J. K. Lewis, R. E. Whitaker, and R. Legeckis, A model for the analysis of drifter data with an application to a warm core ring in the Gulf of Mexico, *J. Geophys. Res., 89*, 3417–3424, 1984.

Leben, R. R., G. H. Born, and B. R. Engebreth, Operational altimeter data processing for mesoscale monitoring, *Mar. Geodesy, 25*, 3–18, 2002.

McClain C. R., M. L. Cleave, G. C. Feldman, W. W. Gregg, S. B. Hooker, and N. Kuring, Science quality SeaWiFS data for global biosphere research. *Sea Tech. 39*:10–16, 1998.

Melo-Gonzalez, N., F. E. Muller-Karger, S. Cerdeira-Estrada, R. Perez de los Reyes, I. Victoria del Rio, P. Cardenas-Perez, and I. Mitrani-Arenal, Near-surface phytoplankton distribution in the western Intra-Americas Sea: The influence of El Nino and weather events, *J. Geophys. Res., 105*, 14029–14043, 2000.

Morey, S. L., W. W. Schroeder, J. J. O'Brien, and J. Zavala-Hidalgo, The annual cycle of riverine influence in the eastern Gulf of Mexico basin, *Geophys. Res. Lett., 30(16)*, 1867–1870, 2003.

Morey, S. L., J. Zavala-Hidalgo, and J. J. O'Brien, The seasonal variability of continental shelf circulation in the northern and western Gulf of Mexico from a high-resolution numerical model, contributed paper to this volume.

Mullin, K. D., and G. L. Fulling, Abundance of cetaceans in the oceanic northern Gulf of Mexico, 1996–2001. *Mar. Mammal Sci. 20(4)*, 787–807, 2004.

Muller-Karger, F. E., J. J. Walsh, R. H. Evans, M. B. Meyers, On the seasonal phytoplankton concentration and sea surface temperature cycles of the Gulf of Mexico as determined by satellites, *J. Geophys. Res., 96*, 12645–12665, 1991.

O'Reilly, J. E., S. Maritorena, D. A. Siegel, M. C. O'Brien, D. Toole, F. P. Chavez, P. Strutton, G. F. Cota, S. B. Hooker, C. R. McClain, K. L. Carder, F. E. Muller-Karger, L. Harding, A. Magnuson, D. Phinney, G. F. Moore, J. Aiken, K. R. Arrigo, R. Letelier, and M. Culver, *in* S. B. Hooker, and E. R. Firestone (Eds.), *Ocean Chlorophyll-a Algorithms for SeaWiFS, OC2 and OC4: version 4. SeaWiFS Postlaunch Calibration and Validation Analyses, Part 3. NASA Tech. Memo. 2000-206892, Vol. 11*, 9–23. NASA Goddard Space Flight Center, Greenbelt, Maryland, 2000.

Qian, Y., A. E. Jochens, M. C. Kennicutt II, and D. C. Biggs, Spatial and temporal variability of phytoplankton biomass and community structure over the continental margin of the northeast Gulf of Mexico based on pigment analysis, *Contl. Shelf Res., 23*, 1–17, 2003.

Sindlinger, L. R., D. C. Biggs, and S. F. DiMarco, Temporal and spatial variability of ADCP backscatter on a continental slope, *Contl. Shelf Res., 25*, 259–275, 2005.

Sturges, W., and R. Leben, Frequency of ring separation from the Loop Current in the Gulf of Mexico: A revised estimate, *J. Phys. Oceanogr., 30*, 1814–1819, 2000.

Thode, A., D. K. Mellinger, S. Stienessen, A. Martinez, and K. Mullin, Depth-dependent acoustic features of diving sperm whales (*Physeter macrocephalus*) in the Gulf of Mexico, *J. Acous. Soc. Am., 112*, 308–321, 2002.

Vidal, V. M., F. V. Vidal, and J. M. Perez-Molero, Collision of a Loop Current anticyclonic ring against the continental shelf-slope of the western Gulf of Mexico, *J. Geophys. Res., 97*, 2155–2172, 1992.

Vidal, V. M., F. V. Vidal, A. H. Hernandez, E. Meza, and J. M. Perez-Molero, Baroclinic flows, transports, and kinematic properties in a cyclonic-anticyclonic ring triad in the Gulf of Mexico. *J. Geophys. Res., 99*, 7571–7597, 1994.

Vukovitch, F. M., and B. W. Crissman, Aspects of warm rings in the Gulf of Mexico, *J. Geophys. Res., 91*, 2645–2660, 1986.

Walker, N. D., G. S. Fargion, L. J. Rouse, and D. C. Biggs, The great flood of summer 1993: Mississippi River discharge studied, *EOS Trans. AGU, 75,* 409–415, 1994.

Welsh, S. E., and M. Inoue, Loop Current rings and the deep circulation in the Gulf of Mexico, *J. Geophys. Res., 105*, 16951–16959, 2000.

Wiseman, W. J., Jr., and W. Sturges, Physical oceanography of the Gulf of Mexico: Processes that regulate its biology. pp. 77–92 in: Kumpf, H., K. Steidinger, K., and K. Sherman, (Eds.), *The Gulf of Mexico Large Marine Ecosystem.* Blackwell Science, Inc., Malden, MA, 1999.

Wiebe, P. H., D. G. Mountain, T. K. Stanton, C. H. Greene, G. Lough, S. Kartvedt, J. Dawson, and N. Copley, Acoustical study of the spatial distribution of plankton on Georges Bank and the relationship between volume backscattering strength and the taxonomic composition of the plankton, *Deep-Sea Res. II 43*, 1971–2001, 1996.

Zimmerman, R. A., and D. C. Biggs, Patterns of distribution of sound-scattering zooplankton in warm- and cold-core eddies in the Gulf of Mexico, from a narrowband acoustic Doppler current profiler study, *J. Geophys. Res., 104*, 5251–5262, 1999.

Douglas C. Biggs, Steven F. DiMarco, Matthew K. Howard, and Ann E. Jochens, Department of Oceanography, Texas A&M University, College Station, TX 77843-3146

Chuanmin Hu, and Frank E. Muller-Karger, College of Marine Science, University of South Florida, St Petersburg, FL 33701

Robert R. Leben, Colorado Center for Astrodynamics Research, University of Colorado, Boulder, CO 80309

Keith D. Mullin, Southeast Fisheries Science Center, National Marine Fisheries Service, Pascagoula, MS 39568

Assessment of Data Assimilative Ocean Models in the Gulf of Mexico Using Ocean Color

Eric P. Chassignet[1], Harley E. Hurlburt[2], Ole Martin Smedstad[3],
Charlie N. Barron[2], Dong S. Ko[2], Robert C. Rhodes[2], Jay F. Shriver[2],
Alan J. Wallcraft[2], and Robert A. Arnone[2]

This paper illustrates the value of SeaWiFS ocean color imagery in assessing the ability of three data-assimilative ocean models (configured in five prediction systems) to map mesoscale variability in the Gulf of Mexico (i.e., the Loop Current and associated warm and cold eddies) and in helping to diagnose specific strengths and weaknesses of the systems. In addition, the study clearly illustrates that biological responses of the surface waters are strongly linked to the physical events and processes.

1. INTRODUCTION

An accurate depiction of currents and associated features such as mesoscale eddies, fronts, and jets, is essential to any ocean forecasting system. The ability to use high-resolution ocean models for data assimilation and forecasting is critical to the development of these nowcast/forecast systems. Applications include assimilation and synthesis of global satellite surface data, ocean prediction, optimum track ship routing, search and rescue, antisubmarine warfare and surveillance, tactical planning, high resolution boundary conditions that are essential for even higher resolution coastal models, sea surface temperature for long range weather prediction, inputs to shipboard environmental products, environmental simulation and synthetic environments, observing system simulations, ocean research, pollution and tracer tracking, inputs to water quality assessment, and so forth.

Since most high horizontal resolution observational data (i.e., altimetry and sea surface temperature) are assimilated in these systems, there are very few independent observa-

tions with sufficient coverage to be useful in assessing the ability of ocean prediction models to map mesoscale ocean features. In this paper, we assess the performance of several data assimilative ocean model systems against independent ocean color measurements in the Gulf of Mexico. The near-surface biological and optical properties, derived from ocean color imagery, provide an excellent evaluation method for assessing the capability of data-assimilative numerical models to define the spatial and temporal mesoscale variability. Although these properties are non-conservative and are influenced by biological processes, these optical gradients are an independent data set with the ability to track and monitor mesoscale features such as warm and cold core rings, coastal filaments and the Loop Current position. Feature tracking using a time series of these independent observations in the Gulf of Mexico is a very effective way to compare different ocean prediction systems.

Because of the logistics associated with routine analysis and prediction, most large-scale ocean prediction systems are located within government centers. Recently, the Global Ocean Data Assimilation Experiment (GODAE) has provided an international framework for such systems to be evaluated and inter-compared. The rationale for restricting ourselves to the systems compared in this paper (see section 3 for details) is that they all use the same atmospheric forcing and assimilate the same observational data. This allows us to evaluate the performance of the systems independently of the choice made for external inputs (atmospheric forcing

[1]RSMAS/MPO, University of Miami, Miami, Florida
[2]Naval Research Laboratory, Stennis Space Center, Mississippi
[3]Planning Systems Inc., Stennis Space Center, Mississippi

Circulation in the Gulf of Mexico: Observations and Models
Geophysical Monograph Series 161
Copyright 2005 by the American Geophysical Union.
10.1029/161GM07

or ocean data input). Another example of a Gulf of Mexico data assimilative system is discussed by Kantha et al. in this volume.

2. OCEAN COLOR

The data set used to evaluate the forecasting systems consists of satellite measurements of ocean color data obtained by the Sea-viewing Wide Field-of-view Sensor (SeaWiFS) in orbit on the OrbView-2 (formerly SeaStar) platform. Ocean color satellites such as SeaWiFS and MODIS provide new capability to quantitatively monitor biological and optical properties of the near surface ocean [*Arnone and Parsons*, 2004]. Accurate sensor calibration and atmospheric correction are required in order to detect the subtle changes in ocean color which are used to uncouple the in-water components.

The apparent color of the ocean, which can range from deep blue to varying shades of green and ruddy brown, is primarily determined by the concentration of substances and particles in the euphotic (lighted) zone of the upper ocean. Living phytoplankton (which contain chlorophyll and associated photosynthetic pigments), inorganic sediments, detritus (particulate organic matter), and dissolved organic matter all contribute to the color of the ocean. When visible light from the sun illuminates the ocean surface, it is subject to several optical effects. Foremost among these effects are light absorption and scattering. Spectral absorption selectively removes some wavelengths of light while allowing the transmission of other wavelengths. In open ocean waters absorption is primarily due to the photosynthetic pigments (chlorophyll) present in phytoplankton, whereas in coastal waters additional absorption processes occur from the absorption of detritus and colored dissolved organic matter (CDOM). Following spectral absorption, backscattered light (resulting from organic and/or inorganic particulate matter suspended in the water) produces a modified light radiating from the ocean surface, the "water-leaving radiance," i.e., the ocean color. Spaceborne radiometers such as SeaWiFS measure the spectral radiance intensity at specific wavelengths. These measurements are used to determine quantitative constituents in the water column that interact with visible light, such as chlorophyll. Additional components of the absorption and backscattering light field can be uncoupled from the ocean color signatures, but are beyond the scope of this paper. We used the chlorophyll component in this study because it is the dominant contributor to the ocean color in open ocean processes.

The chlorophyll products from SeaWiFS represent the integrated near-surface concentrations within the first attenuation length of visible light (490 nm). For Loop Current waters, this is approximately 20 meters, whereas in turbid coastal areas, this is reduced to as little as two to three meters. This product is different from a satellite SST product that only senses the skin temperature and responds to diurnal heating. Ocean color products can be used to detect ocean features throughout the year, which is especially useful during the summer months when the SST in the Gulf of Mexico is nearly isothermal and cannot be used for feature detection. However, chlorophyll concentrations can change rapidly through phytoplankton growth and decay and through vertical movements of phytoplankton into or out of view of the satellite.

The SeaWiFS mission is a public-private partnership between NASA and Orbital Sciences Corporation. SeaWiFS was launched in August 1997 and acquires approximately 15 pole-to-pole orbital swaths (~ 2800 km wide) of data per day, and approximately 90% of the ocean surface is scanned every two days. The SeaWiFS ocean data set used in this effort was collected and processed daily for the Gulf of Mexico at the Naval Research Laboratory (NRL) receiving site using the Automatic Processing System developed for the Naval Oceanographic Office to provide operational products. Atmospheric correction and in-water chlorophyll algorithms (OC4) were applied [*Arnone et al.*, 1998; *O'Reilly et al.*, 1998] as extensions of the NASA algorithms to address the coastal waters. Note that the OC4 algorithm uses a simple ratio of spectral channels at 443, 488, 520 and 550 nm of the water-leaving radiance to determine chlorophyll. This simple algorithm is closely associated with total absorption coefficient, which in the open ocean is closely associated with chlorophyll concentrations [*Lee et al.*, 1998]. However, for other locations, such as in coastal areas, the algorithm significantly overestimates chlorophyll, and the values are closely associated with the total absorption coefficient rather than chlorophyll.

Because the majority of visible imagery is limited by cloud cover, the latest pixel composite is created daily to represent the optimum coverage of ocean features within the last 7 days. The compositing technique provides the most recent image of ocean conditions each day. A seven-day period effectively removes cloud cover and still retains consistency in feature tracking to monitor the slower mesoscale ocean features for comparison with the model results.

Ocean color can change rapidly in response to 1) open ocean biological processes occurring at daily time scales and 2) subduction of surface chlorophyll below the satellite detection depth. Phytoplankton concentrations can double within a day provided optimum light and nutrient availability. More traditionally, chlorophyll response in upwelling regions occurs on time scales of 4–7 days as the bloom matures, and blooms with a continuous nutrient supply can remain observable for months. A similar response occurs in frontal boundaries where a steady increase in chlorophyll is observed with constant frontal upwelling. Similarly, blooms can degrade rapidly in 4–7 days as nutrients are rapidly depleted. However, these daily color changes mainly influence the magnitude of the chlorophyll concentration, whereas the location of the fronts and eddies is driven primarily by the circulation. Therefore, the use of the seven day composite provides a reasonable location of the mesoscale features which are identified by the models.

3. THE DATA ASSIMILATIVE OCEAN MODELS

The three data assimilative ocean models (configured into five prediction systems) discussed in this paper are designed for nowcasting and forecasting of the mesoscale, large scale, and upper ocean. Two of the prediction systems use the NRL Layered Ocean Model (NLOM) [Hurlburt and Thompson, 1980; Wallcraft et al., 2003] and two use the Navy Coastal Ocean Model (NCOM) [Barron et al., 2004, 2006, Kara et al, 2005], all four developed at the Naval Research Laboratory, Stennis Space Center. The final system is based on HYCOM (HYbrid Coordinate Ocean Model) [Bleck, 2002; Chassignet et al., 2003; Halliwell, 2004] which has been developed by a community effort (http://hycom.rsmas.miami.edu) supported by the National Ocean Partnership Program (NOPP). The rationale for restricting ourselves to these systems is that they all use the same atmospheric forcing from the Fleet Numerical Meteorology and Oceanography Center (FNMOC) Navy Operational Global Atmospheric Prediction System (NOGAPS) [Hogan and Rosmond, 1991; Rosmond et al., 2002], and assimilate the same observational data, the altimeter sea surface height (SSH) as well as the operational 1/8° MODAS [Modular Ocean Data Assimilation System; Fox et al., 2002] sea surface temperature (SST) analyses of AVHRR (Advanced Very High Resolution Radiometer) data. This allows us to evaluate the performance of the systems independently of the choice made for external inputs (atmospheric forcing or ocean data input).

3.1 The Numerical Models

NLOM was designed for use in a first generation global ocean prediction system by allowing high horizontal resolution at much lower computational cost than other models [Hurlburt, 1984]. NLOM has been configured globally from 72°S to 65°N at 1/16° and 1/32° (~7 and 3.5 km mid-latitude, respectively) resolution. However, NLOM has only 7 layers in the vertical, including the mixed layer, and excludes the Arctic and most shallow water (i.e., the lateral boundaries usually follow the 200 m isobath.).

Global NCOM was designed to complement NLOM by covering these regions and using much higher vertical resolution, but with coarser 1/8° (15–16 km mid-latitude) horizontal resolution. NCOM has a mixed σ-z vertical coordinate [Barron et al., 2005] where terrain-following σ coordinates are used when the bottom depth is <137 m in the current global configuration. This configuration has 40 levels in the vertical with vertical resolution ~1 m very near the surface for fine resolution of the mixed layer. Global NCOM encompasses the open ocean to 5 m depth in a curvilinear global model grid with 1/8° latitudinal grid spacing at 45°N. The grid extends from 80°S to a complete Arctic cap with grid singularities mapped into Canada and Russia. A regional configuration of NCOM is discussed in section 5.

Recent model comparison exercises performed in Europe (DYnamics of North Atlantic MOdels—DYNAMO) [Willebrand et al., 2001] and in the U.S. (Data Assimilation and Model Evaluation Experiment—DAMÉE) [Chassignet et al., 2000] have shown that no single vertical coordinate (depth, density, or terrain-following sigma) can by itself be optimal everywhere in the ocean. Isopycnal (density tracking) layers are best in the deep stratified ocean, z-levels (constant fixed depths) are best used to provide high vertical resolution near the surface within the mixed layer, and sigma-levels (terrain-following) are often the best choice in shallow coastal regions. HYCOM combines all three approaches by choosing the optimal distribution at every time step. The hybrid coordinate is one that is isopycnal in the open, stratified ocean, but that makes a dynamically smooth transition (via the layered continuity equation) to terrain-following coordinates in shallow coastal regions, and to pressure coordinates in the mixed layer and/or unstratified seas. The capability of assigning additional coordinate surfaces to the oceanic mixed layer in HYCOM provides the option of implementing sophisticated vertical mixing turbulence closure schemes (see Halliwell [2004] for a review). HYCOM has been configured globally, on basin scales, and regionally.

The 1/12° (~7 km mid-latitude resolution) North Atlantic HYCOM system is configured from 28°S to 70°N, including the Mediterranean Sea (see Chassignet et al. [2005] for details).The vertical resolution consists of 26 hybrid layers, with the top layer typically at its minimum thickness of 3 m (i.e., in fixed coordinate mode to provide near-surface values). In coastal waters, there are 15 sigma-levels, and the coastline is at the 10 m isobath. The northern and southern boundaries are treated as closed, but are outfitted with 3° buffer zones in which temperature, salinity and pressure are linearly relaxed toward their seasonally varying climatological values.

3.2 Data Assimilation

The 1/16° and 1/32° global NLOM-based systems [*Smedstad et al.*, 2003; *Shriver et al.*, 2005] assimilated track data from available real-time satellite altimeter observations (GFO and JASON) using an optimal interpolation (OI) deviation SSH analysis with the model field as the first guess and mesoscale covariances calculated from Topex-Poseidon and ERS-2 altimeter data [*Jacobs et al.*, 2001]. A statistical inference technique [*Hurlburt et al.*, 1990] updates all layers of the model based on the analysed SSH deviations, including geostrophic updates of the velocity field outside an equatorial band. The global model is then updated to produce a nowcast using slow insertion to further reduce gravity wave generation. In addition, model deviations from full field MODAS SST analyses are assimilated in the form of heat fluxes.

The 1/8° global NCOM system [*Barron et al.*, 2004, 2005] uses relaxation (i.e., nudging) to assimilate 3-D synthetic temperature and salinity fields generated from MODAS SST analyses and the 1/16° NLOM SSH fields using the MODAS system algorithms and statistics [*Fox et al.*, 2002]. The nudging in NCOM is weak from below the surface to the base of the mixed layer to allow greater model impact on the mixed layer.

The data assimilation in the current 1/12° North Atlantic HYCOM system consists of assimilating daily operational MODAS SSH analyses of available real-time satellite altimeter observations (GFO and JASON) [*Chassignet et al.*, 2005]. The Cooper and Haines [1996] technique is used to project the surface information to the interior of the ocean. A relaxation to the MODAS SST analysis is also included.

All systems provide nowcasts and forecasts at least once a week, with forecasts for 30 days from the global NLOM systems, 5 days from the global NCOM system, and 14 days from the 1/12° Atlantic HYCOM system.

3.3 The Mean Sea Surface Height

In order to assimilate the anomalies determined from the satellite altimeter data into the numerical model, it is necessary to know the oceanic mean SSH over the time period of the altimeter observations. Unfortunately, the geoid is not known accurately on the mesoscale. Several satellite missions are either underway or planned to try to determine a more accurate geoid, but until the measurements become accurate to within a few centimeters on scales down to approximately 30 km, one has to define a mean oceanic SSH. At the scales of interest (tens of kilometres), it is necessary to have the mean of major ocean currents and associated SSH fronts sharply defined. This is not feasible from coarse hydrographic climatologies (~1° horizontal resolution) and

the approach taken by the above systems has been to use a model mean SSH. This requires a fully eddy-resolving ocean model which is consistent with hydrographic climatologies and with fronts in the correct position.

For the NLOM-based system, the mean SSH from a global NLOM 1/16° run without data assimilation was compared to the mean dynamic height calculated from available temperature and salinity measurements and then modified with data over the period of the satellites. To help determine the position of the fronts, mean frontal positions from satellite IR plus the variability from satellite altimeter data were also used. Following this effort to accurately determine the mean position of the front, the model mean was modified by moving the SSH features in an elastic way (rubber-sheeting) [*Hord*, 1982; *Clarke*, 1990; *Carnes at al.*, 1996] (see Smedstad et al. [2003] for an illustration). The 1/8° global NCOM uses the same mean SSH and assimilates the 1/16° NLOM SSH fields. The HYCOM-based forecasting system uses a different model mean, one generated by a previous 1/12° North Atlantic simulation performed with the Miami Isopycnic Coordinate Ocean Model (MICOM) (see Chassignet and Garraffo [2001] for a discussion).

4. COMPARISON TO OCEAN COLOR

Plates 1–4 compare SSH analyses from the three ocean data assimilative systems which overlie ocean color imagery from SeaWiFS in the Gulf of Mexico. Each Plate is for a different date (July 9, July 19, July 26, and August 8, 2003) and contains results from the 1/16° global NLOM (8 km resolution in the Gulf of Mexico), the 1/32° global NLOM (4 km resolution), the 1/8° global NCOM (17 km resolution), and the 1/12° Atlantic HYCOM (8 km resolution). These dates correspond to a Loop Current shedding event as well as further interactions between the newly formed ring and the Loop Current. These snapshots clearly illustrate how ocean color has the potential to characterize the performance of each system, but do not allow for a quantitative assessment of the systems. The SeaWiFS imagery is a composite of the most recent cloud-free pixel over the preceding 6 days and is colored prismatically from low (violet) to high (red). Both bright and dark areas of ocean color provide useful information. Dark areas of low chlorophyll (< .1 mg/m^3) usually correspond to Caribbean water from outside the Gulf of Mexico and can be seen inside the Loop Current and shed eddies. When the model is in good agreement with SeaWiFS imagery, the SSH contours of the Loop Current and shed eddies capture the dark areas quite well. The mismatches show up clearly and the four-way comparison is very effective in bringing them out. One has to be careful when performing the evaluation, since the overlain SSH contours

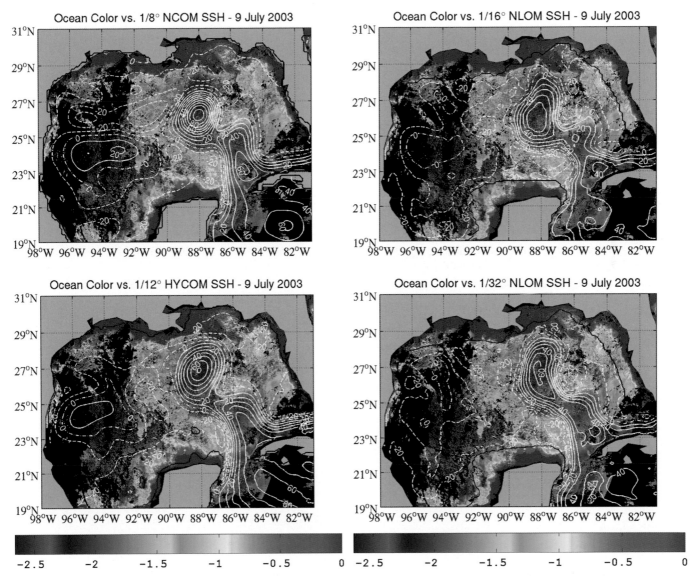

Plate 1. SSH from four ocean data assimilative systems on 09 July 2003 (contour interval of 10 cm and normalized over the displayed area), overlying ocean color imagery from SeaWiFS (natural log of the concentration in mg m^{-3}). The model land-sea boundary is defined by the black contour line.

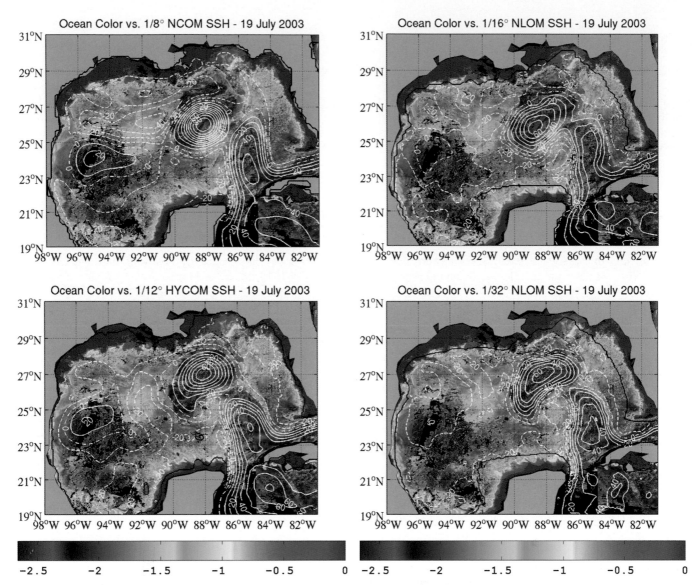

Plate 2. SSH from four ocean data assimilative systems on 19 July 2003 (contour interval of 10 cm and normalized over the displayed area), overlying ocean color imagery from SeaWiFS (natural log of the concentration in mg m^{-3}). The model land-sea boundary is defined by the black contour line.

can occasionally obscure some of the features seen in the imagery. The bright areas of high chlorophyll ($>.7mg/m^3$), usually generated along the coast in the Gulf of Mexico, tend to be advected into plumes by strong currents but sometimes are also present in the center of cyclonic (counterclockwise) eddies.

09 July 2003: Plate 1 clearly illustrates the final process leading to the separation of a Loop Current ring. Specifically, there is a bright patch of high chlorophyll ($~.5mg/m^3$) that is being advected across the Loop Current as the eddy separates. The four models differ from each other significantly. The NLOM 1/16° does a very good job of capturing the complex pattern of the Loop Current, especially southeast of the bright chlorophyll patch. The soon-to-be shed eddy, however, does not extend as far north as the SeaWiFS imagery suggest. The NLOM 1/32° does a better job in that regard, but the westward front of the Loop Current does not extend sufficiently southwest. The wedge represented by the bright chlorophyll patch is also not well defined in NLOM 1/32°. Close examination of NLOM 1/16° and NLOM 1/32° also show that the westward extent of the eastern front of the Loop Current closest to the Florida Keys may be slightly exaggerated in NLOM 1/16°. Despite the fact that NCOM 1/8° assimilates the SSH fields of NLOM 1/16°, the two nowcasts differ significantly. NCOM 1/8° clearly misses a large portion of the northward extent of the Loop Current and shows a Loop Current ring already well defined. Part of the bright patch of chlorophyll can actually be seen inside the newly formed ring, a clear indication of the misplacement of the frontal features. The wedge represented by the bright chlorophyll patch is perhaps best represented by HYCOM 1/12° but, as in NLOM 1/32°, the western front of the Loop Current does not extend sufficiently far south.

19 July 2003: This snapshot (Plate 2) is especially illustrative, since the SeaWiFS imagery clearly shows a separated Loop Current ring. Furthermore, the irregular shape of the ring is quite challenging for the models. For that specific day, NLOM 1/32° agrees best with the color measurements. Of particular note is the accurate representation of the shape and position of the ring and of the Loop Current as depicted by the low chlorophyll water (deep purple). The HYCOM 1/12° also does quite well at representing these features, but the ring does not extend far enough to the southwest and the Loop Current does not extend far enough to the north. The Loop Current is pretty well represented in both NLOM 1/16° and NCOM 1/8° (except at the neck in NLOM 1/16°), but both systems fail to place the newly formed ring properly. The location of the circular ring in NCOM 1/8° (already formed on July 09, 2003) shows a large mismatch between the model results and the chlorophyll observations. Additionally, only NLOM 1/32° and HYCOM 1/12° capture

the small cyclonic eddy depicted by high chlorophyll on the south side of the Loop Current ring. HYCOM 1/12° is the most successful in depicting a large anticyclonic eddy in the western Gulf with NCOM 1/8° in second place. NLOM 1/32° is the least successful.

26 July 2003: The Loop Current ring that showed as clearly separated on July 19, 2003 is still separated, but showing signs of being recaptured by the Loop Current (Plate 3). NCOM 1/8° shows very little change from 09 July 2003 and fails completely to move the ring northeastward (actually moves it slightly southwestward). The ring's shape remains persistently circular and the eastward position of the Loop Current is misplaced. NLOM 1/16° does show a reattachment of the ring, but the position of both the ring and Loop Current do not agree with the chlorophyll measurements. Again, NLOM 1/32° shows the best agreement with the SeaWiFS imagery, with the eastward edge of the eddy clearly delineated by a long plume of high chlorophyll originating from the Louisiana coast. NLOM 1/32°, however, fails to place the eastward frontal position of the Loop Current properly between Cuba and the Florida Keys. HYCOM 1/12° is the only system that accurately positions that specific front. HYCOM 1/12° also places the reattached eddy reasonably well, but is missing some of its lateral extent, especially the eastern area next to the high chlorophyll plume. NLOM 1/32° and HYCOM 1/12° continue to represent the cyclonic eddy depicted by high chlorophyll on the south side of the Loop Current ring. This area of high chlorophyll is overlain by the NCOM 1/8° Loop Current ring and is bisected by the southern side of the NLOM 1/16° Loop Current ring.

08 August 2003: With most of the ring reattached, the Loop Current is now elongated and extends farther to the west and to the north (Plate 4). A fragment of the Loop Current ring remains detached west of the reattached portion (small dark area), but is not depicted by any of the systems. In NLOM 1/16°, the ring did not remain attached and moved westward of 90°W. The Loop Current position is reasonably well represented east of 88°W. The NCOM 1/8° SSH field shows very little agreement between the chlorophyll measurements and the circular ring formed on 09 July 2003, now located farther to the west. In addition, the Loop Current does not extend far enough to the north. Both HYCOM 1/12° and NLOM 1/32° do a good job at capturing the full northwestward extent of the Loop Current. There are some small differences in the representation of the recaptured Loop Current ring. The ring in HYCOM 1/12° is broader and shows more closed SSH contours, but both have nearly the same the ring center location. Both HYCOM 1/12° and NLOM 1/32° fail to correctly place the eastward frontal position of the Loop Current, which is well delineated by high chlorophyll in the observations.

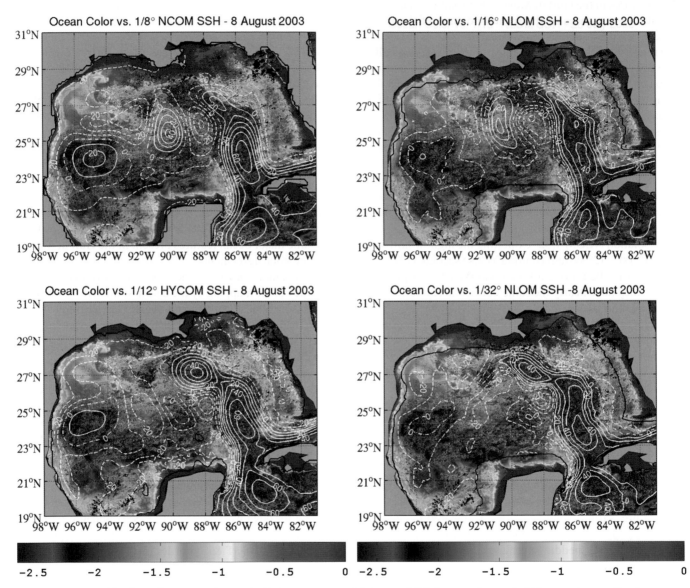

Plate 4. SSH from four ocean data assimilative systems on 08 August 2003 (contour interval of 10 cm and normalized over the displayed area), overlying ocean color imagery from SeaWiFS (natural log of the concentration in mg m^{-3}). The model land-sea boundary is defined by the black contour line.

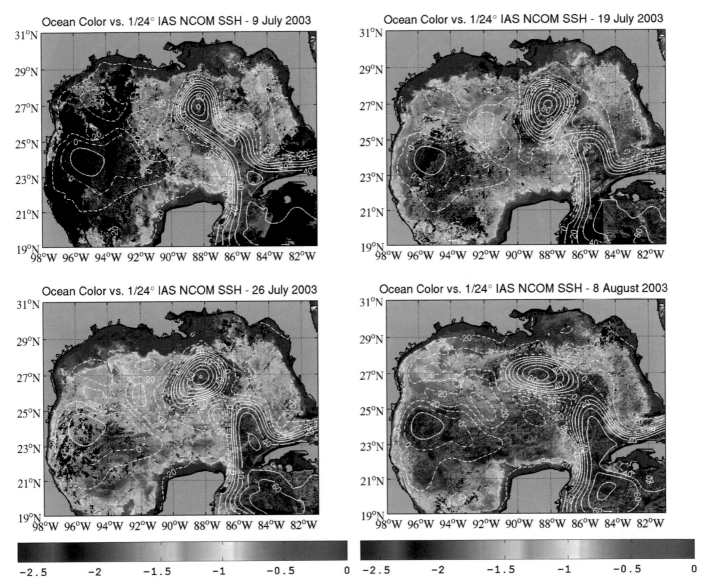

Plate 5. SSH from the 1/24° NCOM data assimilative system on 09, 19, 26 July, and 08 August 2003 (contour interval of 10 cm and normalized over the displayed area), overlying ocean color imagery from SeaWiFS (natural log of the concentration in mg m⁻³). The model land-sea boundary is defined by the black contour line.

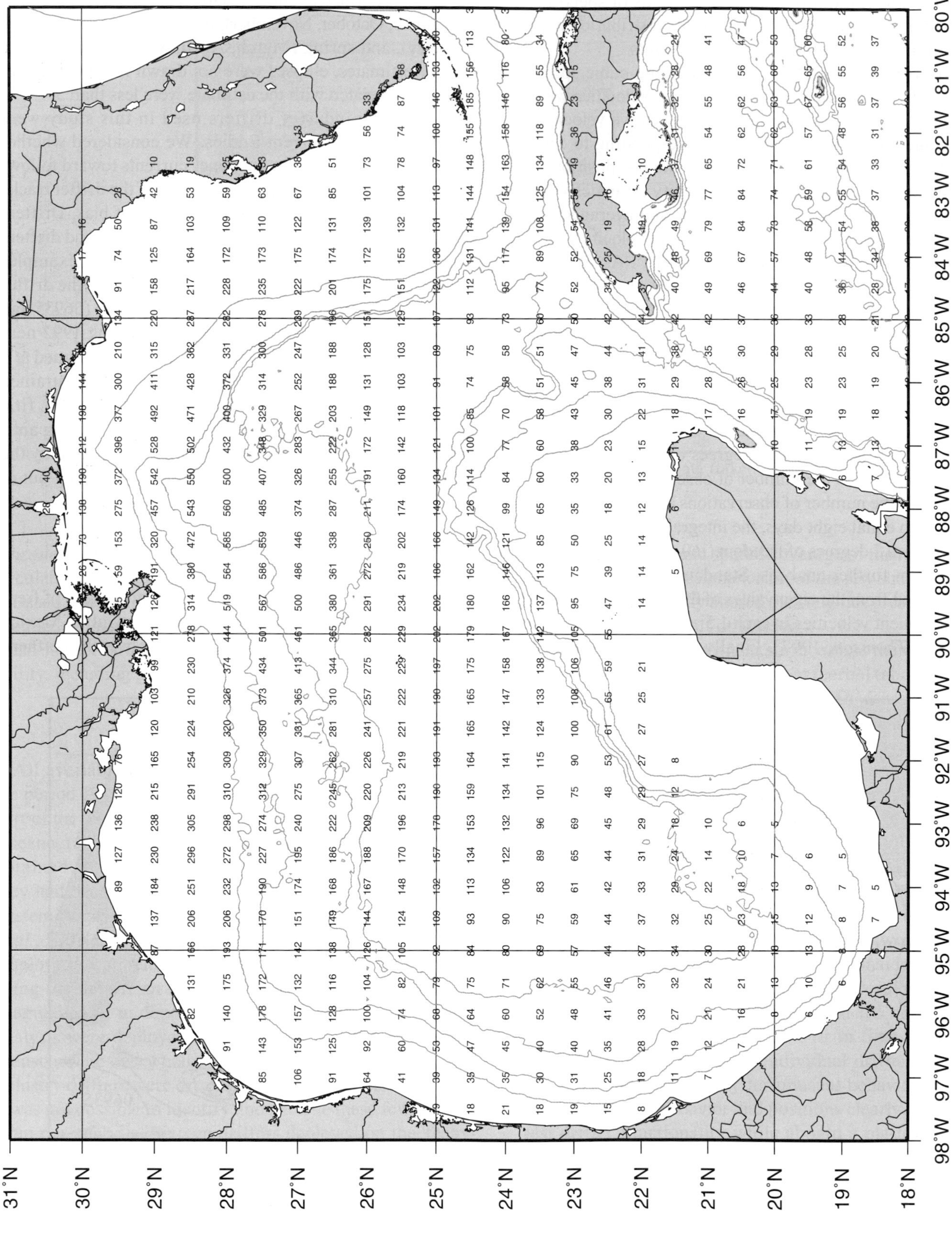

Figure 3. Degrees of freedom associated with each overlapping 1.5° x 1.5° bin for drifter observations of the Gulf of Mexico and Cayman Basin. Degrees of freedom are based on number of observations divided by 32 (integral time scale for the Gulf of Mexico).

deepwater region generally exceed 100 and up to nearly 600; numbers in the southern Gulf are smaller, less than 100. Numbers, vector tails and variance ellipses are plotted at the center of the 0.5° intervals; any datum or number that appears to be over land represents the information associated with the oceanic portion of the box, e.g., western Cuba.

The corresponding fields of vector velocity and standard deviation ellipses are shown in Figures 4 and 5, respectively. Seen is a westward intensification of the Yucatan Current in the Cayman basin with strong (mean exceeding 60 cm·s⁻¹) inflow to the Gulf through the Yucatan Channel. Standard deviation ellipses are elongated in the direction of mean flow in this current regime. The expected outflow just west of Cape San Antonio, Cuba, [*Sheinbaum et al.*, 2002; *Nowlin et al.*, 2001] may have been masked by the 1.5° smoothing (it was seen in less smoothed [0.5° bins] vector fields).

Energetic westward flow (15–20 cm·s⁻¹) is seen in the zonal band between 22° and 24°N extending to the Mexican shelf. The zonal band is likely the combined effects of local wind forcing, which is predominately westward, and the accumulation of westward currents along the southern limbs of westward migrating LCEs. South of that flow a cyclonic circulation is seen clearly in the Bay of Campeche as found by *Vazquez* [1993]. *Vazquez et al.* [2005, this volume] further explore this wind driven, cyclonic feature in the Bay of Campeche and attribute the circulation pattern to the wind stress curl over the region.

Evidence appears for an elongated anticyclonic feature in the west-central Gulf, centered about 24–25°N, 92.5°W, as reported by *Nowlin and McLellan* [1967] and studied in some detail by *Sturges and Blaha* [1975] and *Sturges* [1993]. The northern boundary of this feature, presumably over the northern slope, is not well defined. This may be considered as evidence of the migration of Loop Current Eddies across the region [*Vukavich and Crissman*, 1986; *Hamilton et al.*, 1999] and the current variability due to the episodic interaction of Loop Current Eddies with the continental slope. Such interactions frequently result in the spin up of smaller scale cyclones and anticyclones [*Hamilton et al.*, 1992] which when averaged can result in small mean currents with large variances. Within this large anticyclone there is evidence for a smaller anticyclonic circulation over the central deep Gulf, centered near 25.5°N, 90°W.

The outflowing limb of the Loop Current leading into the Straits of Florida shows large speeds and standard deviations. Standard deviations are larger in the eastern part of the deep water Gulf, where the mean currents are greatest. The lack of closure between inflowing and outflowing limbs of the Loop Current in this long-term mean representation

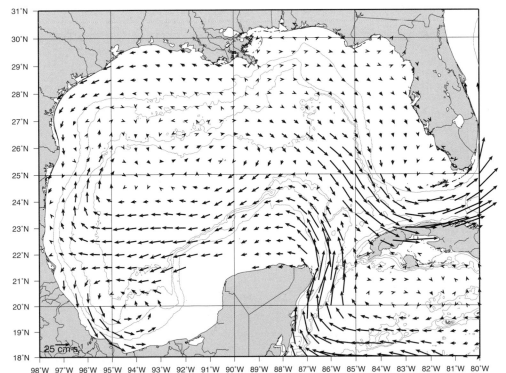

Figure 4. Near-surface velocity estimates for each overlapping 1.5° x 1.5° bin based on averaging all drifter velocity estimates in that bin for the period 1989–1999. Shown are the 200-, 1000-, 2000-, and 3000-m isobaths. Amount of overlap among bins is 1°.

is likely due to the separation and westward migration of eddies from that current.

The mean circulation pattern over the northern continental shelves show marked contrast between the eastern and western regions. The mean flow over the Texas-Louisiana Shelf is coherent and downcoast (cyclonic or in the direction of propagation of a Kelvin wave in the Northern Hemisphere) in agreement with the *Cochrane and Kelly* [1986] scheme of circulation of this region. Standard deviations here are slightly smaller than those found in the deep-water regions. The mean velocity vectors over the Mississippi-Alabama and northern West Florida Shelves are small (< 5 cm·s⁻¹) with no discernable spatial pattern. South of Tampa on the West Florida Shelf, currents remain small but tend to flow southward near the shelf edge indicating the frequent influence of the Loop Current [Hetland et al., 1999; *He and Weisberg*, 2003]. Standard deviations on the West Florida Shelf are among the smallest in the Gulf, but are considerably larger than the mean currents.

Seasonal patterns of circulation and current variability were prepared and studied. We show in Figures 6–9 the average vectors for spring (March–May), summer (June–August), fall (September–November), and winter (December–February) seasons. The numbers of drifter velocity estimates in each bin used to produce the averages in Figures 6 thru 9, respectively,

are roughly one-fourth the numbers seen in Figure 3 indicating an almost even split between seasons.. An exception to this is in the Bay of Campeche where spatial coverage was best in winter and sparse in all other seasons. For the seasonal fields, vectors were drawn if the degrees of freedom associated with the vector were equal to or greater than one. With this information the reader can estimate the standard error of the estimate of seasonal sample vectors, assuming the data are random realizations from a Gaussian estimate. The individual standard deviation ellipses by season are all similar and are merely small sample size versions of the ellipses based on all seasons (Figure 5) and are not shown.

Inflow and outflow of the Loop Current is clearly seen in both seasons, but only for the winter season do the average vectors illustrate the classic picture of a continuous Loop Current through the eastern Gulf (Figure 9). Note also that the small anticyclonic feature seen centered at 25.5°N, 90°W in the mean and summer fields has weakened and moved westward (center near 91.5°W) in the winter field. The Gulf interior during fall and spring appears to consist of a more complex set of several smaller cyclones and anticyclones. For the period 1989–1999 that included this drifter set, *Sturges and Leben* [2000] estimated five eddy separations from the Loop Current in summer and fall, four in spring, but none in winter. Therefore, the long-term

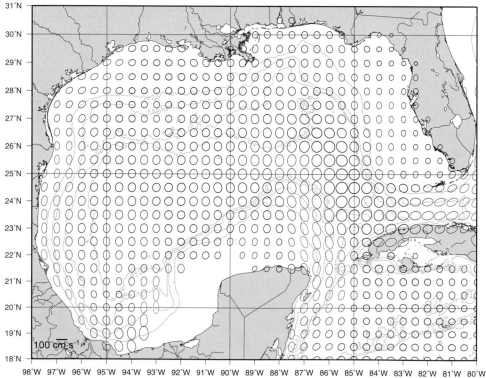

Figure 5. Variance ellipses corresponding to the long-term near-surface velocity estimates shown in Figure 4.

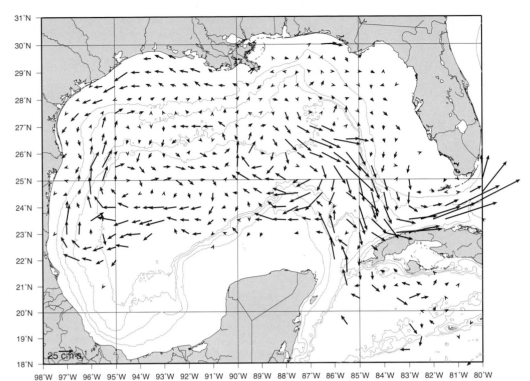

Figure 6. Near-surface velocity estimates for each overlapping 1.5°x 1.5° bin based on averaging all springtime (March through May) drifter velocity estimates in that bin for the period 1989-1999. Shown are the 200-, 1000-, 2000-, and 3000-m isobaths. Amount of overlap among bins is 1°.

mean picture and the seasonal pictures for spring, summer, and fall do not show the classic continuous Loop Current because of eddy separations, whereas the winter picture does, because of the absence of eddy separations during that season for the period studied. Because there was no deep penetration of the Loop Current into the Gulf during the winters considered, the northern limb of the Loop Current in Figure 9 is seen at about 26.5°N.

The strong westward flow across the Gulf in the band 22°–24°N is seen in the summer (exemplified by Figure 7) but is broader, further south, and less energetic in winter (exemplified by Figure 9), even though the numbers of velocity estimates are similar. In winter and spring the circulation regimes of the eastern and western Gulf seem more separated than in summer and fall. This might be related to the annual cycle of wind-stress curl over the Gulf or to the fact that there were more Loop Current eddy separations in summer and fall (five in each season) for the period examined, but only four in spring and none in winter, which shows the most separation between eastern and western circulation regimes.

The western boundary current between 95°W and 96.5°W at 24°N is strongest in summer (Figure 7) and weakest in winter (Figure 9). This tends to support the wind-driven

signal proposed by *Sturges* [1993], who found, based on an analysis of ship drift data, strongest currents in July and weakest in October. They found that the major driving for this current to be the wind stress curl, which peaks in May and local Ekman pumping, which peaks in July. The boundary current strengths for summer and winter are significantly different at the 95 percent confidence level. The western Gulf of Mexico is also a region frequently impacted by Loop Current Eddies that have migrated across the Gulf and interact with the western shelves [*Brooks and Legeckis*, 1982; *Brooks*, 1984; *Kirwan et al.*, 1984].

The cyclone in the Bay of Campeche is apparent in fall and strongest in winter pictures. The seasonal variability of that wind driven feature is discussed in *Vazquez et al.* [2005, this volume]. However, we note that since most drifters analyzed in this study were deployed in U.S. waters, the relative abundance of winter drifter data available in the Bay of Campeche is related to the weakening of the western boundary current during winter.

Downcoast (cyclonic) coastal flow occurs along the northwestern shelf region of the Gulf for nonsummer (exemplified by winter Figure 6). In summer, the shelf flow derived from drifter data is much weaker, in agreement with the SCULP drifter data [*Herring et al.*, 1999]. Current meter and hydrographic data on the other hand show significant upcoast flow in the summers of

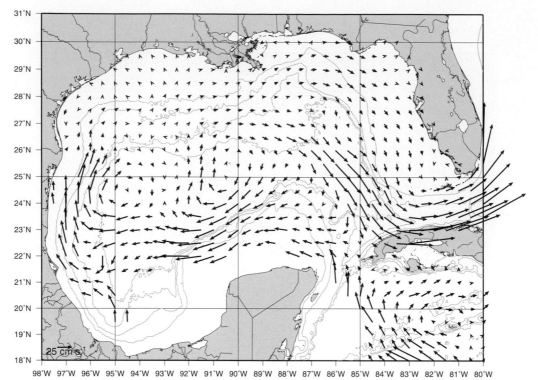

Figure 7. Near-surface velocity estimates for each overlapping 1.5°x 1.5° bin based on averaging all summertime (June through August) drifter velocity estimates in that bin for the period 1989-1999. Shown are the 200-, 1000-, 2000-, and 3000-m isobaths. Amount of overlap among bins is 1°.

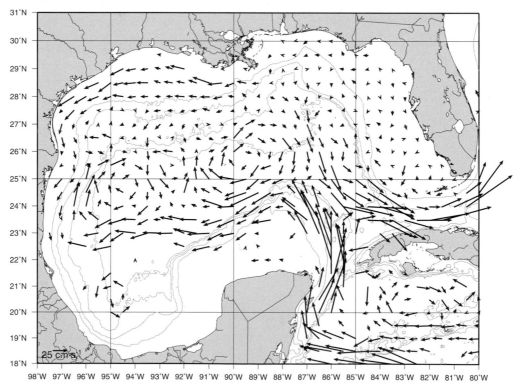

Figure 8. Near-surface velocity estimates for each overlapping 1.5°x 1.5° bin based on averaging all falltime (September through November) drifter velocity estimates in that bin for the period 1989-1999. Shown are the 200-, 1000-, 2000-, and 3000-m isobaths. Amount of overlap among bins is 1°.

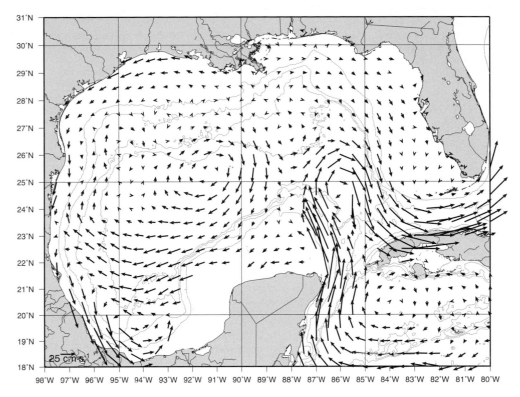

Figure 9. Near-surface velocity estimates for each overlapping 1.5°x 1.5° bin based on averaging all wintertime (December through February) drifter velocity estimates in that bin for the period 1989-1999. Shown are the 200-, 1000-, 2000-, and 3000-m isobaths. Amount of overlap among bins is 1°.

1992–1994 (See *Nowlin et al.* this volume). The drifter-derived results for summer and winter are based on comparable sample sizes. The difference may be due to several factors, which include: the relatively few number of shelf drifters used in this study, the spatial averaging of the drifter data into 1.5° bins, and interannual variability of the coastal currents.

In the northeastern shelf region, the summer and winter patterns of the vector fields are similar to the mean patterns indicating no apparent seasonality of the upper circulation in this region. However, we note the currents over the outer West Florida Shelf and slope north of 27°N during winter and summer and more coherent and organized than during the spring and fall months.

In summary, the mean circulation of the upper waters of the Gulf of Mexico is mainly driven by the combined effects of the Loop Current system in the eastern Gulf and by basin and regional scale wind patterns. In the western Gulf of Mexico, the finding of a mean anticyclonic flow pattern is consistent with previous findings from hydrography [Nowlin, 1972] and ship drift [Sturges, 1993]. Typically, the periodic shedding of LCEs, their westward migration, and their interaction with topography and other eddies tends to perturb the mean anticyclonic flow, particularly along

the northern slopes. The wind-driven patterns are mainly manifest in the cyclonic circulaion of the Bay of Campeche and along the western boundary. Both wind-driven features are seasonally dependent. The westward zonal flow in the western Gulf between 22°N and 24°N is broadest in winter but most intense and narrow in summer.

Acknowledgments. We thank Todd Ooten, Ou Wang, and Yue Fang (TAMU) for preliminary processing of the drifter data. This work was partially supported by the Minerals Management Service under Contract Nos. 1435-01-98-CT-30910. The remarks and comments of Dr. Alexis Lugo-Fernandez (MMS) were invaluable to this research. We thank Drs. M. K Howard and A. E Jochens (TAMU) for their valuable comments regarding this research.

REFERENCES

Brooks, D. A. (1984), Current and hydrographic variability in the northwestern Gulf of Mexico, *J. Geophys. Res., 89*(C5), 8022–8032.

Brooks, D. A., and R. V. Legeckis (1982), A ship and satellite view of hydrographic features in the western Gulf of Mexico, *J. Geophys. Res., 87*, 4195–4206.

Cochrane, J. D., and F. J. Kelly (1986), Low-frequency circulation on the Texas-Louisiana continental shelf, *J. Geophys. Res., 91*(C9), 10,645–10,659.

Dietrich, G. (1937), Die Lage Meeresoberfläche im Druckfeld Ozean und Atmosphäre mit besonderer Berücksichtigung des westlichen Nordatlantischen Ozeans und des Golfs von Mexiko. Veröffentlichungen des Institut für Meereskunde an der Universität Berlin, Heft 33, 5–52.

DiMarco, S. F., M. K. Howard, and R. O. Reid (2000), Seasonal variation of wind-driven diurnal current cycling on the Texas–Louisiana shelf, *Geophys. Res. Lett., 27*(7), 1017–1020.

Emery, W. J., and R. E. Thomson (1997), *Data Analysis Methods in Physical Oceanography*, 650 pp., Pergamon, New York.

Hamilton, P. (1992), Lower continental slope cyclonic eddies in the central Gulf of Mexico, *J. Geophys. Res., 97,* 2185–2200.

Hamilton, P., G. S. Fargion, and D. C. Biggs (1999), Loop Current eddy paths in the western Gulf of Mexico, *J. Phys. Oceanogr., 29,* 1180–1207.

He, R., and R. H. Weisberg (2003), A Loop Current intrusion case study on the West Florida Shelf, *J. Phys. Oceanogr., 33,* 465–477.

Hetland, R. D., Y. Hsueh, R. R. Leben, and P. P. Niiler, (1999). A Loop Current-Induced Jet Along the Edge of the West Florida Shelf, *Geophys. Res. Letters,* **26**(15), 2239–2242.

Herring, H. J., M. Inoue, G. L. Mellor, C. N. K. Mooers, P. P. Niiler, L.-Y. Oey, R. C. Patchen, F. M. Vukovich, and W. J. Wiseman, Jr. (1999), Coastal Ocean Modelling Program for the Gulf of Mexico, Final Report, Volume 2., 194 pp., Prepared for Minerals Management Service Environmental Division, Technical Analysis Group. Report No. 115.2, New Orleans, Louisiana.

Kirwan, A. D., Jr., W. J. Merrell, Jr., J. K. Lewis, R. E. Whitaker (1984), Lagrangian observations of an anti-cyclonic ring in the western Gulf of Mexico, *J. Geophys. Res., 89*(C3), 3417–3424.

Nowlin, W. D., Jr., and H. J. McLellan (1967), A characterization of the Gulf of Mexico waters in winter, *J. Mar. Res., 25,* 59.

Nowlin W. D., Jr., (1972). Winter circulation patterns and property distributions, in *Contributions on the Physical Oceanography of the Gulf of Mexico*, edited by L. R. A. Capurro and J. L. Reid, Texas A&M University Oceanographic Studies, Vol. 2, pp. 3–53, Gulf Press, Houston, Texas.

Nowlin, W. D, Jr., A. E. Jochens, S. F. DiMarco, R. O. Reid, M. K. Howard (2001), Deepwater physical oceanography reanalysis and synthesis of historical data: Synthesis Report, OCS Study MMS 2001-064, 528 pp., U.S. Dept. of the Interior, Minerals Management Service, Gulf of Mexico OCS Region, New Orleans, Louisiana.

Nowlin, W. D., A. E. Jochens, S. F. DiMarco, R. O. Reid, and M. K. Howard (2005), Low Frequency circulation over the Texas-Louisiana continental shelf, (this volume).

Ohlmann, J.C., and P. P. Niiler (2005). Circulation over the continental shelves of the northern Gulf of Mexico. *Progress in Oceanography* **64**, 45–81.

Poulain, P.-M., and P. P. Niiler (1989), Statistical analysis of the surface circulation of the California Current system using satellite-tracker drifters, *J. Phys. Oceanogr., 19,* 1588–1603.

Sheinbaum, J., J. Candela, A. Badan, and J. Ochoa (2002), Flow structure and transport in the Yucatan Channel, *Geophys. Res. Lett., 29*(3), 10.1029/2001GL013990.

Sturges, W., (1993). The annual cycle of the western boundary current in the Gulf of Mexico, *J. Geophys. Res.,* 98 (C10): 18053–18068.

Sturges, W., and J. P. Blaha, (1975). A western boundary current in the Gulf of Mexico, *Science,* 192 (4237): 367–369.

Sturges, W., and R. Leben (2000), Frequency of ring separations from the Loop Current in the Gulf of Mexico: a revised estimate, *J. Phys. Oceanogr., 30,* 1814–1819.

Vazquez De la Cerda, A. M., (1993). Bay of Campeche Cyclone, Ph.D. Dissertation. College Station, Texas: Texas A&M University, 91 pp.

Vazquez, A., R. O. Reid, S. F DiMarco, A. E. Jochens (2005), Bay of Campeche Circulation: an Update (this volume).

Vukovich, F. M., and B. W. Crissman (1986), Aspects of warm rings in the Gulf of Mexico, *J. Geophys. Res., 91*(C2), 2645–2660.

Texas A&M University, Department of Oceanography, 3146 TAMU, College Station, TX 77843-3146

Gulf of Mexico and Atlantic Coast Sea Level Change

Bruce C. Douglas

Laboratory for Coastal Research, Florida International University

Twentieth-century relative sea level rise shows considerable variability along the U.S. East and Gulf coasts. Local rates of rise lie in the range of about 1.5 to more than 4 mm per year for records from Key West, Florida, to New York City. Rates of sea level rise in the Gulf of Mexico can be much higher. In Texas and Louisiana, long-term water levels are rising up to about 10 mm per year. This is having disastrous consequences in the form of wetlands loss in the region, estimated to be as much as 65 km^2 per year in the Mississippi Delta area of Louisiana alone. Beach erosion is also significant along both the Gulf and Atlantic coasts, resulting in ever-increasing exposure of fixed structures to the damaging impacts of storms. The especially high rates of sea level rise in Louisiana and Texas are a result of their particular geomorphology, and anthropogenic alterations in the form of sediment diversion and withdrawal of underground fluids. The average long-term local rate of sea level rise on the rest of the U.S. East and Gulf coasts when corrected for glacial isostatic adjustment is about 2 mm per year, in conformity with 20th century global sea level rise. U.S. East and Gulf coast tide gauge records also have regionally coherent low frequency (decadal and longer) variations that need to be understood because of their impact on wetlands loss, and to enable accurate determination of long-term trends of sea level rise.

1. INTRODUCTION

Sea level change is receiving considerable attention these days, and for good reasons. As a practical matter, about 100 million people live in the immediate severely flood-prone coastal zone within one meter elevation above mean sea level (Nicholls and Leatherman, 1995). It has been further estimated that about 450 million persons live within 10 meters elevation above MSL (Small and Nicholls, 2003). An increasing water level leads to more rapid beach erosion thereby increasing the exposure of coastal populations to the damaging effects of storms, and contaminates groundwater and raises water tables. In addition, coastal wetland flora do not have an unlimited ability to adjust to rising sea levels, so

that increased rates of sea level rise in the 21st century will accelerate the present alarming rates of coastal wetland loss. Finally, sea level rise is an important indicator of climate change. Recent results (Miller and Douglas, 2004) indicate that ocean thermal expansion and melting of small glaciers cannot account for all of the observed global sea level rise in the 20th century; Greenland and/or Antarctica or some other source of fresh water must also be contributing an amount up to a sea level equivalent of about 1 mm per year.

On the U.S. East and Gulf coasts, beach erosion and wetland loss are ubiquitous. There are a few beach areas in the US that experience accretion of sand, e.g., the immediate area updrift of barrier island tidal inlets, but such are the exception. Galgano (1998) has estimated that about 86% of US East coast seaward barrier beaches are eroding. There are several reasons for this, including coastal development, changes of sediment supply, and rapidly rising sea levels (Boesch et al., 1994; Zhang, et al, 2004; Morton et al.,

Circulation in the Gulf of Mexico: Observations and Models
Geophysical Monograph Series 161
Copyright 2005 by the American Geophysical Union.
10.1029/161GM09

2004). Erosion and loss of low-lying coastal wetlands is especially severe in regions of the Gulf coast (the Mississippi Delta and Texas) where land subsidence gives an effective rate of long-term sea level rise of as much as five times the 20th century average global rate of nearly 2 mm per year (Douglas, 1991, 1997; Peltier, 2001) Wetlands loss in coastal Louisiana has in the second half of the 20th century occurred at a rate of about 65 km[2] per year (Boesch, et al., 1994) albeit with some evidence of a reduction in recent decades. This dramatic loss occurs because of the high rate of relative sea level rise, the small slope of the coastal plain (<1:1000), and lack of sediments to replace those that are compacting and producing the land subsidence. Under these conditions, the gently sloping land is inundated and plants are lost. Since Louisiana coastal wetlands amount to 40% of the national wetland area, the loss there is responsible for about 80% of the total U.S. annual wetland loss (see Boesch, et al., 1994, and S.J. Williams (no date) at http://marine.usgs.gov/fact-sheets/LAwetlands/lawetlands.html).

In this chapter, the primary attention will be given to the character of sea level rise, and its measurements. There is a good reason for this. It is that the measurement and impact of sea level change, rather like what has been asserted about politics, "is local." That is to say, local conditions can drastically alter the long term apparent sea level rise (SLR) at a site from the so-called eustatic or global mean value of about 2 mm per year. Indeed, in some areas (Fennoscandia and Hudson Bay), sea level is falling locally by up to a cm per year because of *uplift* of the land, whereas coastal Louisiana and Texas experience their rising sea levels of up to a cm per year from land *subsidence*. Sea level measurements made by tide gauges are thus referred to as relative, because they are made with respect to a nearby local land reference which may itself change in the vertical direction, rather than to an absolute reference such as the mass center of the earth. We shall see that untangling local land reference, regional, and global components of relative SLR is a complex task, but can yield important insights into geophysical and climate processes.

The best way to appreciate the nature of relative sea level change at a site is just to examine a sea level record. Since our interest here is primarily the Gulf of Mexico, we will consider a site there. Figure 1 shows a plot of annual means (the dots) of relative sea level (RSL) recorded by a tide gauge at Pensacola, Florida. Annual means are shown in order to suppress the seasonal cycle which has an amplitude of about 90 mm. In this paper we will not consider the familiar semi-daily or other ocean tides. A comprehensive presentation of the theory of tides can be found in Pugh (1987, 2004). These data were obtained from the Permanent Service for Mean Sea Level (see www.pol.ac.uk/psmsl) based at the Proudman

Oceanographic Laboratory in the United Kingdom. The record is long, extending from 1924 to the present, and the location is an excellent one. It is geologically stable and not close to any significant river influence. In addition, the site is carefully maintained by the National Ocean Service (NOS) of NOAA, including annual surveys of the pier-mounted tide gauge relative to a stable benchmark on land. These surveys are done to determine if the gauge mount has settled, and if reinstallations for maintenance or technology upgrades have been done correctly. The PSMSL refers to gauge records for which such a complete geodetic history is available as Revised Local Reference (RLR) sites. The PSMSL states that with only a few exceptions, RLR sites are the ones that should be used for scientific applications.

Figure 1 shows that RSL is rising at Pensacola, in fact at 2.1 ± 0.2 mm per year, just a little more than the global mean value. We shall see later that this is not merely fortuitous. Any statistically significant acceleration of RSL is obscured by the large interannual variations in this and the other records in this study. Woodworth (1990) and Douglas (1992) in their large scale studies did not find evidence of an acceleration of GSL in the 20th century.

It is also obvious that there is a lot of year-to-year fluctuation of RSL at Pensacola. The standard deviation of the annual means around the upward trend is 35 mm, a little less than 1/2 of the seasonal cycle. Approximately this level of fluctuation is seen in all long tide gauge records of annual means, and is mostly due to local meteorological forcing (Tsimplis and Woodworth, 1994). But there is more to see in the Pensacola record than high-frequency noise; the smooth curve shown is the result of plotting a 5-year so-called boxcar filter (i.e., a running average that moves one year at a time) of the annual means of RSL, and shows that there are low-frequency variations of sea level as well. These interdecadal

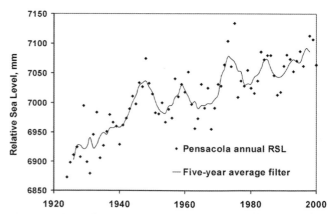

Figure 1. Annual means of Relative Sea Level at Pensacola, FL. Note the large low frequency fluctuations around the upward trend.

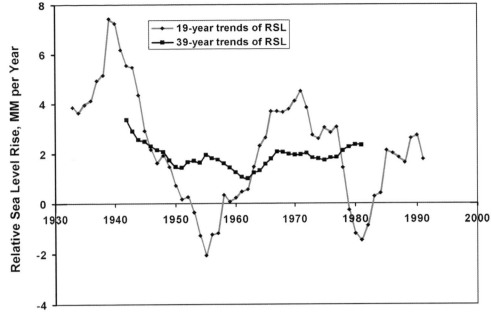

Figure 2. 19- and 39-year trends of relative sea level rise at Pensacola, Florida. The 19-year trends are highly variable, the 39-year trends much less so.

changes fluctuate from peak to trough up to 100 mm, and are the reason why long records are needed to determine the long-term trend of sea level rise at a site. Pugh (1987) noted that 10 year records of relative sea level have variable trends, with not even the sign of the long-term trend being determined. Records twice as long are also unreliable. To illustrate this, Figure 2 presents values for 19- and 39-year running trends of RSL at Pensacola. The values for each year in the figure are at the mid-points of the trend computation span. Note that an arbitrarily selected 19-year trend can vary from -2 to nearly 8 mm per year due to the influence of the low-frequency variations of sea level. Emery and Aubrey (1991) asserted that 20-year records of RSL rise are adequate to determine a long-term trend of sea level at a site, but this and numerous other counterexamples vitiate that notion. The 39-year trends shown in Figure 2 are more stable in value, having a smaller amount of variation. This sort of analysis was what convinced Douglas (1991) that about 50 years is required in most cases to determine a reasonably reliable value of the long-term trend of relative SLR. For some sites, such as Buenos Aires, Argentina which is affected by ENSO-associated rainfalls, much longer records are required (Douglas, 2001).

There are not many tide gauges on the Gulf coast with records longer than 50 years, so we will bend the rules somewhat and include records from 37 years on up for consideration. The example of Pensacola suggests that in this region, records of this length can give at least a good approximation

to the actual long term trend of relative sea level. There are 11 tide gauges distributed around the Gulf of Mexico from Key West, Florida, to Vera Cruz, Mexico with data records 37 years or longer. In this study we shall also consider additional tide gauge records along the eastern Atlantic Coast of the United states from New York to south Florida (Table 1). The gauge locations in Table 1 are listed along the east coast from the north, southward to Key West, FL, and then counterclockwise around the Gulf of Mexico. This arrangement is especially useful in evaluating the impact of glacial isostatic adjust (GIA) on sea level trends, as will be seen below.

Figure 3 shows the locations of tide gauges used in this study. They are distributed over a very large area. What do these gauge sites have to do with one another, especially the gauge at Cristobal in Panama? The answer is to be found in the following figures, which sort out the effects of glacial isostatic adjustment (GIA), local land subsidence, and the everywhere-present and highly correlated low frequency variations of relative sea level. Figure 4 presents the rates of SLR for these sites. Obviously something very interesting is happening on both the mid-Atlantic, and Gulf Coasts.

2. ANALYSIS OF EAST AND GULF COAST RELATIVE SEA LEVEL TRENDS.

In Figure 4, the solid black columns are the raw values of sea level trends in mm per year taken from the PSMSL web site. The gauges shown follow the coastline down the US East

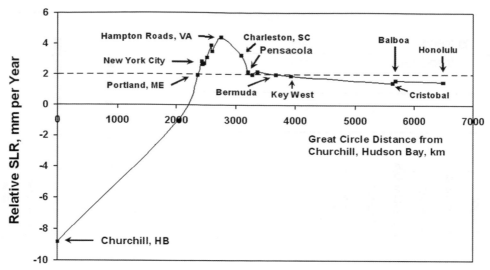

Figure 5. Observed relative sea level trends as a function of distance from Hudson Bay, Canada, the area of the greatest ice load at the peak of the last glacial maximum.

the city of New Orleans, Louisiana that exist because of the very high rate of subsidence-induced SLR that exists in this region. They analyzed a long sequence of repeated leveling surveys, and emphasize the severe threat to New Orleans from storm surges.

The Texas coast has a very wide range of rates of SLR ranging from about 3.4 to more than 10 mm per year. Freeport, TX is obviously a SLR "hot spot", since it has such a high rate of SLR compared to other nearby locations. But none of the high rates of SLR on the Texas coast are caused by GIA (which is small in this region). They reflect varying rates of local subsidence (Emery and Aubrey, 1991).

High rates of sea level rise are associated with rapid beach erosion because storm waves can reach farther up the beach as sea level increases. Figure 7 shows the consequences of the rapid sea level rise on the beach at Galveston, TX. This is a modern photo (courtesy of S.P. Leatherman) of the Galveston Island sea wall at high tide. The sea wall was constructed after the hurricane disaster of 1900, and originally had a wide beach in front of it. It has served the City well for a century, but the relentless increase of sea level has led to the virtual disappearance of the beach through a rapid erosion of the sand. Unlike The Mississippi Delta area of Louisiana, the Texas Gulf coast may be subsiding and eroding more from extraction of underground fluids than sediment compaction and starvation (Emery and Aubrey, 1991).

A simplified rule of thumb for the relation of beach erosion to sea level rise proposed decades ago by Bruun (1962) suggests that beaches that are not armored or near spits and inlets will erode shoreward about two orders of magnitude faster than the rate of rise of sea level. Zhang, et al. (2004) compared barrier island sandy beach erosion rates to local

sea level rise on the U.S. East coast, and found evidence that supports the Bruun theory in at least a general way. Of course this simple rule cannot be used to predict the fate of the beach at an arbitrarily selected specific site, but it does demonstrate the huge leverage that sea level rise has in promoting beach erosion.

3. LOW-FREQUENCY VARIATIONS OF RELATIVE SEA LEVEL.

The example of Pensacola shows that the rate of SLR can depart from its long-term average value by a large factor for an extended time. Recall that 19-year trends of sea level at Pensacola ranged from -2 to nearly 8 mm per year (Figure 2.) These large and enduring variations, which on the Mississippi Delta and the Texas coastline can effectively double the already high rate of relative SLR from an average of about 10 mm per year, may play an important role in exacerbating the huge wetlands loss. Rapidly rising sea level can cause drowning of wetland plants if the high rate is sustained for an extended time. Boesch et al (1994) say that "…these episodes of rapid sea level rise could be a significant factor in wetlands loss,…and should be investigated further."

What causes the large low frequency variations of sea level? How extensive are they geographically? The answers to these questions are interesting indeed. Maul and Hanson (1991) have discussed the interdecadal variations of sea level in the southeast U.S. and noted the high correlation of sea level records there at low frequency. Their analysis used data from 1930–1986, while the present analysis has the luxury of 15 years more data and in addition adds several more records (Vera Cruz and Cristobal) to the analysis. We start by

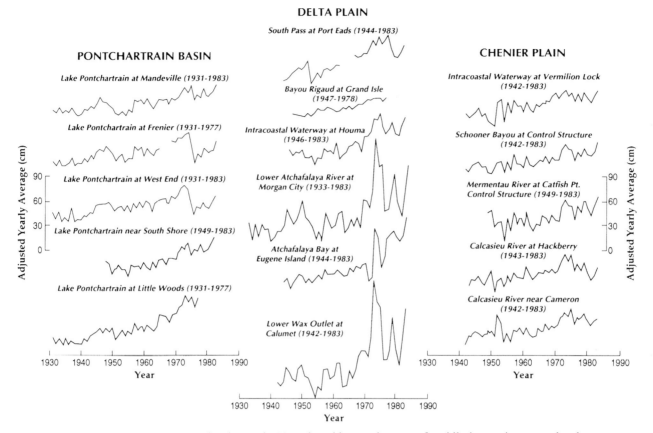

Figure 6. Mississippi Delta water level records. Note the widespread extent of rapidly increasing water levels over a broad area. The Delta Plain is south of the Pontchartrain Basin, and the Chenier Plain is in southwestern coastal Louisiana. From Boesch, et al. (2001).

examining in Plate 1 five-year smoothed (boxcar filter) and detrended tide gauge records for a representative sample of U.S. Southeast and Gulf Coast sites. Each record was detrended over its entire length to give a zero mean for the sake of clarity in the figure. These low-pass filtered sea level records have only about 2/3 of the variation of the annual means from which they were computed. The records are shown from Charleston southward to Key West, and then up and around the Gulf Mexico and on down to Cristobal, in Panama. They are all on the shoreward side of the Gulf Stream.

Low frequency variations of Gulf Coast sea level records are obviously similar in phase and amplitude to each other and those of the southeast coast, a result that is probably unexpected by most readers. It is necessary to go clear around to Cristobal, Panama for the low frequency correlation with Gulf and Atlantic coast gauge records to begin to disappear. The correlation of Cristobal and Charleston when detrended over their common length is only $R^2 = 0.18$ ($p < .001$). Equally as striking as the correlations are the big synoptic anomalies that occurred about 1920, 1950, and

Figure 7. Modern photo of the Galveston seawall by S.P. Leatherman. The once-wide beach has eroded away since its construction after the devastating hurricane of 1900.

gauges at Guantanamo, Cuba, and Magueyes Island, Puerto Rico. The former site is at the east end of Cuba, and the latter is located at the extreme west end of Puerto Rico, so they are not far apart. Compositing was accomplished by correcting each for GIA using the values computed by Peltier (2001), and calculating and applying the offset (10.1 mm) over their overlap period of 1960–1971. Plate 2 shows this composite record along with that for Pensacola over their common time interval. Both series have been smoothed with a running 3 year boxcar filter. The results are surprising in that one might have anticipated that Cuba and Puerto Rico were on the "offshore", i.e., eastward side of the Gulf Stream, so that the Pensacola and the composite Cuba/Puerto Rico series would be 180 degrees out-of-phase. Apparently at Cuba and Puerto Rico, the Gulf Stream, for all practical purposes, is westward of the Caribbean, except for a relatively small amount of transport that takes place on the outside.

4. CONCLUSIONS

The rapid rates of long-term sea level rise in Louisiana and Texas are due to local conditions including interference with sediment supplies (especially in Louisiana) and by withdrawal of underground fluids in both places. The consequences of the high rates are well documented (Boesch et al., 1994; Morton, et al., 2004), and mitigation is necessary to avoid severe social and ecological consequences. Long-term sea level rise on the rest of the Gulf coast and the south U.S. East Coast is remarkably consistent when corrected for glacial isostatic adjustment, and is in good agreement with more geographically extensive analyses that give a global rate of sea level rise of about 2 mm per year for the 20th century. The Gulf of Mexico and U.S. Atlantic coast also experience decadal and longer variations of up to about 100 mm peak-to-trough which may endure long enough to exacerbate wetland damage. These low-frequency variations appear to have their origin in wind forcing on the Atlantic Ocean. A common interdecadal signal is seen even as far around as Vera Cruz, Mexico, fading away but still detectible at Panama.

Such regionally coherent low-frequency variations of sea level are seen everywhere there are enough tide gauges to show it. These variations have a character that depends on whether they are on eastern or western ocean boundaries. Chelton and Davis (1982) demonstrated the effect on the West Coast of the North America due to ENSO events. Douglas (1991, 1997, 2001) grouped a global set of long tide gauge records into regions based on the coherence of their low-frequency signal components and then averaged the regional rates to avoid overweighting areas with many sites. More recently Woodworth (2003) has discussed the coherence of variations in the Mediterranean to as far away as Newlyn near Land's End in the UK that arise from atmospheric pressure changes. Miller and Douglas (2004) have shown that Hawaii's decadal sea level variations are similar to those on the U.S. West coast, and that sea level records from Cascais in Portugal and Tenerife in the Canary Islands are also coherent at low frequencies. Thus for the purpose of determining global sea level rise for the entire 20th century, there are fewer statistically independent tide gauge locations than has been supposed. It is clear that any issue involving sea level variation must be carefully posed because of the complex and red spectrum of sea level.

Acknowledgements. This investigation was supported by the U.S. Minerals Management Service and the NASA Solid Earth and Natural Hazards program. I am grateful to W. Sturges for his help in interpreting sea level records, and to L. Miller, R. Nicholls, and P. Woodworth for their helpful comments on the manuscript.

REFERENCES

Boesch, D.F., Josslyn, M.N., Mehta, A.J., Morris, J.T., Nuttle, W.K., Simestad, C.A., and D.j. Swift, Scientific Assessment of Coastal Wetland Loss, Restoration and Management in Louisiana, *J. Coastal Res., Special Issue no. 20,*. 103 pp., May, 1994.

Bruun, P., Sea-Level Rise as a Cause of Shore Erosion, *American Society of Civil Engineers Proceedings, Journal of Waterways and Harbors Division 88*, pp. 117–130, 1962.

Burkett, D.R., Zilkoski, D.B., and D.A. Hart, Sea Level Rise and Subsidence: Implications for Flooding in New Orleans, Louisiana, *USGS Open File Report 03-308*, pp.63–70, 2001.

Chelton, D.B., and R.E. Davis, Monthly Mean Sea-Level Variability along the West Coast of North America, *J. Phys. Oceanog. 12*, August 1982.

Church, J.A., White, N.J., Coleman, R., Lambeck, K., and J.X. Mitrovica, Estimates of the Regional Distribution of Sea Level Rose over the 1050–2000 Period, *J. Climate 17*, pp. 2609–2624, July, 2004.

Douglas, B.C., Global Sea Level Rise, *J. Geophys. Res. 96*, no. c4, April 15, 1991.

Douglas, B.C., Global Sea Level Acceleration, *J. Geophys. Res. 97*, no. c8, August 15, 1992.

Douglas, B.C., Global Sea Rise: A Redetermination, *Surveys in Geophys. 18*, pp. 279–292, 1997.

Douglas, B.C., 2001, Sea Level Change in the Era of the Recording Tide Gauge, Chapter 3 in *Sea Level Rise: History and Consequences*, Douglas, B.C., Kearney, M.S., and S.P. Leatherman (eds.), chapter 4, pp. 65–93, Academic Press, San Diego, 2001.

Emery, K.O., And D.G. Aubrey, *Sea Levels, Land Levels, and Tide Gauges*, Springer Verlag,, New York, 237 pp., 1991.

Galgano, F., *Geomorphic Analysis of Modes of Shoreline Behavior and the Influence of Tidal Inlets on Coastal Configuration*, Ph.D. dissertation, Dept. of Geography, University of Maryland, College Park, MD, 484 pp., 1998.

Hong, B.G., Sturges, W. and A. J. Clarke, Sea level on the U.S. east coast: decadal variability caused by open-ocean wind-curl forcing. *J. Phys. Oceanogr., 30*, pp. 2008–2098, 2000

Maul, G.A., and K. Hanson, Interannual coherence Between North Atlantic Atmospheric Surface Pressure and Composite Southern U.S.A. Sea Level, *Geophys. Res. Letters, 18*, no. 4, pp 653–656, April 1991.

Miller, L, and B.C. Douglas, Mass and Volume Contributions to 20th-Century Global Sea Level Rise, *Nature, 428*, pp. 406–409, March 25, 2004.

Morton, R.A., Miller, T.A., and L.J. Moore, National Assessment of Shoreline Change: Part 1: Historical Shoreline Changes and Associated

Coastal Land Loss Along the U.S. Gulf of Mexico. *U.S. Geological Survey Open-file Report 2004–1043*, 45pp., 2004.

Nicholls, R.J., and S.P. Leatherman, The Implication of Accelerated Sea Level Rise for Developing Countries: A Discussion, *J. Coastal Res., Special Issue no. 14*, pp. 303–323, 1995.

Peltier, W.R., Global Glacial Isostatic Adjustment and Modern Instrumental Records of Relative Sea Level History, in *Sea Level Rise: History and Consequences*, Douglas, B.C., Kearney, M.S., and S.P. Leatherman (eds.), chapter 4, pp. 65–93, Academic Press, San Diego, 2001.

Peltier, W.R., Global Sea Level Rise and Glacial Isostatic Adjustment: An Analysis of Data from the East Coast of North America, *Geophys. Res. Lett., 23*, no.7, pp 717–200, April 1, 1996.

Pugh, D.T., *Tides, Surges, and Mean Sea Level*, 472 pp., John Wiley and Sons, New York, 1987.

Pugh, D.T., *Changing Sea Levels: Effects of Tides, Weather, and Climate*, 280 pp., Cambridge University Press, Cambridge, UK, 2004

Small, C., and R.J. Nicholls, A Global Analysis of Human Settlement in Coastal Zones, *J. Coastal Res. 19*, no. 3, pp. 584–599, 2003

Sturges, W., and B.G. Hong, Decadal Variability of Sea Level, in *Sea Level Rise: History and Consequences*, Douglas, B.C., Kearney, M.S., and S.P. Leatherman (eds.), chapter 7, pp. 165–180, Academic Press, San Diego, 2001.

Tsimplis, M.N. and Woodworth, P.L. The global distribution of the seasonal sea level cycle calculated from coastal tide gauge data. J. *Geophys. Res., 99* (C8), pp. 16031–16039, 1994.

Woodworth, P.L., A search for Accelerations in Records of European Mean Sea Level, *International J. Climatology, 10*, pp. 129-143, 1990.

Woodworth, P.L., Some Comments on the Long Sea Level Records from the Northern Mediterranean, *J. Coastal Res. 19*, no. 1, pp. 212–217, winter 2003.

Zhang, K., Douglas, B.C., and S.P. Leatherman, Global Warming and Coastal Erosion, *Climatic Change 64* (1–2), pp. 41–58, May 2004.

Bruce C. Douglas, Florida International University, Laboratory for Coastal Research, 6544 Wiscasset Road, Bethesda, MD 20816-2113 (bruced7082@aol.com)

Eddies and Jets Over the Slope of the Northeast Gulf of Mexico

Peter Hamilton

Science Applications International Corporation, Raleigh, North Carolina

Thomas N. Lee

Rosenstiel School of Marine and Atmospheric Science, University of Miami, Miami, Florida

The eddy circulation over the northeastern Gulf of Mexico slope and the DeSoto canyon has been characterized using data from a 13 mooring array and 7 hydrographic surveys over a period of two years. The presence of cyclones and anticyclones over the lower Mississippi-Alabama slope is strongly influenced by the positions of cyclonic frontal eddies on the Loop Current or a Loop Current warm eddy to the south of the study region. Both anticyclones and cyclones feed an eastward surface jet on the upper slope that can follow the rim of the canyon or meander across the slope. Since this eastward jet often transports part of the discharge of the Mississippi delta, it provides a mechanism to transfer nutrient-rich, river-derived water to offshore or, on occasion, to the outer west Florida shelf. A westward countercurrent, between 200 and 500 m depth against the Alabama slope, often accompanies this surface jet. This usually occurs in conjunction with a cyclone over the slope of the west Florida terrace. The eastward jet and its countercurrent are replaced by westward cyclonic flow if a Loop Current frontal eddy extends northwards onto the middle and upper slope. A time scale for the dominant eddies is of order 100 days, with eddy diameters and swirl speeds of order 100-150 km and 50 cm/s, respectively. Over the head of the canyon smaller scale cyclones and anticyclones occur with shorter time scales (~ 20 days), and their interaction with the Alabama slope eddies can cause strong cross-slope currents, which are an effective mechanism for the transport of material on and off the shelf. The predominance of the eastward jet and anticyclonic currents promotes upwelling at the head of the canyon and along the shelf break of the west Florida shelf, where the water is ~ 2° C colder on average than further west.

1. INTRODUCTION

The upper-layer waters in the deep basin of the eastern Gulf of Mexico are dominated by the Loop Current (LC) and its shedding, at irregular intervals, of large vigorous anticyclones, with diameters ~ 300 to 400 km, and swirl velocities ~ 100 to 150 cm/s. These LC eddies (LCE) then translate, at speeds of ~ 3 to 6 km/day, into the western Gulf [*Vukovich et al.,* 1979; *Elliott* 1982; *Kirwan et al.,* 1984; *Glenn and Ebbesmeyer,* 1993]. Over the last quarter of a century, it has become increasingly clear that many other eddy-like flows, both cyclonic and anticyclonic, with a range of length and time scales, are present in the upper layer (above ~ 1000 m). Satellite imagery has shown cyclonic

Circulation in the Gulf of Mexico: Observations and Models
Geophysical Monograph Series 161
Copyright 2005 by the American Geophysical Union.
10.1029/161GM010

frontal eddies propagating clockwise around the LC front and shed LC anticyclones [*Vukovich*, 1986; *Paluszkiewicz, et al.*, 1983; *Fratantoni et al.*, 1998; see also the satellite SST images on the CD accompanying this book], similar in many ways to the meanders of the Gulf Stream front in the western Atlantic [*Lee and Mayer*, 1977; *Bane et al.*, 1981; *Lee et al.*, 1995]. At the western side of the basin, the collision of the westward-translating warm LC eddies with the steep Mexican continental slope were observed to generate complex eddy circulations, including the generation of companion cyclones [*Merrell and Morrison*, 1981; *Brooks*, 1984; *Lewis et al.*, 1989; *Vidal et al.*, 1992; *Vidal et al.*, 1994]. These processes have been modeled numerically [*Smith*, 1986], and theoretical mechanisms proposed for interactions of warm anticyclones with a wall [*Shi and Nof*, 1993]. The formation of a cyclone-anticyclone pair of eddies next to the Mexican slope often produces a strong offshore-directed jet that causes cold plumes of water to be ejected from the shelf. These plumes are often observed in satellite imagery around 25 to 26°N during the winter and spring, when surface temperature contrasts between the Texas shelf and the deep Gulf are large [*Brooks and Legeckis*, 1982].

High-resolution hydrographic surveys of the northern slope by ship and aircraft [*Hamilton et al.*, 2002; *Berger et al.*, 1996], along with extensive surface drifter deployments, showed fields of cyclones and anticyclones with diameters ranging from ~ 40 to 150 km. Swirl velocities were ~ 50 cm/s, and rotation periods were ~ 10 to 20 days, rather longer than the typical 7 to 10 days of LC anticyclones in the central Gulf. Further, apart from major LC anticyclones, satellite altimeter maps of sea level anomalies (SLA) have shown that complex fields of mesoscale cyclones and anticyclones are ubiquitous over most of the deep waters of the basin [*Leben*, this volume]. There is some evidence that cyclones can be advected on and off the Louisiana–Texas slope by the swirl velocities of the leading and trailing edges of westward translating major LC eddies [*Hamilton*, 1992; *Hamilton et al.*, 2002]. However it is not clear at present how cyclones are generated in the center of the Gulf, away from direct influences of the western boundary and LC frontal eddies. Though the hydrographic surveys provided snapshots of the northern slope eddy fields and the drifters provided some continuity, it was nonetheless difficult to track how eddies evolved between surveys. Thus little is known about the origin of eddies over the Louisiana–Texas slope, how they interact with shelf and deepwater circulations, and their eventual dissipation. Limited evidence suggests that some cyclones on the slope can remain relatively stationary for periods longer than three to six months [*Hamilton*, 1992]. Louisiana–Texas slope eddies provide effective mechanisms for material to be transported from the shelf to the deep basin and vice-versa [*Hamilton et al.*, 2002; *Ohlmann et al.*, 2001].

The first moored array on the northern Gulf slope that was more than just a single transect [e.g., *Hamilton*, 1992], was deployed between the 100 m and 1200 m isobaths east of the Mississippi delta as part of the DeSoto Canyon Eddy Intrusion study [*Hamilton et al.*, 2000]. The 13 moorings (Figure 1) were deployed for two years (March 1997 to April 1999), and as part of the experiment seven (7) CTD surveys of the northeastern slope region were performed at 4-month intervals. Thus this study allows examination of the time evolution of the current and temperature fields over the complex topography of the slope. The DeSoto canyon can be characterized as the almost perpendicular intersection of two different slopes: the rough Mississippi–Alabama slope and the broader west Florida slope (the Florida Terrace), which becomes very steep below 1000 m (the Florida Escarpment). The region around transect D will be called the head of the canyon. Because of its position, the DeSoto canyon is more likely to be directly and indirectly influenced by a northward-extending LC or LCE and its associated frontal eddies than the slope to the west of the delta. The LC rarely intrudes over the northeastern slope, but LC extrusions of warm water, advected by frontal or peripheral cyclones, can occasionally intrude almost as far as the coast of the Florida Panhandle [*Huh et al.*, 1981]. During the time period of this study, however, the mean position of the northern edge of the LC front is ~ 26.5 to 27°N, 150 to 200 km south of the study region. LC northward extensions have dominant return periods near 11 and 6 months with a mean of about 9 months [*Sturges and Leben*, 2000]. Thus there can be quite long periods when the northern edge of the LC or a LCE is south of 25°N and other smaller scale (~ 150 km diameter) cyclones and anticyclones can occupy the region between the LC and the DeSoto canyon. Examples of this situation are given by *Muller-Karger* [2000] (see his Figure 7). However there are few in-situ observations of the deep waters of the eastern Gulf when the LC is absent.

Wang et al. [2003] used the near-surface currents from this study to examine whether a numerical circulation model and geostrophic currents derived from SLA could generate similar spatial flow patterns. The model assimilated both SLA and satellite-derived sea surface temperature (SST). Two eddy modes were extracted using singular value decomposition (SVD) analysis (SVD extends empirical orthogonal function (EOF) or principal component analysis in that it extracts related temporal and spatial patterns from two independent data sets, such as observed and modeled surface currents [*Bretherton et al.*, 1992]). The major (first) pattern for both the model with observations and SLA with observations showed that a cyclonic (anticylonic) eddy in the DeSoto canyon could be associated with a northward (southward) movement of the LC front. Thus a coupling of eddy flows over the northeastern

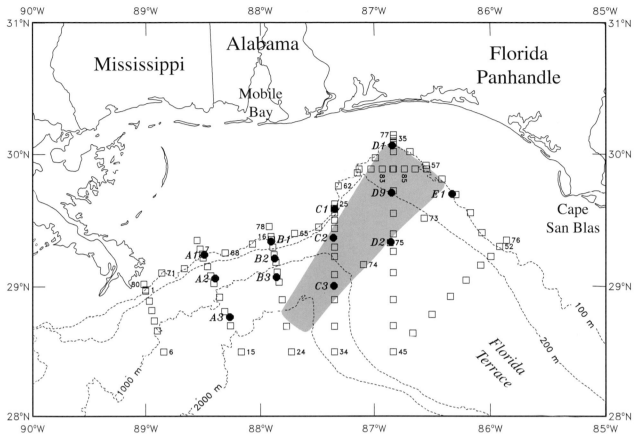

Figure 1. Map showing standard hydrographic grid (excluding stations 81–87) and all 13 current meter mooring locations (solid circles). Light gray shading indicates the region of the DeSoto Canyon.

slope with fluctuations of the north–south position of the LC front was demonstrated even though the LC (or a LCE) did not directly intrude over the slope. This paper complements *Wang et al.* [2003] in that it provides a more comprehensive description of the current and temperature variability, as well as considering flows below the 100-m surface layer.

2. OBSERVATIONS

Figure 1 shows the positions of the moorings, which are laid out on 5 cross-slope transects (A to E) with the 1, 2, and 3 moorings being on the shelf break (100 m), 500 and 1300 m isobaths. All the moorings were equipped with an upward-looking 300 kHz ADCP at 80 or 90 m and conventional current meters below this at varying spacing. The current meters and ADCPs were supplemented with temperature and conductivity/temperature (Micro- and SeaCats) instruments. An exception to these moorings was D9, equipped with a single narrowband 150 kHz ADCP at 180 m, that was placed to improve the horizontal coverage towards the head of the canyon. Details of the moored array are given in Table 1.

The moorings were deployed for a period of two years (March 1997 to April 1999) with rotations every 4 months. Data return was about 96%, and the only mooring that had substantial data loss was B1 during the second year because of fishing activity. Quality assurance, gap filling and 40-hour low pass (HLP) filtering of the time series records were the same as discussed in *Hamilton* [1990]. The principal analysis methods were spectra and frequency domain EOFs, also described in *Hamilton* [1990].

On the deployment, recovery, and rotation cruises, a standard grid hydrographic survey was performed. The CTD stations are given in Figure 1 and coverage was over a little larger region than the moorings. CTD profiles were collected to a maximum depth of 1200 m. On the December 1998 cruise (PE9923), a pump plumbing problem resulted in unreliable conductivity data. These data were discarded; replacement salinity data were generated from temperature through the T/S relation from the November 1997 cruise, using the technique developed for XBT data that is discussed in *Hamilton et al.* [2003]. This gives reasonable results for dynamic height because of the tight T/S relation for tem-

Table 1. Moorings and Instruments

Mooring / Water Depth (m)	Instrument Depth (m)	Measurement Type	Instrument Type
A1	16	C, T	SeaCat (Deployments 5 & 6, only)
100	20	T, V	MK2 (Deployment 6, only)
	80	T, V-Profile	300 kHz ADCP – up
	94	C, T, V	S4
A2	62	T	Hugrun
500	90	T, V-Profile	300-kHz ADCP – up
	150	T	Hugrun
	200	T, V	RCM-7
	250	T	Hugrun
	300	T, V	RCM-7
	490	T, V	S4/MK2
A3	80	T, V-Profile	300-kHz ADCP – up
1310	500	T, V	MK2/RCM-8
	1300	T, V	RCM-8/MK2
B1	20	C, T, P	SeaCat (Deployment 1, only)
100	62	T, V	RCM-7/MK2
	80	T, V-Profile	300 kHz ADCP – up
	82	C, T, P	MicroCat
	95	T, V	S4
B2	62	T	Hugrun
500	90	T, V-Profile	300-kHz ADCP – up
	150	T	Hugrun
	200	T, V	RCM-7
	250	T	Hugrun
	300	T, V	RCM-7
	490	T, V	MK2
B3	80	T, V-Profile	300-kHz ADCP – up
1300	500	T, V	MK2/RCM-8
	1290	T, V	RCM-8/MK2
C1	16	C, T, P	SeaCat (Deployments 5 & 6, only)
100	20	C, T, P	MicroCat (Deployments 1–4)
	80	T, V-Profile	300 kHz ADCP – up
	82	C, T	MicroCat
	95	T, V	RCM-7/S4
C2	62	T	Hugrun
500	90	T, V-Profile	300-kHz ADCP – up
	150	T	Hugrun
	200	T, V	RCM-7
	250	T	Hugrun
	300	T, V	RCM-7
	490	T, V	MK2
C3	80	T, V-Profile	300-kHz ADCP – up
1300	500	T, V	MK2/RCM-8
	1290	T, V	RCM-8/MK2
D1	62	C, T, P	SeaCat (Deployments 5 & 6, only)
100	80	T, V-Profile	300 kHz ADCP – up (Deployments.2–6)
	95	T, V	MK2/S4 (Deployments 2–6)
	99	T, V-Profile	300kHz ADCP (Deployment 1, only)

Table 1. (continued)

Mooring / Water Depth (m)	Instrument Depth (m)	Measurement Type	Instrument Type
D9	180	T, V-Profile	150kHz ADCP – up
200			
D2	62	T	Hugrun
500	90	T, V-Profile	300-kHz ADCP – up
	150	T	Hugrun
	200	T, V	RCM-7
	250	T	Hugrun
	300	T, V	MK2/RCM-7
	490	T, V	MK2
E1	20	C, T, P	MicroCat (Except Deployment 3)
100	62	C, T, P	MicroCat (Deployment 3, Only)
	80	T, V-Profile	300kHz ADCP – up
	82	C, T	MicroCat (Deployments 1–4)
	95	T, V	RCM-7/S4

peratures < 16° C. For dynamic heights, *Csanady*'s [1979] method for integrating the bottom density profile across the isobaths was used. If the bottom σ_t isolines are approximately parallel to the isobaths so that integration along transects is independent of position, then the method is considered accurate. This is a reasonable assumption for most of the upper slope. Dynamic heights and other hydrographic variables are gridded, for mapping, using statistical interpolation [*Pedder*, 1993]. Geostrophic velocities and vorticity are calculated using straightforward finite difference formulae. More details of both the moored array and the hydrographic cruises can be found in *Hamilton et al.* [2000].

3. SLOPE CIRCULATION PATTERNS FROM HYDROGRAPHY

Data from the seven hydrographic surveys allow high-resolution snapshots of the upper layer circulation over the slope. Patterns observed during these cruises are similar to mean circulations obtained from the moored instrumentation during a number of different periods (1 to 3 months long) of sustained flows that are discussed in the next section. *Wang et al.* [2003] also noted the predominance of long periods (> 15 days) in the upper-layer currents. Thus geostrophic flows, derived from the CTD surveys, reveal some of the details of the slope eddy field. Plate 1 shows the near surface geostrophic flows (relative to 1000 m) and relative vorticity (ζ) for each of the seven surveys. These geostrophic velocities compare well with the 5-day averaged, 40-HLP currents from the moorings, centered on the cruise periods, which are overlaid on the maps. The near-surface velocity maps show complex eddy circulations with diameters of order the width of the slope (~ 50 to 100 km). Smaller scale features are sometimes present on the upper slope and at the head of the canyon.

Five of the seven maps (Plate 1(a, b, c, e & f)) show predominantly eastward flows along the upper slope, south of Mobile Bay. These eastward jet-like flows had differing configurations depending on the sense of rotation of circulation patterns further south. In March 1997 (Plate 1(a)), the eastward flow, south of Mobile Bay, diverged from the shelf break to bypass the head of DeSoto canyon. A weak cyclone-anticyclone pair of small eddies occupied the canyon. A SST image for March 24 1997 shows a streamer of cold shelf water that moved southeast along the northern edge of the jet as it turned away from the shelf break (SST images, mentioned in the text, are provided on the accompanying CD). This image shows two moderately sized warm eddies centered at 28°N, 88.7°W (named LC eddy D) and 28°N, 87°W, respectively. The survey documented the northern part of this latter circulation. The northern front of the LC was south of 26°N at this time. Similar circulations occurred during the November 1997 and August 1998 surveys (Plate 1(c & e)). In the former, the eastward flow turned sharply to the south at approximately 87°W. In the latter, the flow was part of a vigorous anticyclone to the south, and the canyon circulation was cyclonic. An elongated warm anomaly formed parallel to the west Florida escarpment in the wake of the detachment of eddy F (~ 400 km diameter) from the LC in April 1998 (see Figure 5.2-10a in *Berger et al.* [2000], also included on the accompanying CD). This anomaly, which was detected in the SLA maps, remained for most of the year and was probably a direct contributor to the strong anticyclonic flows observed in the August survey. A couple of surface drifters released in August 1998 made a number of circuits of this anticyclone [*Hamilton et al.*, 2000] and confirmed that its center was at ~ 28.5°N, 87.5°W with a diameter of ~ 150 to 200 km and rotation period of ~ 15 to 20 days.

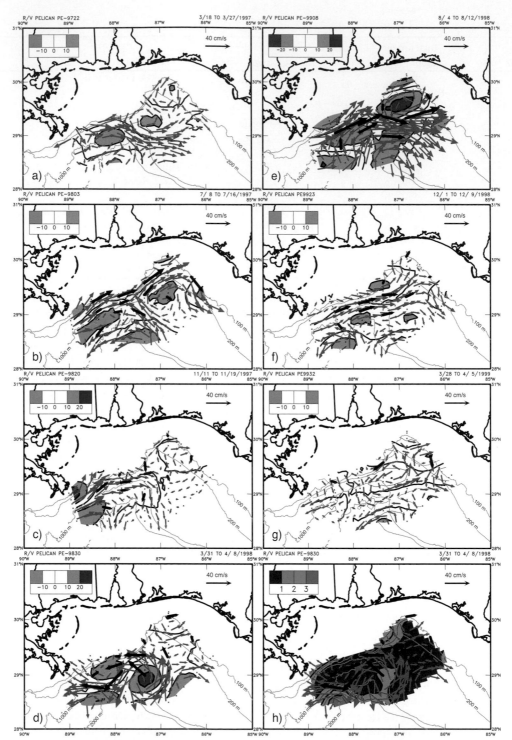

Plate 1. Relative vorticity (10^{-6} s^{-1}) contoured for each hydrographic survey (a–g) and overlaid with geostrophic velocity vectors (pink arrows) calculated from dynamic height fields at 6 m relative to 1000 m. The heavy black arrows emanating from the dots are the 5-day average, 40-HLP current vectors from the ADCP's using near surface measurements (6 to 12 m depth). The average is centered on the mid-point of the survey. The final plot (h) shows the Ertel potential vorticity (10^{-9} m^{-1}s^{-1}) between the surface (6 m) and 26 σ_t for the April 1998 survey (PE9830), corresponding to (d). The relative vorticity for the PV calculation was calculated for the 25.6 σ_t surface, which was approximately mid-way between the upper and lower surfaces.

The remaining two surveys that show eastward flows (Plate 1(b & f)) are similar in that the jet-like flow roughly followed the shelf break around to the eastern side of the canyon, and cyclonic flows were present over the deep water south of Mobile Bay. In July 1997, the cyclone and anticyclone on the west Florida slope appears to have generated strong flows toward the shelf break that enhanced flow around the head of the canyon. In December, however, the interaction of the deepwater cyclone and a weaker west Florida slope anticyclone (Plate 1(f)) did not generate onshore flow that could transport warm salty water to the outer shelf.

The map from March 1998 (Plate 1(d)) shows flow to the west that is driven by a compact, vigorous slope cyclone in the center of the study area. There was another region of cyclonic flow to the west that appears in the altimetry as a cold frontal eddy on the northern periphery of eddy F, which was in the process of detaching from the LC. In the SST image for April 4, 1998, the two cyclones can be identified clearly. There was also an anticyclone over the west Florida Terrace that interacted with the western cyclone to promote onshore flow toward mooring C1. The March 1999 survey, however, shows a situation with weak, confused flows consisting of small-scale cyclones and anticyclones along the upper slope on both sides of the canyon (Plate 1(g)).

The relative vorticity maps (Plate 1) show most of the surveys had similar amounts of cyclonic and anticyclonic flows even when the flows are predominantly eastward or westward. The relative vorticity magnitudes have maximum values of ~ $0.3f$, where the Coriolis parameter is $0.7 \ 10^{-4} \ \mathrm{s}^{-1}$. The spacing of the cyclones and anticyclones is ~ 100 km and is similar to the dominant scales of the near surface relative vorticity found on the northern slope west of the delta, where direct influence of the LC and LCE's is rare (see Figures 2b, 5b and 10f in *Hamilton et al.* [2002]). Therefore, in terms of the space and time scales of slope cyclones and anticyclones, there appears to be no significant differences between the eastern and western sides of the delta. Maps of the upper layer potential vorticity were estimated for each survey using the depth of the lower pycnocline represented by the 26 σ_t surface, and the relative vorticity on an isopycnal that was approximately at the middle depth of this upper layer. The Ertel potential vorticity [*Pollard and Regier*, 1992] is defined by:

$$PV = (f + \zeta) \ (1/\rho) \ \partial\rho/\partial z \qquad (1)$$

The PV for the March 1998 vigorous cyclone (Plate 1(d)) is given in Plate 1(h). Except for the shelf break regions where low surface densities, derived from the Mississippi outflow, cause higher values, the PV is fairly uniform over most of the slope. This is also the case for the other surveys even though

the depth of the 26 σ_t surface can vary as much as 60 m between the centers of cyclones and anticyclones. Therefore, one of the effects of the upper-layer eddy flows appears to homogenize the PV over the slope. Strong PV gradients, normal to the isobaths, would be expected to inhibit cross isobath flows that are commonly observed on the northern Gulf of Mexico slope.

For two of the surveys (March 1997 and August 1998), the geostrophic currents at 300 to 500 m depth are observed, in agreement with the current meter data, to be in the opposite direction to the upper slope eastward surface flows (Plate 2(a & e)). These two cases of counter currents at depth, as well as a weaker example in April 1999, have in common a cyclonic eddy, in the canyon, to the east of where a strong jet turns anticyclonically away from the upper slope. It appears that the westward cyclonic flow, which is against the northern slope of the eastern part of the canyon, penetrates below the eastward surface jet to produce a counter flow at depth. These counter flows are most strongly observed at section B, and the geostrophic currents (relative to 1000 m) and density fields for this transect are given in Plate 2. This section is due south of Mobile Bay and extends into offshore waters deeper than 2000 m. Observations from March 1997 and August 1998 (Plate 2(a & e)) most clearly show the surface jet and the slope counter current below 200 to 300 m. The core of the westward flow was at a depth of about 300 and 400 m and had a vertical extent of 500 to 600 m for March 1997 and August 1998, respectively. Further offshore the observed eastward flow had a subsurface maximum at depth that was part of an anticyclone over deep water (Plate 1(a & e)). The last survey in April 1999 (Plate 2(g)) documented a weak two-layer structure with the westward current maximum at the shallow depth of 200 m. These sections suggest that the upper slope eastward surface currents and their counter currents adjacent to the slope are somewhat distinct from the anticyclonic flows further offshore.

In contrast, the purely cyclonic case (Plate 2(d)) showed no reversals with depth. The velocity maximum, over the upper slope, was westward and subsurface at about 100 m as expected for this type of eddy [*Hamilton et al.*, 2002]. The remaining sections for July 1997, November 1997 and December 1998 (Plate 2(b, c, & f)) all have cyclonic westward flow offshore in deep water, but eastward, strongly sheared, jet-like flows over the upper slope. Again the surface geostrophic velocity fields (Plate 1) show that the eastward upper slope currents were separate from the westward flows offshore. Further east on the slope this eastward jet took various paths, including following the canyon rim (Plate 1(b)), depending on the configuration of the eddies. Occurences of a jet-like flows on the upper slope of the northeast Gulf make this slope dynamically different from that of the central

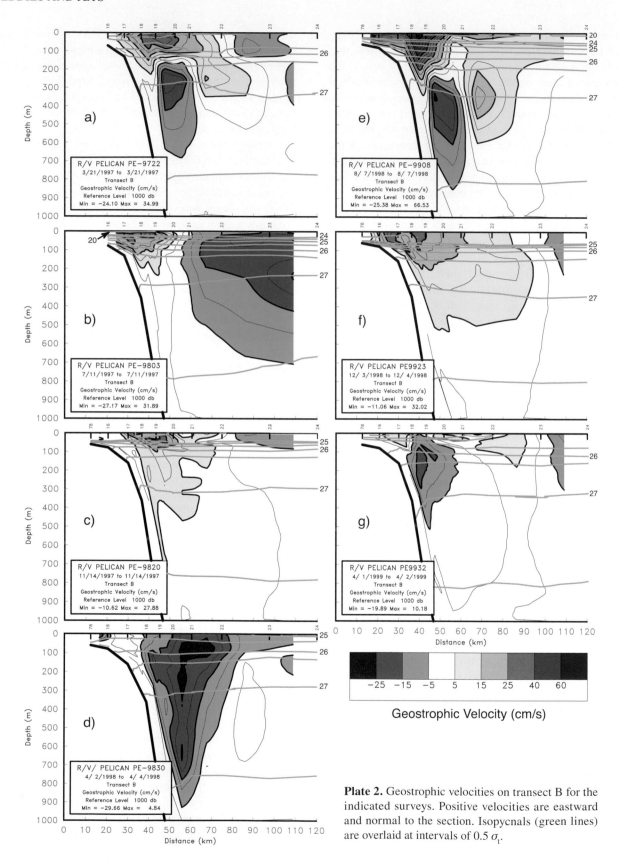

Plate 2. Geostrophic velocities on transect B for the indicated surveys. Positive velocities are eastward and normal to the section. Isopycnals (green lines) are overlaid at intervals of 0.5 σ_t.

and northwestern Gulf in that there, upper slope flows seem to be dominated by small scale eddies [*Ohlman et al.,* 2001; *Hamilton et al.,* 2002].

Consequences of these circulation patterns included the transport of low salinity water from the shelf to the slope, and the transport of warm, salty offshore water toward the shelf break. Figure 2a shows the near-surface salinity for July 1997. The eastward shelf break flow (Plate 1(b)), assisted by an eastward flow on the shelf driven by the prevailing westerly winds, distributed brackish water from the Mississippi delta along the canyon rim. This could be an important mechanism by which nutrient-laden, river-derived water was input to the shelf on the east side of the canyon. There are no major river discharges into the shelf on the east side of the DeSoto canyon. The northward flow between the two counter-rotating eddies (Plate 1(b)) also caused a warm salty intrusion to approach the shelf break.

Eastward flows that prevail along the shelf break in the western portion of the study area together with the occurrence of a cyclone around the head of the canyon also cause isotherms to be upwelled over the upper slope. Figure 2b shows contours of the temperature at 90 m for the August 1998 survey. Plate 1(e) gives the surface geostrophic velocity vectors, and the 5-day mean currents at 70 m are also overlaid on Figure 2b. Cooler water was present on the northern edge of the eastward current that, in this case, bypasses the canyon head and roughly flows along the path of the subsurface temperature front. However, the weaker cyclone in the canyon also raised the isotherms to produce water colder than 17°C, and some of this water intrudes over the head of the canyon. Of the 7 surveys, 4 show water over the eastern part of the canyon rim (100 m isobath) colder by 1° or 2 °C at the bottom than is the case further west near the delta.

4. LONG PERIOD EVENTS

Snapshots of the slope circulation raise questions on the persistence of various flow patterns and their frequency of occurrence. Movies of daily averaged currents and temperature fields for the surface 100 m layer, as well as at 500 m, are included on the CD accompanying this book. They give a good idea of the spatial and time variability of the slope eddy field. Similarly, the movies of SLA [*Leben,* this volume], derived from satellite altimetry, show the time variability of the larger scale eddy features over the whole of the Gulf basin. Characteristic events and their time scales are examined using the moored array data with emphasis on flows that have some degree of persistence over periods of a few months. The middle slope (500 m) moorings were the most heavily instrumented and allow both the temperature and velocity structure to be documented for the upper 300

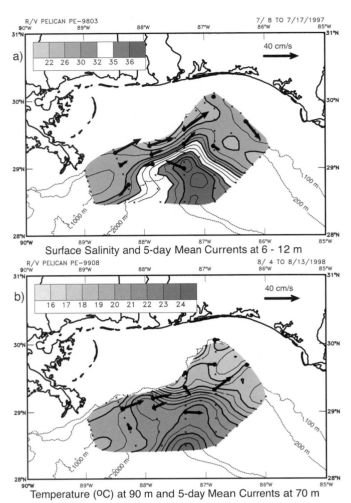

Figure 2. (a) Surface salinity from July 1997 (PE-9803) overlaid with near surface 5-day mean current vectors, and (b) temperature at 90 m from August 1998 (PE-9908) overlaid with 5-day mean current vectors at 70 m.

m of the water column. Plate 3 shows the 7-day low pass (DLP) temperature and velocity profiles at C2 over the two years of the study. The records from the 500 m moorings at C2 are similar to the data from the other three 500 m moorings. Basic flow patterns set up for periods of several months and three long-period events have been selected for further analysis. The isotherms at different depths had similar displacements, indicating that the whole of the upper water column warmed and cooled in response to the advection by large-scale flows. A small amount of seasonal warming and cooling was evident at 62 m, and the first year of the study (March 1997 to March 1998) was notably cooler than the second year, when the mean depth of the 15°C isotherm was about 50 m deeper. The difference in the flows is that eastward, upper layer currents predominated in the first year whereas there were more sustained periods of westward flow

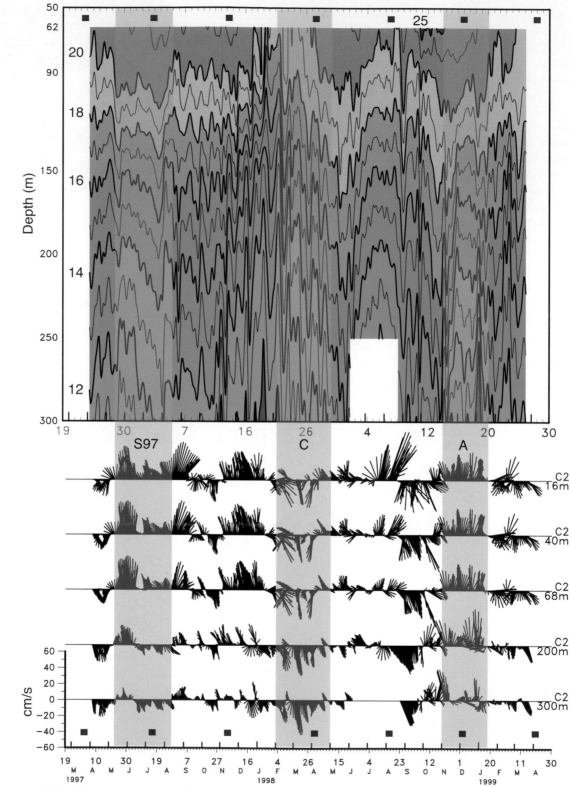

Plate 3. 7-DLP current vectors and isotherm depths (upper panel) for the upper 300 m of mooring C2. Up vectors are directed eastward along the isobaths (75°T). The times of the 7 hydrographic surveys are given by the red squares.

in the second year. Generally, eastward and westward flow events raised and lowered the 15°C isotherm, respectively, which is consistent with C2 being on the northern side of the anticyclones and cyclones. The cyclonic flow episodes were also effective in advecting warmer water from the deep eastern basin towards the northeastern slope.

As shown in Plate 3, the first period of sustained flows occurred in the summer of 1997 (denoted by S97) with eastward flow above 200 m and weaker westward currents at depth. The July 1997 survey occurred in the middle of this period (Plate 1(b), Plate 2(b), and Figure 2a) and documented westward flow further offshore resulting from the northern edge of a cyclone. Between 1 February and 30 April 1998 (denoted by C for cyclonic), flow was westward at all depths with speeds at 300 m equaling or slightly exceeding those at the surface. The early April 1998 survey showed a vigorous cyclone over the western part of the canyon (Plates 1(d) and 2(d)). Both velocities and temperatures had energetic short time-scale (~ 10 to 20 days) fluctuations that rarely reversed the direction of the prevailing westward currents. Similar periods with all flows at C2 toward the west included August

and September 1998 and February and March 1999. The short-period fluctuations in the latter interval were not as energetic as during C. The last event (denoted A for anticyclonic) was also characterized by along-slope flows toward the east. The difference from the S97 period is that flows decreased little with depth and no undercurrent was apparent. Observations from the December 1998 survey (Plates 1(f) and 2(f)) show an anticyclonic circulation extending vertically through the water column over the central part of the canyon. The other two sustained anticyclonic episodes, in November and December 1997 and July 1998, have more in common with S97 in that counter currents develop at depth that were caught by the July 1998 survey (Plate 2(e)) but not by the November 1997 survey (Plate 2(c)), which occurred early in this event.

The means and standard deviations for the three long-period events and the complete two-year period of the study are given, for the surface layer (40 m) and lower layers (200 and 500 m), in Figures 3 and 4, respectively. The 40-m depth currents were representative of the upper 100 m of the water column as measured by the ADCPs. The 16-m level was

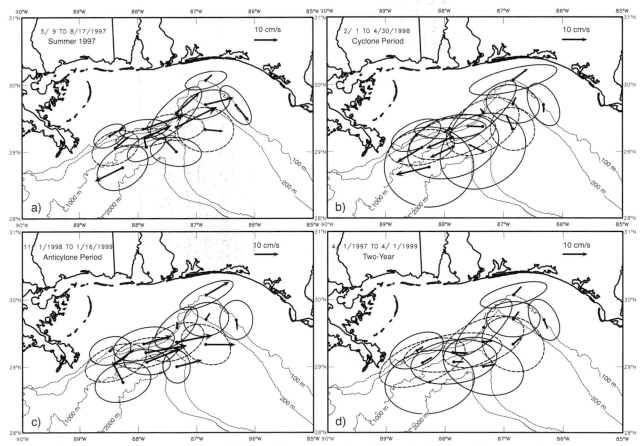

Figure 3. Mean currents and standard deviation ellipses at 40-m depth, from 40-HLP data, for the periods: (a) Summer 1997 (S97); (b) Cyclonic flow (C); (c) Anticyclonic flow (A); and (d) Complete two-year records.

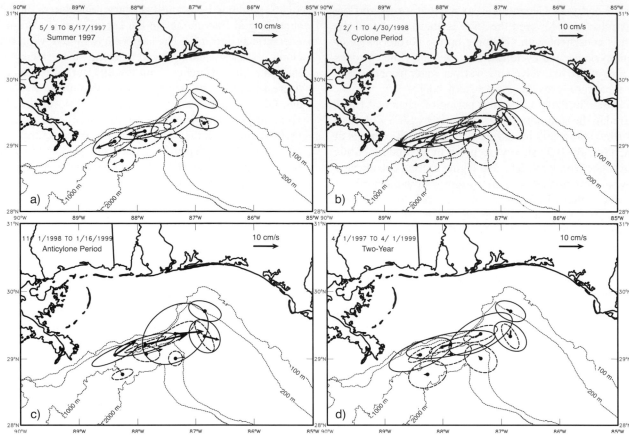

Figure 4. Mean currents and standard deviation ellipses at 200-m (thick arrows and solid ellipses) and 500-m (thin arrows and dashed ellipses) depths, from 40-HLP data, for the periods: (a) Summer 1997 (S97); (b) Cyclonic flow (C); (c) Anticyclonic flow (A); and (d) Complete two-year records.

more energetic than the 70-m level, but the differences were not substantial. Similarly, for the lower-layer currents, there was very little difference between the 200- and 300-m levels, and the bottom level on the 1300-m moorings had very small means and standard deviations compared to the 500-m data [*Hamilton et al.,* 2000]. For the summer 1997 (S97) period, the 40-m currents show eastward flow following the trend of the upper slope isobaths with westward flow further offshore (Figure 3a). The mean currents at A3, B3, and C3 indicate that some of this westward flow was being entrained into, or acting as a source for, the upper slope jet. At depths of 200 and 500 m (Figure 4a), the flow was westward with the maximum (~ 10 cm/s) being found at 300 m. Though in general the means were comparable to the standard deviations, it is noteworthy that the cross-slope variances were a substantial fraction of the along-isobath variances, particularly offshore and in the canyon. The fluctuation magnitudes showed no marked variation across the array at any given depth level; however, there was a strong decrease with increasing depth. The cyclonic flow on the southern edge of the array was

related to the presence of a LC frontal eddy (LCFE) in the region between 28°N and 29°N [*Hamilton et al.,* 2000].

The mean velocities and standard deviation ellipses for the anticyclonic flow period, A, are similar in many respects to S97 (Figures 3c and 4c). The mean velocities in the upper 100 m were eastward with northward inflow occurring on the western edge. The main current bypassed the head of the canyon but a subsidiary loop maintained the anticyclonic flow along the rim. At 200 m the means were eastward, and at B2 they exceeded those in the upper 100 m. At 500 m, however, a narrow region of westward flow (at B2 and C2) existed next to the slope. The eastward mean flow (at B3 and C3) over the 1200-m isobath appears to have been blocked by the west Florida slope and was partly returned to the west via a deep cyclonic circulation in the canyon (i.e., 500-m mean flow at D2 is northward). Thus the circulation was similar to, but less energetic than, that occurring during the S97 interval, except that there was no cyclonic circulation offshore and the undercurrent was restricted to a narrow region close to the bottom at the 500-m isobath. This undercurrent was evidently too close to the slope to be seen in the geostrophic velocity section (Plate 2(f)). The anticyclonic circulation implies uplifted isotherms on the outer edges of the eddy. The water at 200 m rapidly cooled at the beginning of this event (Plate 3). This particular anticyclone was not directly associated with the LC. SLA maps and SST images show the LC mostly south of 25°N at this time.

The opposite situation, with the slope being dominated by westward flows, is given for period C in Figures 3b and 4b. Mean flows at 40 m were toward the west with a magnitude of approximately about 10 cm/s over the Alabama slope with an anticyclonic circulation over the head of the canyon. The smallest westward velocities were at the shelf break. The standard deviation ellipses, in the upper 100 m, show that the kinetic energy increased from the shelf break toward deeper water. The energy levels were also greater in the western part of the array than for S97 or A (Figure 3). In the canyon, upper layer mean flows were northward, thus, cyclonic flows in deep water can generate onshore velocities along the axis of the canyon (transect D). At 200 m, the westward means exceeded 25 cm/s in the center of the slope (B2) and approached 10 cm/s at 500 m. This period is characterized by a rapid succession of cyclonic frontal eddies propagating around a well extended LC or LC eddy F, which at this time was in the process of forming and detaching from the LC. This explains the predominance of cyclonic flow and the presence of short period fluctuations during February to April 1998. The intrusion of LCFEs onto the slope during this period caused the steady deepening of the 15°C isotherm at mid-slope locations (Plate 3).

The differences in the surface-layer mean flow patterns between the cyclonic (C) and anticyclonic periods (S97) (Figures 3a and 3b), are not that large, particularly around the head of the canyon. Indeed, the means in Figure 3a could be generated by displacing the cyclonic flows over the Alabama slope in Figure 3b off to the southwest, thus allowing northward flow over the lower slope that turns to the east to feed the upper slope jet. The eastward turn is consistent with PV conservation for flows over shoaling topography. This interpretation emphasizes a fundamental connection between the observed slope circulations, whether cyclonic or anticyclonic, and the locations of the deep-water cyclones or LCFEs further south.

Wang et al. [2003] performed a SVD analysis of the upper-layer currents after filtering to remove periods shorter than 15 days and geostrophic currents derived from satellite altimetry. They state that the first two SVD patterns were very similar to the first two EOF modes of the observed, near-surface currents. The mean currents at 40 m (Figure 3) for the S97, C and A periods have strong similarities to the first two patterns from the SVD analysis. The anticyclone

and cyclone period means correspond to positive and negative amplitudes of the first SVD pattern, and the S97 and the canyon flows of C are similar to the second SVD pattern. The long periodicities of the events in Plate 3 make this correspondence reasonable, and indicate that the geostrophic flows in Plate 1 are representative of eddy patterns over the slope and canyon.

Figures 3d and 4d give the means and standard deviations for the currents, at the given depths, for the complete two years of the observations. The standard deviations greatly exceeded the means at all depths. Unlike the event statistics, however, the two-year records contained both eastward and westward flow events. The means were quite robust, and subdividing the record into two separate 12-month intervals produced very similar statistics. Upper layer means were eastward and tended to follow the isobaths at the surface. Mean flows at D9 had an up-slope component above about 120 m. At 200 m and below, the flow was all to the west with slightly larger magnitudes at the 500 m isobath than at the 1300 m isobath. Therefore the counter current at depth was a distinct feature of the mean slope circulation.

The anticyclonic and cyclonic nature of the upper- and lower-layer mean flows, respectively, was also reflected in the two-year mean temperature fields at 90 m and 500 m (not shown). The 90 m temperatures had a strong cross-slope gradient with the warmest water on the offshore side of the Alabama slope. The water at the head of the canyon was more than 2°C colder. Thus persistent anticyclonic flows caused upwelling at the head of the canyon to a greater degree than elsewhere along the shelf break. At 500 m, the mean cross-slope temperature gradient, though small (~ 0.2 °C between A3 and A2), had the opposite sign consistent with the westward flows found in the lower layer.

5. LONG PERIOD VARIABILITY

A frequency domain EOF analysis [*Preisendorfer*, 1988] was performed using all available velocity records (both components with means removed) that spanned the two-years of observations. Because there was high coherence of the records in the upper 100 m, three levels (16, 40, and 68 m) were used from the 90 m ADCPs, and four levels (37, 72, 98 and 168 m) from the D9 ADCP. The kinetic energy spectra showed that motions could be divided into two frequency bands: > 50 days and 15 to 50 days [*Hamilton et al.*, 2000]. Using the criteria of *North et al.* [1982] that eigenvalues do not overlap when their errors, as determined from the degrees of freedom, are taken into account, only the first mode of the lower frequency band (0-0.02 cpd), was significant; however, this accounts for 57.7% of the total variance, Figures 5a and 5b show the EOF amplitudes and phases (in

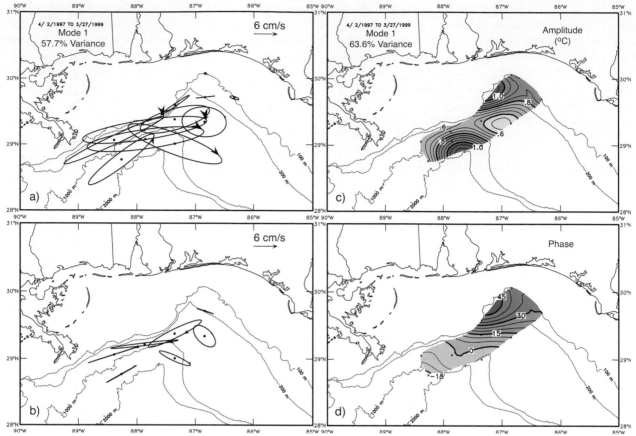

Figure 5. Left hand panels: Mode 1 EOF currents for the 724 to 50-day period band fluctuations. (a) is the subset for 40 m, and b) for 200 m (C2, D2 & D9), 300 m (A2 and B2), and 500 m (A3 and C3). Right hand panels: Mode 1 EOF for the 90 m temperatures for the 724 to 50-day period band fluctuations. (c) Amplitude (°C), and (d) phase (degrees), where positive angles lead.

the form of hodographs) at selected upper and lower layer depths for mode 1. The fluctuations did not have a great deal of variability with depth, but had maximum amplitudes near the surface. For this first mode, the fluctuations followed the trend of the isobaths and were most energetic along the Alabama slope, but had little or no influence on the canyon and shelf break moorings C1, D1, D9, and E1 and thus represent large scale, low-frequency (~ 100-day period) eastward and westward flows along the slope. This is very similar to the first SVD spatial pattern of *Wang et al.* [2003], which had dominant periods of 100 to 200 days. An advantage of frequency domain analysis is that propagation of coherent signals is part of the mode. Phase changes from east to west were small but show some interesting features. Along the 500 m isobath, the east side led the west at both the 40 m and deeper depths, but for the 1300 m moorings the signals were essentially in phase. This implies offshore and westward propagation of phase that in turn suggests surface trapped topographic waves.

A similar EOF analysis, using the same 0–0.02 cpd frequency band, was performed for all the temperature records at 90 m. Only the first mode was significant, accounting for 63.6% of the total normalized variance. The amplitude and phase of this temperature mode are given in Figures 5c and 5d. The amplitudes of the fluctuations were similar across the array, with maxima on the western rim of the canyon and the offshore side of the Alabama slope. Again, the phase indicates offshore and westward propagation of the coherent signals. A least squares fit of wave numbers to the array of phases obtained from the velocity and temperature modes [*Hamilton*, 1990] are given in Table 2, where the (k, l) axes are aligned with the general trend of the northern slope. Estimates of wavelengths range from 250 to 600 km with the wave vector directed approximately due south. The along-slope wavelength (k^{-1}) has the approximate east–west extent of the northern slope (~ 1000 km) that could be consistent with a first normal mode in a closed basin [*Pedlosky,* 1982;

Table 2. Wave Numbers From EOF Mode 1 Phases (0.01036 cpd)

Variables	Along-Isobath Wave Number k (km^{-1})	Cross-Isobath Wave Number l (km^{-1})	Wavelength (km)	Direction of Phase Propagation (°T)
U, V-cmpts at 40 m	-0.000846	-0.002456	385	179
U, V-cmpts at 200 to 500 m	-0.001204	-0.004198	229	176
Temperature at 90 m	-0.000560	-0.001538	611	180

Section 3.25]. Therefore this large spatial scale westward propagation could be an indication that the slope acts as a topographic waveguide for long period (~ 100 day) disturbances. The time and length scales are appropriate to LC- or LC-eddy induced motions as found for the similar SVD pattern discussed in *Wang et al.* [2003]. The steep west Florida slope has been proposed as a wave guide for northward propagating pressure disturbances generated by the impingement of the LC on the west Florida shelf break [*Hetland et al.*, 1999]. The disturbance is manifested as a southward flowing shelf-edge jet, and the signal is observed in satellite altimeter tracks. Such signals propagating into the DeSoto canyon could also be contributing to the observed long-period variability.

For the 50- to 15-day period (0.02–0.065 cpd) motions, only the first two modes were significant, and they accounted for less than half the variance. This implies that a majority of the energy was contained in small-scale motions that were not coherent across the array. The fluctuations were also less energetic than for the low frequency band, discussed above. The EOFs for both modes are given in Figure 6, and in both cases the velocity fluctuations were essentially in phase, both horizontally and with depth. Therefore there is little evidence of propagating signals.

For mode 1, the fluctuations consisted of two similar-sized (~ 100 km diameter) eddies with the same sign, one over the Alabama slope with a center over or seaward of the 2000 m isobath, and the other, weaker, circulation over the head of the canyon with a center between D9 and D2 (Figure 6a). The sign of the fluctuations (i.e., two anticyclones or two cyclones) alternate at ~ 10 day intervals. At depth, only the slope anticyclonic-cyclonic circulation was present (Figure 6b). Mode 2 (Figures 6c and 6d) reversed the sign of the slope circulation such that there was cyclonic flow when there was anticyclonic flow over the head of the canyon and vice-versa. The flows were also more perpendicular to the isobaths on the lower slope (1300 m moorings) than for the first mode. Because the second mode

accounted for less of the total variance, the occurrence of counter-rotating eddies was less prevalent than eddies of the same sign. This second mode resembles the second spatial pattern of *Wang et al.* [2003] for the combined current observations and altimetry SVD analysis, which also had eddies of opposite sign over the head of the canyon and the Alabama slope. However, most of the energy in their mode was at long periods similar to their first SVD pattern. The flow patterns for both EOF modes indicate that the two eddies were interacting through flows along the upper Alabama slope. The implication of this analysis is that slope and canyon eddy circulations were relatively short-lived (~ 10 to 25 days) and quite complex.

6. DIVERGENCE AND RELATIVE VORTICITY

The arrangement of the moored array allowed resolution of both north-south and east-west spatial gradients of velocity. The divergence (D) and relative vorticity (ζ) are given by:

$$D = \partial u/\partial x + \partial v/\partial y, \text{ and } \zeta = \partial v/\partial x - \partial u/\partial y, \quad (2)$$

respectively, where (u,v) and (x,y) refer to the geographic coordinate system. Direct calculation of these quantities is expected to be noisy. Therefore, following *Chereskin et al.* [2000], least square planes were fitted to arrays of 4 to 6 velocity components at a given depth level. Thus

$$u(x,y,t) = u_o + x\ \partial u/\partial x + y\ \partial u/\partial y + HOT, \text{ etc.,} \quad (3)$$

where (x,y) were measured from the center position of the array. In the least square fit, the velocity components were weighted by their standard deviations. The resulting standard deviations of the gradient terms in the plane model were of the order 0.05 to 0.09f. Standard least square methods [*Press et al.*, 1992; Chapter 15], rather than the matrix inversion method described by *Chereskin et al.* [2000] were used for these calculations. D and ζ were estimated at various depths for B3, using

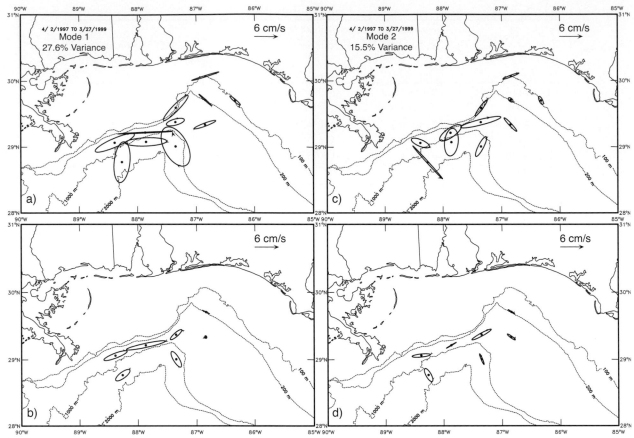

Figure 6. Modes 1 (RHS a and b) and 2 (LHS c and d) EOF currents for the 50 to 15-day period band fluctuations. (a) and (c) are the subset for 40 m, and (b) and (d) for 200 m (C2, D2 & D9), 300 m (A2 and B2), and 500 m (A3 and C3).

A2, A3, B2, B3, C2, and C3; C2, using B1, C1, C2, D2, and C3; and D9, using C1, D1, D2, D9, and E1 (see Figure 1).

The time series of divergence and relative vorticity at 40 and 500 m depths are given in Figure 7. Calculations were also made at 70 m depth, but these were almost identical to the 40 m level. Substantial fluctuations of D and ζ occurred at both the 40 m level for B3 and C2. Relative vorticity was also coherent with depth in the upper 100 m and also between the upper levels of B3 and C2. The divergences at the upper levels of B3 and C2 were weakly positively correlated (R = 0.25 and 0.35, respectively) with temperature records at the same positions. The 80 m temperatures at B3 were, however, negatively correlated (R ≈ -0.48) with the vorticity. The same relation does not hold at C2. The implication is that only a small part of the temperature signal could be attributed to local vertical velocities. At B3, there was a stronger relationship to whether a cold or warm eddy (positive or negative vorticity, respectively) was present on the lower slope. The same relation did not hold for C2, because it was often situated in the outer parts of different eddy flows.

It is clear from Figure 7 that the vorticity fluctuations at B3 were closely related to the along-slope currents. Thus eastward flow corresponded to negative vorticity and vice-versa. The large negative anomaly in July and August 1998 corresponded to the intense anticyclonic circulation observed over the slope (see Plate 1(e)). Even though the vorticity at C2 was similar to that at B3, it was much less closely related to the C2 currents. Again, this most likely occurred because of the position of C2 on the edges of the major deep-water eddy circulations (see Plate 1). The small magnitude of the vorticity fluctuations at D9 is somewhat surprising given that flows following the rim of the canyon should have considerable curvature. However, major flow events often bypassed the head of the canyon, and the eddy scales were often smaller than the separation between the moorings used for these calculations.

The spectra of D and ζ show that most of the energy was at periods longer than 50 days (Figure 8a). The peak at ~ 100 days is similar to those found in the velocity spectra and suggests that the major vorticity fluctuations were associated

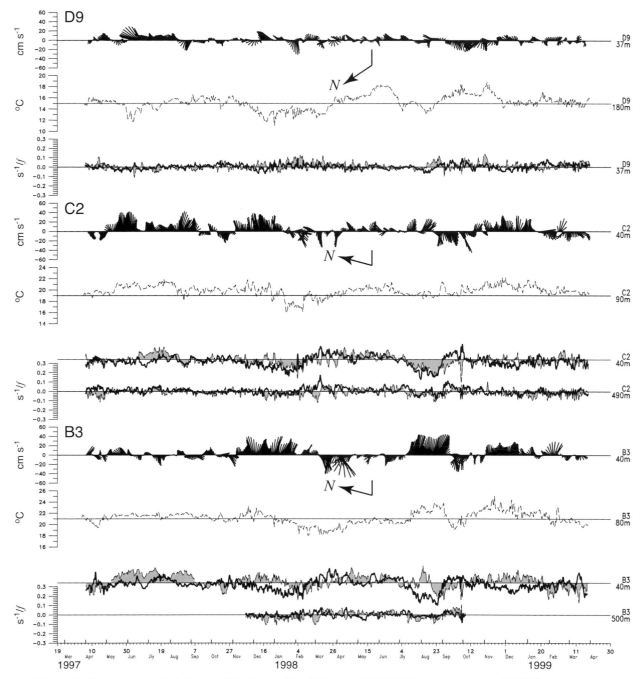

Figure 7. Divergence and relative vorticity (scaled by f) along with 40-HLP temperature and 7-DLP current vectors for the 3 indicated moorings at the depths given on the RHS of the time axis. For each mooring the records from the top are current vectors, temperature, and relative vorticity (heavy line) and divergence (dashed line with shading) at 40 m and 500 m (except D9).

with the long period propagating slope waves that are probably related to LC and LC eddy effects. Figure 8 also shows the coherence squared and phase differences of the most coherent records with D and ζ at B3 and C2 (40 m depth). At

long periods, the vorticity was coherent and 180° out-of-phase with the along-slope current at B3. This confirms the visual correlations of the time series in Figure 7. However, D at B3 was coherent and in-phase with the along-slope current at

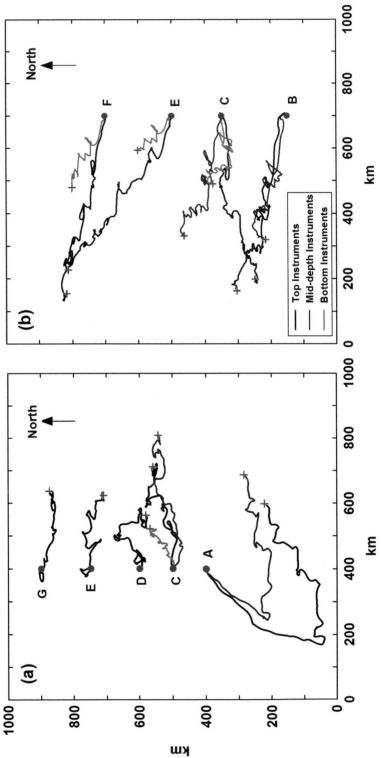

Plate 1. Progressive vectors diagrams for (a) the upcoast and (b) downcoast flow regime; depths are color-coded and offset for clarity; the solid red circles mark the initial position; the red pluses mark the final position.

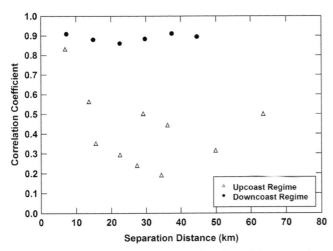

Figure 3. Correlation coefficients as a function of the cross-shore separation distance for the upcoast (open triangles) and downcoast (solid circles) flow regimes; the coefficients are computed only for the top instruments.

summer coefficient does not display a monotonic decrease with increasing separation distance as often observed in other coastal regions [*Kundu and Allen*, 1976; *Dever*, 1997]. However, inspection of the correlation values and their errors suggest that the summer cross-shore spatial length scale could be between 30 km and 50 km.

The distinct difference in spatial length scales of the LTCC currents between seasonal regimes may result from differences in the scale and strength of forcing mechanisms such as winds [*Gutiérrez de Velasco and Winant*, 1996; *Wang et al.*, 1998]. During the downcoast regime, the spatial length scale is much larger than 60 km, meaning that it is greater than the width of the LTCC (48 km as estimated by *Murray et al.* [1998]). This large spatial scale probably results from coherence imposed on the surface layer by the large and energetic synoptic scale weather systems (cold fronts) impacting the northern Gulf of Mexico in fall, winter, and spring. These systems usually induce strong vertical mixing and drive flows inside and outside of the LTCC with similar efficiency [*Wiseman et al.*, 1986; *Wiseman and Kelly*, 1995; *Murray et al.*, 1998]. In summer, more complicated spatial structure of the coastal current, its weaker and more variable speed, smaller scale forcing generated by less organized and weaker winds and/or well-developed stratification could be partly responsible for the much shorter cross-shelf spatial length scale.

There is also a difference between values of the correlation coefficients of the upcoast and downcoast regimes as a function of vertical separation (results not shown). The correlation coefficients are high (> 0.8) in both regimes for adjacent instruments and drop to approximately 0.54 for the summer

season and to 0.75 for the fall/winter time period for larger vertical separations (9 m or more). The weaker correlation of the summer currents at the separation distance of 9 m or more could be related to summer stratification. The well-developed summer pycnocline is able to isolate effectively flow observed in the lower layer of the water column from direct forcing such as winds, for example. Consequently, the flow in the lower layer is usually very weak, disorganized and often opposed to the flow observed above the pycnocline [*Crout et al.*, 1983; *Murray et al.*, 1998]. The relationship between currents at different depths strengthens significantly due to weaker stratification in fall/winter months and deeper influence of wind forcing.

VELOCITY CEOFS

To extract variability that is coherent across the mooring array and quantify major patterns in the LTCC currents, we subjected the data to complex empirical orthogonal function analysis (CEOF) [*Kundu and Allen*, 1976]. For each flow regime, the velocity CEOFs were computed by forming a matrix containing demeaned complex velocities obtained from the low-frequency along- and cross-shore current components for all available depths in a given season. Complex eigenvectors estimated from the associated covariance matrix were then converted into amplitudes and veering angles for each statistically significant mode (Table 3). The amplitudes of each mode were also scaled by the square root of the corresponding eigenvalue, and the veering angles were relative to the Btop and Atop currents for the downcoast and upcoast flows, respectively.

For the fall/winter months, only the first CEOF mode is statistically significant when tested with an algorithm proposed by *North et al.* [1982]. It explains 81.1% of the current variability, and its spatial structure shows very little variation in veering angles in the Cameron cross-section (Table 3). The spatial distribution of the amplitudes and veering angles suggests that the primary downcoast flow pattern is in a form of a surface-intensified coastal jet. The first mode is also highly correlated with the alongshore wind stress with a squared correlation coefficient (R^2) of 0.73 that is significant at the 95% confidence level. This high correlation clearly shows that the alongshore wind stress is an important driving mechanism of the LTCC fluctuations in the downcoast flow regime. Additionally, dominance of the first mode and its high correlation with the wind stress suggest that this mode represents a predominant barotropic response of the inner shelf current to the strong fall/winter winds and storms. Plate 2 demonstrates how swiftly the currents respond to changes in the alongshore wind stress forcing. This plate shows daily values of wind stress (Plate

Table 3. Amplitudes[1] (cm/s) and veering angles[2] (deg) of the CEOF modes.

Downcoast Regime			Upcoast Regime				
Mooring	Amp 1	Angle 1	Mooring	Amp 1	Dir 1	Amp 2	Dir 2
Btop	13.1	0	Atop	12.4	0	6.9	0
Bmid	12.1	356	Amid	10.4	3	5.1	27
Ctop	8.7	356	Ctop	5.8	341	8.8	163
Cmid	11.1	358	Cmid	6.1	331	6.5	167
Cbot	8.4	344	Cbot	2.3	279	1.5	154
Etop	10.6	352	Dtop	6.8	349	6.2	172
Ebot	5.2	3	Etop	4.1	357	0.6	255
Ftop	14.0	8	Gtop	5.3	357	0.9	159
Fbot	8.5	2					

[1] Amp 1 and Dir 1 are amplitudes and veering angles of mode 1; Amp 2 and Dir 2 are amplitudes and veering angles of mode 2. [2] Veering angles are relative to currents measured at Btop and Atop moorings for the downcoast and upcoast regimes, respectively.

2a) and combined mean flow and mode 1 for each mooring on 6 and 9 October 1996 (Plates 2b and 2c, respectively). On 6 October, a wind stress of 0.11 Pa was directed downcoast as were the currents across the entire Cameron section. On 7 October, this downcoast stress started diminishing and on 8 October, changed direction to upcoast. The currents followed these changes in wind stress. On 9 October, the downcoast flow stopped and at most locations, the flow began to move upcoast as shown in Plate 2c. Similar behavior of the currents in the fall/winter/spring months on the Louisiana-Texas inner shelf, i.e., their swift response to the wind stress variations, is reported by *Murray et al.*, [1998].

For the upcoast current data, CEOF modes 1 and 2 are the only statistically significant modes. Mode 1 is not as dominant as in the downcoast flow, but it can explain 46% of the flow variance, while mode 2 accounts for 25% of the variance. The amplitudes of these modes are smaller and the veering angles vary more (especially those of mode 2) than those of mode 1 estimated from the downcoast data. The angles of summer mode 2 imply two current reversals occurring (a) offshore (between moorings A and C) and (b) near the coast (between moorings E and G) (Table 3). Additionally, the spatial structure of the amplitudes and veering angles of mode 1 suggest that one of the current patterns in summer could be a jet confined to the upper 10 to 14 m of the water column. This conclusion agrees with summer current observations collected on the Louisiana inner shelf and presented by *Murray et al.*, [1998]. The spatial structure of the second mode could represent more complex summer flow also described by *Murray et al.*, [1998]. Their ADCP, CTD measurements, and SCULP drifters [*Johnson and Niiler*, 1994; *Ohlmann et al.*, 2001] showed that in summer 1994, there were three distinct zones in the region where our moorings were later deployed namely (a) a coherent eastward flow of waters with salinity of 30 psu and higher, (b) an interior zone of slow, disorganized flow of intermediate salinity

waters (20–30 psu), and (c) a near coastal westward flow of low salinity waters (less 20 psu).

In contrast to the high correlation between the currents and alongshore wind stress in the downcoast regime, the summer current fluctuations are not very correlated with this forcing. The squared correlation coefficient between summer CEOF mode 1 and the alongshore wind stress is low and its magnitude is 0.31 (R^2 greater than 0.12 is significant at the 95% confidence level which was computed from the algorithm proposed by *Sciremammano* [1979]), while there is no correlation between mode 2 and this component of the wind stress ($R^2 = 0.04$). The complex spatial structure of the coastal current, its highly variable speed, and/or well-developed stratification could be partly responsible for the low correlation between the current and wind stress in summer.

A WIND-DRIVEN MODEL

To further clarify the role of local wind stress as a major force controlling current variability within the LTCC off Cameron, LA, a wind-driven model, previously used successfully to study variability of currents within the Amazon River coastal plume [*Lentz*, 1995], is employed. This model assumes that current variability is caused solely by wind stress. Thus the input of energy from the wind into the plume layer is balanced only by a constant linear friction at the base of the plume and the temporal acceleration of the alongshore flow. This alongshore momentum balance is expressed as followed:

$$\frac{\partial u}{\partial t} + \frac{ru}{h} = \frac{\tau_{sx}}{\rho h} \quad (1)$$

where u is the alongshore current, ρ is the reference water density, r is the linear friction coefficient, h is the plume thickness and τ_{xs} is the alongshore wind stress. Integrating (1) in time and solving for u yields

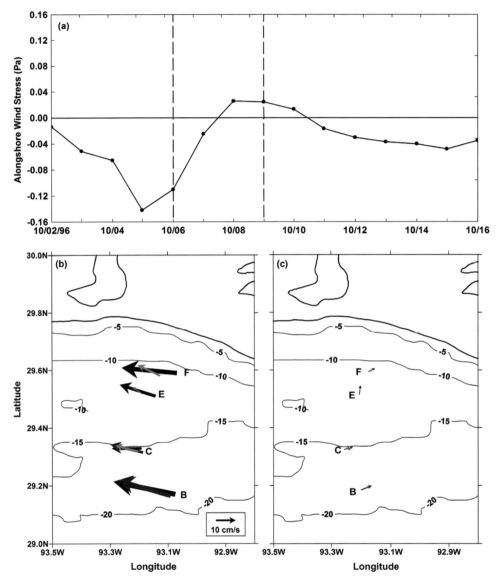

Plate 2. (a) Daily alongshore wind stress time series (thin vertical lines indicate two dates: 6 and 9 October 1996), and maps of daily values of CEOF mode 1 combined with the downcoast means at the mooring locations on (b) 6 October 1996 and (c) 9 October 1996; black, blue, and green arrows represent top, mid-depth, and bottom current meters, respectively. Positive values are upcoast (eastward).

$$u(t) = \int_0^t \frac{\tau_{sx}}{\rho h} \exp(-r(t-t')/h)dt' + u(t=0)\exp(-rt/h) \quad (2)$$

The alongshore wind stress component was computed from the wind data measured at Freshwater Bayou using the approach proposed by *Large and Pond* [1981]. As in *Lentz* [1995], the linear friction coefficient was chosen to yield the best agreement with the observations. For the downcoast flow regime, the alongshore current component was calculated from (2) at three different mooring locations (B, C and F), and then compared with the observations collected by the top meters at each locations. The best agreement between the wind-driven alongshore velocities based on (2) and the observed velocities is found at mooring F (Figure 4a), with h = 12 m and the reference density of 1021 kg/m³. We used h equal to the total water depth in moorings F and G (data from mooring G were used for the upcoast regime computations presented in the next paragraph) because the historical hydrographic data suggested that the entire water column near these mooring locations was filled out with light waters (salinity less then 30 psu). The squared correlation coefficient between the observations and predictions with r = 0.0003 m/s is 0.66 (all correlations presented in this section are significant at the 95% confidence level computed from the algorithm proposed by *Sciremammano* [1979]). Variability of the predicted wind-driven alongshore current is very similar to that observed at F, for instance, events

observed between 3 and 11 October 1996, 22 October and 4 November 1996, and 7 and 19 November 1996. Analogous conclusions are obtained for the other two moorings (mooring C: $R^2 = 0.58$; mooring B: $R^2 = 0.58$). These results reinforce the earlier conclusion that alongshore wind stress, which alone explains at least 58% of the current variance, is a dominant driving force of the current variability observed within the LTCC in the downcoast flow regime.

When the wind stress observed in the summer regime is applied to (2), amplitudes of the alongshore velocities are poorly modeled for almost entire June but they are quite well reproduced for July and August 1996. The model estimates are again compared with subsurface current observations at three different locations (moorings A, C, and G). The best agreement is found at mooring G (Figure 4b), with a squared correlation coefficient of 0.37 between the observed alongshore velocities and estimates, with r = 0.00035 m/s, h = 10 m, and the reference density of 1010 kg/m³. It is obvious that major velocity fluctuations in July and August, particularly events observed between 1 and 6 August are wind driven. However, the predicted variability in June, still in transition from downcoast to upcoast dynamics, is different from the observations, with amplitude differences reaching even 30 cm/s, which is quite large because lower speeds are usually measured within the LTCC in summer. Similar conclusions are found when predictions are compared with subsurface

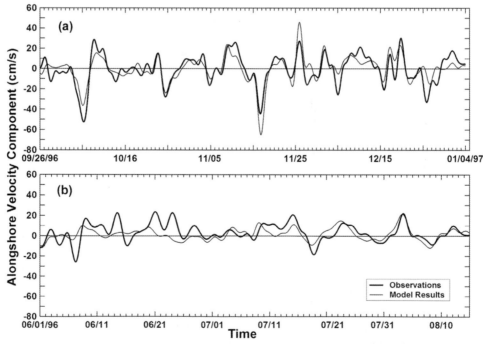

Figure 4. (a) The alongshore 6.5m current observed at mooring F during the downcoast flow regime and (b) the alongshore 5.5m current observed at mooring G during the upcoast flow regime compared to the model results (equation 2); positive values indicate upcoast velocities, while negative values indicate downcoast velocities.

currents at moorings C ($R^2 = 0.34$) and A ($R^2 = 0.32$). These results suggest that alongshore wind stress, explaining by itself as much as 37% of the current variance, is an important driving force of the coastal current in the upcoast flow regime. The unexplained variance of the currents implies that other driving forces, such as barotropic and baroclinic pressure gradients, could be also important in summer.

The friction coefficients used for the simulations of the LTCC currents are in the range of bottom friction coefficients estimated for the Louisiana-Texas inner shelf by *Chuang and Wiseman* [1983]. For both flow regimes, these friction coefficients are an order of magnitude larger than those utilized for the Amazon plume (r = 0.00002 m/s) [*Lentz,* 1995], and this magnitude difference is mainly due to the vertical size of the Amazon plume and Mississippi-Atchafalaya outflow at their respective locations. In the case of the Louisiana inner shelf, the low salinity waters at moorings F and G extend from the surface to the bottom, and the friction coefficient is simply the bottom friction coefficient, while it is the interlayer friction coefficient for the Amazon plume.

CROSS-SECTIONAL TRANSPORT

In order to calculate the transport within the LTCC through the section off Cameron, we first produce a time series of the low-passed alongshore current component at each current

meter. The data are then interpolated onto a grid (1-m depth by 2-km cross-shore distance) at each hour, using linear interpolation. A rigorous definition of the offshore extent of the coastal current is obviously difficult because of its inherent temporal variability; however, the limited salinity data recovered from the moorings suggest the presence of coastal plume water across the section even as far out as mooring A. Thus we make an operational definition of the coastal current transport by integrating across the section between moorings A and G, recognizing that this estimate is a lower bound. It does, however, have the advantage of being consistent with the ship-mounted ADCP transport estimates of *Murray et al.* [1998].

The alongshore transport through the section during the downcoast regime of 1996–1997 is shown in Figure 5a. Note the persistent downcoast transport of -0.1 to -0.15 Sv with a record length mean of -0.06 Sv (a root-mean-squared error of 0.02 Sv). Significant bursts of elevated downcoast transport occur on 6 October, 17 November, and 8 and 13 January. We will examine the spatial characteristics of some of these bursts later. The corresponding transport during the preceding upcoast regime is shown in Figure 5b. As expected, regional forcing conditions in the summer of 1996 cause the coastal current to reverse directions, but it shifts predominantly upcoast by 9 June and remains upcoast for most of the summer. Note, however, the temporal variability

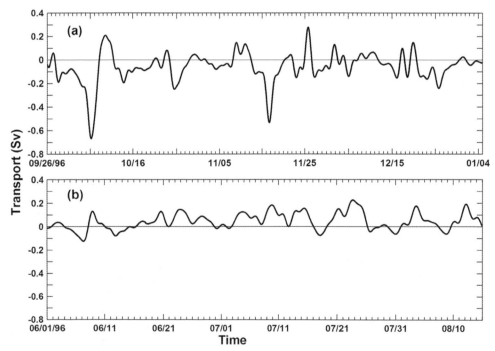

Figure 5. Transport in the LTCC during (a) the downcoast flow regime and (b) upcoast flow regime; positive values indicate the upcoast transport, while negative values indicate downcoast transport.

in the transport. These transport fluctuations, which range from slightly below 0 to maximums of 0.15 to 0.23 Sv, occur roughly on a 5-day time scale. The record length mean for this period is 0.05 Sv with a root-mean-squared error of 0.03 Sv. Our mean seasonal transport results are in surprisingly good agreement with results from a high-resolution (a 1/20° grid in latitude and longitude) NCOM model of the Gulf of Mexico driven with climatological monthly surface fluxes [*Zavala-Hidalgo et al.*, 2003, Figure 2A]. The model results for a section close to our observations, integrated from the 25 m isobath to the coast, show a 10-month transport of the downcoast flow reaching -0.1 Sv and a 2-month transport of the summer upcoast flow reaching approximately 0.04 Sv.

The most energetic transport oscillations during the fall/winter period appear closely related to direct wind driving, as seen during the event between 2 and 11 October (Figure 5a). Ten days of northeasterly winds exert downcoast wind stresses that produce downcoast current speeds in excess of 30 cm/s across the entire section with two zones of intensification (Plates 1b and 2b). A high-speed current (speeds of 60 cm/s) occupies the outer (southern) edge of the section, and a second jet, nearly as strong, occupies the inshore (northern) end of the section. Previous data taken during LATEX B cruises [*Murray et al.*, 1998] suggest that intense vertical mixing, combined with zones of intensified horizontal density gradients (fronts), are associated with these jets in the coastal current. The cessation of the northeasterly winds driving the October 6 intensification of the downcoast flow event is followed by an upcoast flow event.

The time series of transport during the summer regime (Figure 5b) is similarly characterized by fluctuations with several days to week time scales. The strongest upcoast intensification, centered on 23 July, is notable for its rather modest current speeds of 20 cm/s observed along the mooring array. These wind driven upcoast currents are upwelling favorable, leading to offshore transport in the surface layer and a consequent weakening of cross-shore density gradients and reduction of baroclinic alongshore velocities.

DRIVING FORCES OF THE LTCC

To seek further insight into dynamical relationships controlling the transport observed along the transect, we examined multiple and partial coherence between the transport and four forcing mechanisms: (a) alongshore wind stress, (b) cross-shore wind stress, (c) alongshore barotropic pressure gradient (a sea-surface slope), and (d) buoyancy forcing proxied by the river discharge as in *Münchow and Garvine* [1993]. Wind stress was computed from the anemometer at Fresh Water Bayou. The sea-surface slope was calculated from subsurface pressure gauges located at Oyster Bayou

and Freeport for both seasons, and buoyancy forcing is represented by Atchafalaya River discharge.

Figure 6 shows multiple and partial coherence over the 0.05 to 0.6 cpd frequency band between the transport and possible driving mechanisms. For the fall/winter observation period (Figure 6a), 78% to 96% of the total transport fluctuations, shown by the multiple coherence curve, are explained by the four forcing variables. Looking at individual partial coherences, the alongshore wind stress clearly accounts for a majority (at least 65%) of the transport variance. There is also an indication that the cross-shore wind stress might be considered a significant forcing for the transport fluctuations with periods between 3 and 5 days. Partial coherences of the alongshore sea-surface slope and river discharge with the downcoast transport are rarely statistically significant, thus their importance as possible sources of subtidal transport fluctuations is extremely small, if any.

In the summer flow regime, 60% to 92% of the transport fluctuations (Figure 6b) are accounted for by the four possible driving forces over the entire frequency band examined. During this regime, the transport fluctuations are not only coherent with the alongshore wind stress, but also with the alongshore sea-surface slope at frequencies lower than 0.45 cpd. In addition, the cross-shore wind stress could be an important influence on the transport fluctuations with periods between 5 and 10 days, but its importance is clearly secondary compared to the alongshore wind stress and sea-surface slope. Partial coherence between the transport and buoyancy forcing represented by the Atchafalaya discharge only appears statistically significant for frequencies lower than 0.2 cpd, with the percentage of variance explained by this possible driving force reaching about 75% at 0.1 cpd.

SUMMARY AND CONCLUSIONS

Temporal and spatial variability of the currents and volume transport of the Louisiana-Texas Coastal Current were studied with data from a cross-shore transect located south of Cameron, Louisiana. In addition to the current meter data, subsurface pressure, wind, and the Atchafalaya River discharge were also analyzed to better understand the local behavior of the LTCC on the western Louisiana inner shelf. Two subsets of the data, one for the upcoast (summer) and another for the downcoast (fall/winter) flow regimes, were analyzed to determine whether there were any differences in the current characteristics and transport between these two regimes of the LTCC.

In the fall/winter season, the fluctuations of the currents and transport are highly dependent on the wind forcing. The overall mean flow is downcoast (westward) as expected for this part of a year; however, the rotating winds associ-

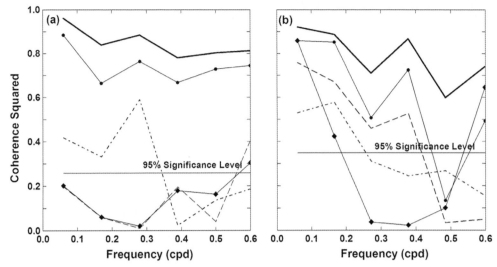

Figure 6. Multiple (thick line) and partial coherence (symbol lines) between the transport observed during (a) the downcoast regime and (b) the upcoast regime and alongshore wind stress (solid line with circles), cross-shore wind stress (dash-dot line), alongshore sea-surface slope (dashed line), and the Atchafalaya River discharge (solid line with diamonds); the lower straight solid line is the 95% significance level.

ated with cold fronts do temporarily reverse its direction to upcoast (eastward) for 18–36 hours. The downcoast flow is clearly polarized in the alongshore direction. High correlations in the vertical and large cross-shore spatial length scale of the observed currents as well as the spatial structure of the amplitudes and veering angles of the dominant velocity CEOF mode (mode 1) strongly suggest a surface-intensified jet-like response to the wind forcing. Based on high correlation of the currents and transport with the alongshore wind stress, we conclude that this wind stress component is the principal driving mechanism of the LTCC off Cameron, LA during the downcoast flow, which is in agreement with the findings of other investigators [*Crout et al.*, 1984; *Cochrane and Kelly*, 1986; *Li et al.*, 1997; *Murray et al.*, 1998; *Cho et al.*, 1998; *Nowlin et al.*, 2005; *Walker*, 2005].

During the upcoast regime, vertical and cross-shore correlation length scales are markedly lower than during the downcoast regime. The mean flow as well as the mean transport is directed upcoast, but again with persistent fluctuations which are two to three times larger than the seasonal means. These fluctuations also become more isotropic with increasing depth and increasing distance off the coast than those observed in the fall/winter season. The current and transport variability is partly related to the alongshore wind stress; however, the correlation between this wind stress and both currents and their transport is weaker. Thus we conclude that other forcing mechanisms such as barotropic and/or baroclinic pressure gradients could be important in summer. Our data indicate that during the analyzed summer

season, in addition to the alongshore wind stress, the alongshore sea-surface slope was also an important driving forcing of the LTCC, which agrees with the conclusions reached by *Murray et al.* [1998].

Acknowledgments. This study was funded by the Minerals Management Service under contract 14-35-001-30724 awarded to Louisiana State University.

REFERENCES

Cho, K., R. O. Reid, and W. D. Nowlin Jr. (1998), Objectively mapped stream function fields on the Texas-Louisiana shelf based on 32 moored current meter data, *J. Geophys. Res., 103*, 10,377–10,390.

Chuang, W. S. and W. J. Wiseman, Jr. (1983), Coastal sea level response to frontal passages on the Louisiana-Texas shelf, *J. Geophys. Res., 88*, 2615–2620.

Cochrane, J. D., and F. J. Kelly (1986), Low-frequency circulation on the Texas-Louisiana shelf, *J. Geophys. Res. 91*, 10,645–10,659.

Crout, R. L., W. J. Wiseman, Jr., and W. S. Chuang (1984), Variability of wind driven currents, west Louisiana inner continental shelf: 1978-1979, *Contrib. Mar. Sci., 27*, 1–11.

Current, C. L. (1996), Spectral model simulation of wind driven subinertial circulation on the inner Louisiana-Texas Shelf, Ph.D. Thesis, 144 pp., Texas A&M University, Collage Station.

Dever, E. P. (1997), Subtidal velocity correlation scales on the northern California shelf, *J. Geophys. Res. 102*, 8555–8571.

Etter, P. C., M. K. Howard, and J. D. Cochrane (2004), Heat and freshwater budgets of the Louisiana – Texas shelf, *J. Geophys. Res., 109*, C02024, doi:10.1029/2003JC001820.

Gutiérrez de Velasco G., and C. D. Winant (1996), Seasonal patterns of wind stress and wind stress curl over the Gulf of Mexico, *J. Geophys. Res., 101*, 18,127-18140.

Jarosz, E. (1997), Summer flow regime on the Louisiana inner shelf, M.S. thesis, pp. 74,Louisiana State University, Baton Rouge.

Johnson, W. R. and P. P. Niiler (1994), SCULP drifter study in the Northern Gulf of Mexico, AGU Fall Meeting, O52C-8, San Francisco.

Kundu, P. H. (1976), A note on the Ekman veering observed near the ocean bottom, *J. Phys. Oceanog., 6*, 238–242.

Kundu, P. H., and J. S. Allen (1976), Some three-dimensional characteristics of low-frequency current fluctuations near Oregon coast, *J. Phys. Oceanogr., 6*, 181–199.

Large, W. G. and S. Pond (1981), Open ocean momentum flux measurements in moderate and strong winds, *J. Phys. Oceanogr., 11*, 324–336.

Lentz, S. J. (1995), The Amazon River plume during AMASSEDS: Subtidal current variability and the importance of wind forcing, *J. Geophys. Res., 100*, 2377-2390.

Li, Y., W. D. Nowlin, Jr., and R. O. Reid (1997), Mean hydrographic fields and their interannual variability over the Texas-Louisiana continental shelf in spring, summer, and fall, *J. Geophys. Res., 102*, 1027-1049.

Murray, S. P., E. Jarosz, and E. T. Weeks (1998), Physical oceanographic observations of the coastal plume, In: Final report: an observational study of the Mississippi-Atchafalaya coastal plume, OCS Study MMS 98-0040, U.S. Dept. of the Interior, MMS, Gulf of Mexico OCS Region, New Orleans, LA.

Murray, S. P., E. Jarosz, and E.T . Weeks (2001), Velocity and transport characteristics of the Louisiana-Texas Coastal Current, OCS Study MMS 2001-093, U.S. Dept. of the Interior, MMS, Gulf of Mexico OCS Region, New Orleans, LA.

Münchow, A. and R. W. Garvine (1993), Buoyancy and wind forcing of a coastal current, *J. Mar. Res., 51*, 293–322.

North, G. R., T. L. Bell, R. F. Cahalan, and F. J. Moeng (1982), Sampling errors in the estimation of empirical orthogonal functions, *Mon. Wea. Rev.,110*, 699–706.

Nowlin, W. D., Jr., A. E. Jochens, R. O. Reid, and S. F. DiMarco (1998), Texas-Louisiana Shelf circulation and transport processes study: Synthesis report, vol.1, Technical report, OCS Study MMS 98-0035, U.S. Dept. of the Interior, MMS, Gulf of Mexico OCS Region, New Orleans, LA.

Nowlin, W. D., Jr., A. E. Jochens, S. F. DiMarco, R. O. Reid, and M. K. Howard (2005), Low-frequency circulation over the Texas-Louisiana continental shelf, in *New Developments in the Circulation of the Gulf of Mexico*, W. Sturges and A. Lugo-Fernandez, Eds., AGU Monograph Series.

Oey, L. Y. (1995), Eddy and wind-forced shelf circulation, *J. Geophys. Res., 100*, 8621–8637.

Ohlmann, J. C., P. P. Niiler, C. A. Fox, and R. R. Leben (2001), Eddy energy and shelf interactions in the Gulf of Mexico, *J. Geophys. Res., 106*, 2605–2620.

Sciremammano, F., Jr. (1979), A suggestion for the presentation of the correlations and their significance levels, *J. Phys. Oceanogr., 9*, 1273–1276.

Smith, N. P. (1975), Seasonal variations in near shore circulation in the northwestern Gulf of Mexico, *Cont. Mar. Sci., 19*, 49–65.

Smith, N. P. (1978), Low-frequency reversals of near shore currents in the northwestern Gulf of Mexico, *Cont. Mar. Sci., 21*, 103–115.

Vastano, A. C., C. N. Barron, Jr., and E.W. Shaar Jr. (1995), Satellite observations of the Texas current, *Cont. Shelf Res. 15*, 729–754.

Walker, N. D. (2005), Seasonal changes in surface temperature, shelf and slope circulation, and coastal upwelling in the northwestern Gulf of Mexico, in *New Developments in the Circulation of the Gulf of Mexico*, W. Sturges and A. Lugo-Fernandez, Eds., AGU Monograph Series.

Wang, W., W. D. Nowlin, Jr., and R. O. Reid (1998), Analyzed surface meteorological fields over the northwestern Gulf of Mexico for 1992–94: mean, seasonal and monthly patterns, *Mon. Wea. Rev. 126*, 2864–2883.

Wiseman, W. J., Jr., R. E. Turner, F. J. Kelly, L. J. Rouse, Jr., and R. F. Shaw (1986), Analysis of biological and chemical associations near a turbid coastal front during winter 1982, *Contributions in Marine Science, 29*, 141–151.

Wiseman, W. J., Jr., and F. J. Kelly (1994), Salinity variability within the Louisiana coastal current during the 1982 flood season, *Estuaries, 17*, 732–739.

Wiseman, W.J., Jr., and F.J. Kelly (1995), Plumes and coastal currents near large river mouths, *Estuaries, 18*, 509–517.

Wiseman, W. J., Jr., N. N. Rabalais, R. E. Turner, S. P. Dinnel, and A. MacNaughton (1997), Seasonal and interannual variability within the Louisiana coastal current: stratification and hypoxia, *J. Mar. Sys., 12*, 237–248.

Zavala-Hidalgo, J., S. L. Morey, and J. J. O'Brien (2003), Seasonal circulation on the western shelf of the Gulf of Mexico using high-resolution numerical model, *J. Geophys. Res., 108*, doi:10.1029/2003JC001879.

E. Jarosz, Oceananography Division (Code 7332), Naval Research Laboratory, Stennis Space Center, MS 39529, USA. (ewa.jarosz@nrlssc.navy.mil)

S. P. Murray, Office of Naval Research, 800 N. Quincy St., Arlington, VA 22217, USA. (MURRAYS@ONR.NAVY.MIL)

Movies of Surface Drifters in the Northern Gulf of Mexico

Walter Johnson

U.S. Minerals Management Service, Herndon VA

The Minerals Management Service sponsored two major near-shore surface drifter projects in the northern Gulf of Mexico in recent years. The first, off the coast of Texas, is called the SCULP experiment here, for Surface CUrrent Lagrangian Program), in keeping with its description at the time. The second, off the western coast of Florida, is called NEGOM, for North East Gulf Of Mexico.

These drifters are genuinely surface drifters, as they are drogued at ~ 1 m deep; the P.I. of the drifter programs was P. Niiler [see *Ohlmann and Niiler*, 2005]. The drogues were tracked via the ARGOS satellite; in near-real time we received the data from Service ARGOS by email and made maps, usually weekly, of the drifter tracks. Additional processing was performed by the Scripps team to suppress tides and inertial motions and to form the final data sets.

We have put two movies on the CD in the back of this monograph. The "Sculp" movie was made from a set of 385 images showing drifter tracks of a week's duration during October 1993–October 1994; wind vectors at half a dozen or more land stations and buoys are shown as well.

The NEGOM movie was made from daily images, February 6, 1996 to May 31, 1997, of weekly surface drifter tracks. The wind vectors are centered at the observation station. It is instructive also to compare the mean monthly files for September and October which are also included. A complete set of monthly mean drifter motions is included in a subdirectory.

As one watches these movies, two results seem obvious. First, the drifters "mostly go back and forth." As wind events pass over the region, shelf waves are set up; their motions carry the drifters for distances of order ~100km; but then the winds reverse and the drifters go in the other direction. A few drifters are blown off the shelf. Second, it required a very energetic wind event to force the drifters around the bend near New Orleans [*Ohlmann and Niiler*, (2001)]. Other than that, drifters stay fairly close to where they were launched.

REFERENCES

Ohlmann, J. C., and P. P. Niiler, A two-dimensional response to a tropical storm on the Gulf of Mexico shelf. *Jour. Mar. Systems*, 29, 87–99, 2001.

Ohlmann, J. C., and P. P. Niiler, Circulation over the Continental Shelf in the Northern Gulf of Mexico, *Progr. Oceanogr* 64, 45–81, 2005.

Circulation in the Gulf of Mexico: Observations and Models
Geophysical Monograph Series 161
This paper is not subject to U.S. copyright. Published in 2005 by the American Geophysical Union.
10.1029/161GM12

Table 1. The characteristics of all ten eddies shed during the 10 years of model simulation

Eddy	Birth day	Death day	Period Days	Life Days	Center Biirth	N. Edge Birth	Max Penetr.	Center at 92W	Size (Deg)
1	570	1000	-	480	26.5/90.4	28.0	28.4	26.2	3
2	720	1340	150	620	26/87.8	27.4	27.6	24.8	3.5
3	1230	1800	510	570	26.4/89.2	28.0	28.0	26.0	3
4	1360	1970	130	610	25.4/87.8	27.4	27.4	23.5	3
5	1450	1800	90	350	26.2/86.4	27.0	27.4	25.8	1.5
6	1830	2160	380	330	26.4/89.2	27.8	27.8	25.6	3
7	2200	2550	370	350	26.4/88.2	28.0	28.0	24.2	3.5
8	2500	3210	300	710	25.6/87.6	27.0	28.0	23.8	4
9	2940	3590	440	650	26/87.2	27.5	27.8	23.6	3.5
10	3330	3650	390	320	25.5/87.8	27.0	27.6	23.4	3

actual dates are meaningless. A total of 10 eddies were shed during this 10-year run (with an 11th eddy about to shed).

Table 1 shows the characteristics of each of these eddies. The model fields were archived every 5 days and used to analyze the Loop Current intrusion, eddy shedding and the movement of the eddies into the western Gulf. Once shed, the center of the LC eddy was taken as the point corresponding to the maximum SSH in the eddy, and the track and the translation speed of the eddy were determined. Fig. 2 shows the tracks of all the 10 eddies shed during the model run. The eddy tracks show a distinct bimodal pattern, with prominent slightly southwestward and southwestward branches. Also, once they reach the western slope, most eddies turn northward. It is of interest to compare the modeled eddy paths in Fig. 2 to observations. *Vukovich and Crissman* (1986) analyzed AVHRR imagery and found that there are 3 common paths taken by warm core eddies once they are free of the LC, the most preferred one being the southwesterly path, while the least common one is the path that takes them parallel to the northern slope. Their study suffers from the fact that eddies cannot be tracked during summer-fall period when the eddy signatures are obscured by a warm mixed layer on the top. *Hamilton et al.* (1999) used drifting buoys to analyze eddy statistics (such as swirl period) in addition to the paths of 10 eddies shed between 1985 and 1995. They also found that the eddy tracks were generally southwesterly (corresponding to the slightly southwesterly tracks in the model), with the eddy centers lying between 22°N and 25°N, when they approach the western slope.

The model results can also be compared with the EJIP database of LCEs from 1966 to 1991, known as GEM Path and Configuration (P&C) analyses (*Evans-Hamilton* 1992) based on AVHRR and ocean color images, synoptic ship cruises, ship-of-opportunity XBT transects, and drifting buoys. In essence the GEM analysis includes all the data of *Vukovich and Crissman* (1986) and *Hamilton et al.* (1999) plus a substantial amount of additional data. Fig. 3 shows the GEM eddy tracks. If the initial portions that correspond to the intruding Loop Current are discarded, these tracks are consistent with the model tracks, with roughly slightly southwestward and southwestward patterns, with eddy centers ending up between 22°N and 26°N, when the eddies approach the western slope. Most sheddings occur when the eddy center is around 26°N/88°W, while one eddy was shed with the center as far south as 25°N.

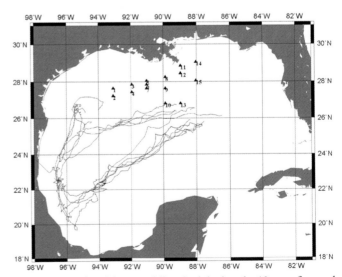

Figure 2. Tracks of the ten eddies shed during the 10-year forward model run. Note the bimodal pattern. Triangles denote monitoring stations.

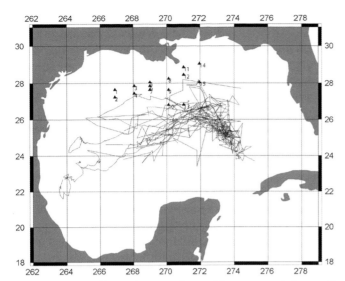

Figure 3. GEM eddy tracks. Note the tracking was carried out well before eddy separation. Triangles denote monitoring stations.

Translational speed of each individual eddy is quite variable, with characteristic "sprints and stalls" observed in the field (*Glenn and Ebbesmeyer* 1993). Modeled eddy translational speeds are typically around 3–5 cm s^{-1}, quite consistent with observations. For example, *Hamilton et al.* (1999) indicate a mean translation speed of around 4.2 cm s^{-1} in the western Gulf for the 10 eddies they tracked using drifter buoys (their Table 6).

The maximum swirl velocity decreases steadily after shedding. Except for the Eddy #5, which was a small, weak eddy of roughly 150 km diameter with an initial maximum swirl velocity of about 0.7 m s^{-1}, the initial maximum swirl velocity is typically 1.5 m s^{-1}, also consistent with observations. The maximum swirl velocity decreases almost linearly from 1.5 m s^{-1} to 0.5 m s^{-1} in roughly 300 days. Hamilton et al. (1993) indicate a mean swirl velocity of nearly 0.5 m s^{-1} at a radius of 65 km for the eddies they tracked.

Model results are summarized in Table 1. These results suggest that the Loop Current variability exhibited by the forward model is reasonable. A total of 10 eddies were shed during the 10-year simulation, with an average shedding period of 306 days (~10 months), but with a minimum of 90 days (~3 months) and a maximum of 510 days (~17 months). The average lifespan of the eddies is 500 days (~17 months), with a minimum of 320 days (~11 months) to a maximum of 710 days (~23 months). The eddy translation speed is highly variable, with a maximum of 15–20 cm s^{-1}, but an average of 2–5 cm s^{-1}.

Overall, the Loop Current variability is not inconsistent with what little is known from existing observations. For example, Table 1 of *Sturges and Leben* (2000), who stud-ied LCE shedding frequency from IR and altimetric data, shows that from mid-1986 to the end of 1994, a total of 10 eddies were shed with an average eddy shedding period of 11 months. If we discard the first 18 months from "spinup" considerations, the model also shed 10 eddies during this same period. The minimum period between successive eddy sheddings is 3 months and the maximum is 17 months in their study of 34 eddy sheddings in the 26-year period from 1973 to 1999 (which also suggests an average shedding period of about 9 months), consistent with the model results.

Most eddies analyzed in the GEM data base are 300–400 km in diameter, except one which was only 150 km, again consistent with model results. The maximum extent of northward penetration of eddies analyzed is also consistent with GEM database, except that for the model eddies the northernmost penetration falls short by about 25 km or so.

Among the other eddy statistics contained in the GEM database, the eddy swirl period, readily obtained by deployment of drifting buoys in the eddy is an important one. The swirl periods in the model can be compared to those in the GEM database. While the GEM eddies appear to have a near-constant swirl period of 12±2 days over a 200 day period, the model eddies show an increase from 12 to 13 days over 200 days. Still, the agreement is remarkably good. Note that drifting buoys appear not to have been deployed in GEM eddies until well after eddy shedding and may not therefore have sampled the high rotation rates prevalent in LCEs just after shedding. On the other hand, *Hamilton et al.* (1999) indicate a mean swirl period of about 10.2–10.8 days (with an rms of 1.6–1.9 days) for the 10 eddies they tracked between 1985 and 1995 (their Table 6). This suggests that the model may be overestimating the swirl periods by a few percentage points.

Model SSH averaged over the entire run period and subtracted from the 5-day SSH fields has been used to derive a map of model SSH variability (Fig. 4). SSH variability can also be obtained from objective analyses of altimetric data along repeat tracks in the Gulf as well as historical CTD data. The observed RMS SSH distribution is similar to that of the forward model.

It is worth noting that this Gulf model does not include the Caribbean Sea. Nor does it include part of the Gulf Stream system in the western Atlantic. The inflow prescribed is monthly averaged value from STACS measurements, with a mean of 28 Sv. Yet, the Loop Current variability is quite realistic and comparable to that from the *Oey et al.* (2003) model, which includes the Caribbean Sea and part of the Gulf Stream system in the western Atlantic. This suggests that while a better replication of Yucatan Channel transport might contribute to better replication of eddy shedding events, and hence might be desirable, it is not essential. We also observe have little dependence on the mean added to the altimeter-derived sea surface height anomalies. Quantitative Loop Current metrics, which I describe in the next section, show more sensitivity to the mean used to estimate the synthetic height. The fidelity of these estimates was evaluated by comparing individual synthetic height maps with coincident satellite imagery. In that preliminary qualitative study, I found that robust metrics could be determined when using the CUPOM data assimilation model mean with referencing of the altimeter data record to a mean computed over the data assimilation time period.

Loop Current Metrics

Having a synthetic estimate of the mean plus dynamic height, it is feasible to objectively track the Loop Current in the GOM SSH maps. In *Hamilton et al.* [2000], I described a novel Loop Current monitoring capability provided by satellite altimetry that uses the 17-cm SSH contour as a proxy for the high velocity core of the Loop Current in the eastern GOM. This proxy allows objective calculation of a variety of Loop Current metrics such as the length, area, volume and circulation associated with the Loop Current penetration into the Gulf, and its maximum northward and westward extent. The metric time series can then be used to objectively identify Loop Current eddy separation periods, the time intervals between eddy separations.

The use of the 17-cm contour was based on our early success tracking Loop Current rings using altimetry [*Leben and Born*, 1993]. Altimetric monitoring of Loop Current rings in the western Gulf of Mexico relies on the small contribution of the mean circulation to the total dynamic topography. In *Berger et al.* [1996], the 17-cm SSH anomaly contour was selected to derive quantitative metrics for the Loop Current eddies because the contour closely matched the location of maximum gradients in the surface topography corresponding to the edge of the high-velocity core of the eddies. This allowed continuous tracking of Loop Current rings during their translation through the western Gulf of Mexico. The 17-cm contour of total dynamic topography (SSH plus mean) also works well as a definition for the outer edge of the high-velocity core of the Loop Current in the eastern Gulf of Mexico (Plate 1).

The procedure for computing the metrics from the SSH fields has been automated by a MATLAB® program that accesses a GOM altimeter data archive and computes the values. Daily values for each metric are computed using the following algorithm:

1. Load the 0.25° gridded SSH field and generate the coordinates of the 17-cm contours within the Gulf.

2. Identify the Loop Current core, which is defined as the continuous 17-cm contour that enters the gulf through the Yucatan Channel and exits through the Florida Straits.

3. Find the maximum west longitude and north latitude coordinates to determine the extent of westward and northward penetration of the Loop Current.

4. Compute the length of the Loop Current by summing the distances between the coordinates on the 17-cm contour.

5. Identify all 0.25° grid cells bounded by the 17-cm contour and compute the total Loop Current area by summing the areas of the individual cells.

6. Estimate the Loop Current volume, assuming a one and a half-layer ocean and a reduced gravity approximation, by evaluating the following area integral over the region bounded by the 17-cm contour:

$$\iint \frac{g}{g'} h \, dx \, dy,$$

where h is the sea surface height; g is the acceleration of gravity; and g' the reduced gravity. (A value of 0.03 m/s^2 was used for g'.)

7. Estimate the Loop Current circulation by the line integral of the geostrophic velocity along the 17-cm contour:

$$\oint \vec{V} \cdot ds = \int u \, dx + \int v \, dy,$$

where u and v are the geostrophic velocity components and dx and dy are the coordinate spacing in the east/west and north/south directions, respectively. The geostrophic velocity components at the midpoint locations are found by bilinear interpolation from the gridded geostrophic velocity components computed from the height field. (The sign convention employed here is such that the anticyclonic vorticity associated with the Loop Current is positive and therefore in positive correlation to the other metrics.)

The most difficult computational aspects of the algorithm are the autonomous tracking of the Loop Current contour in the presence of any other contours (such as those associated with detached Loop Current eddies) and identifying the Loop Current grid cells within the tracking. Otherwise, this is a relatively simple algorithm to implement using existing contouring programs and could be used to compute similar metrics from numerical ocean model experiments. These metric and their statistics would be useful for parametric tuning and/or model skill assessment. The Loop Current tracking contour, however, typically will be different for each model implementation.

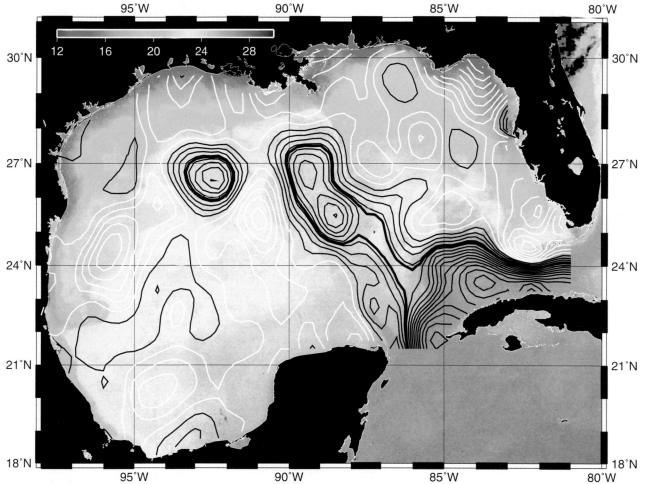

Plate 1. Contoured sea surface height (white is negative, black positive) from the altimetry map for March 13, 2002 overlaid on the nighttime composite SST image from the GOES 8 satellite for the same day (courtesy of Nan Walker, Louisiana State University). The 17-cm contour used to track the Loop Current boundary and eddies is shown by the bold line. The westernmost eddy (Pelagic Eddy or Eddy 13) separated from the Loop Current on February 28th, followed closely by a second eddy (Quick Eddy or Eddy 14) on March 15th, using the objective criteria of breaking of the 17-cm contour.

3. RESULTS

Loop Current statistics [*SAIC*, 1989] and the eddy separation cycle [*Vukovich* 1988; 1995]; *Sturges, 1992, 1994; Vukovich, 1995; Sturges and Leben, 2000*] have been studied in some detail; however, the analysis has been limited by the available observations and by the difficulty in defining a separation event objectively. In this section, I use the continuous 11.5-year altimeter data record and the Loop Current tracking algorithm to describe the Loop Current in much greater detail than possible before tandem altimetry sampling became available starting in 1993.

Loop Current Metrics: Time Series and Statistics

The GOM SSH data archive and Loop Current tracking algorithm were used to compute Loop Current metric values for the time period from 1 January 1993 through 1 July 2004. Time series of Loop Current maximum latitude/longitude extension and length are shown in Plate 2 and area, circulation, and volume in Plate 3. Histograms of the distributions and mean values are shown in the lower panels. Summary statistics for each of the metrics are shown in Table 1.

The total area of open water on the 1/4°grid in the Gulf of Mexico is 1,512,000 km^2. During this time interval, the average area covered by the Loop Current was 147,240 km^2 or approximately 10% of the Gulf. The average length was 1376 km and average circulation was 1,396,200 m^2/s. The total estimated Loop Current volume was 2.26 × 10^{13} m^3. This volume would take over eight days to fill assuming an average inflow of 30 Sv, which is typical of the transport into the Gulf through the Yucatan Channel. The average maximum northward and westward extension of the Loop Current intrusion were 26.2°N and 87.9°W. The maximum and minimum northward extension into the Gulf observed during the time interval were 28.1°N and 24.1°N, respectively. The maximum and minimum westward extension were 93.1°W and 85.8°W. These statistics are quite stable; the Loop Current statistics from this 11.5-year time series differ very little from similar statistics computed from an 8.33-year

altimetric record that was reported in *Hamilton et al.* [2000]. This is further confirmed by the good agreement between the standard deviation of the Loop Current northern boundary reported by *Maul and Vukovich* [1993] with the altimeter-derived result. They report a 102 km standard deviation of the northern boundary location for the time period from 1977 through 1988, compared to 106 km reported here (0.95 degrees) using the altimeter record.

Loop Current statistics have also been derived from 10 years (1976–1985) of monthly frontal analysis maps based on AVHRR thermal images [*SAIC*, 1989]. The summary statistics from that study reported an average area of 210,000 km^2 and 29.75°N and 91.25°W for the maximum northward and westward extensions. The average area derived from thermal imagery was nearly 50% larger than the altimetric estimate reported here. This difference is attributable to the offset between the surface thermal fronts and the location of the Loop Current boundary determined from altimetry. A similar offset of 50 to 100 km was found between the surface frontal analysis maps and subsurface temperature gradients at 125 meters determined from coincident XBT survey transects [*SAIC*, 1989]. Assuming that the mean Loop Current statistics over the two observation intervals are comparable (i.e., stationary), the relative offset can be estimated by the difference of the averages divided by the average length of the Loop Current. This gives an estimate of approximately 45 km for the offset. The integrated effect of this offset is to make estimates of the Loop Current areal extent from thermal imagery significantly larger than values determined using the altimetric Loop Current tracking procedure described herein.

The matrix of correlation coefficients computed from the time series of Loop Current metrics is shown in Table 2. Identifying a "best" metric for monitoring the Loop Current is difficult because of the high correlation between the individual time series. Dynamical considerations may favor the circulation metric, which shows the most consistent growth throughout an intrusion event and the greatest increase just before separation. Loop Current volume and maximum westward extent are the least correlated with the other metrics and

Table 1. Summary statistics for the Loop Current metrics (see Plates 2 and 3) computed from the 1 January 1993 through 1 July 2004 altimetric time series.

	Maximum West Longitude	Maximum North Latitude	Length	Area	Volume	Circulation
Mean	87.9°W	26.2°N	1376 km	147,240 km^2	2.26 × 10^{13} m^3	1,396,200 m^2/sec
Std. Dev.	1.18°	0.95°	365 km	29,295 km^2	0.37 × 10^{13} m^3	338,960 m^2/sec
Maximum	93.1°W	28.1°N	2494 km	213,540 km^2	3.08 × 10^{13} m^3	2,311,200 m^2/sec
Minimum	85.8°W	24.1°N	614 km	55,840 km^2	0.85 × 10^{13} m^3	611,420 m^2/sec

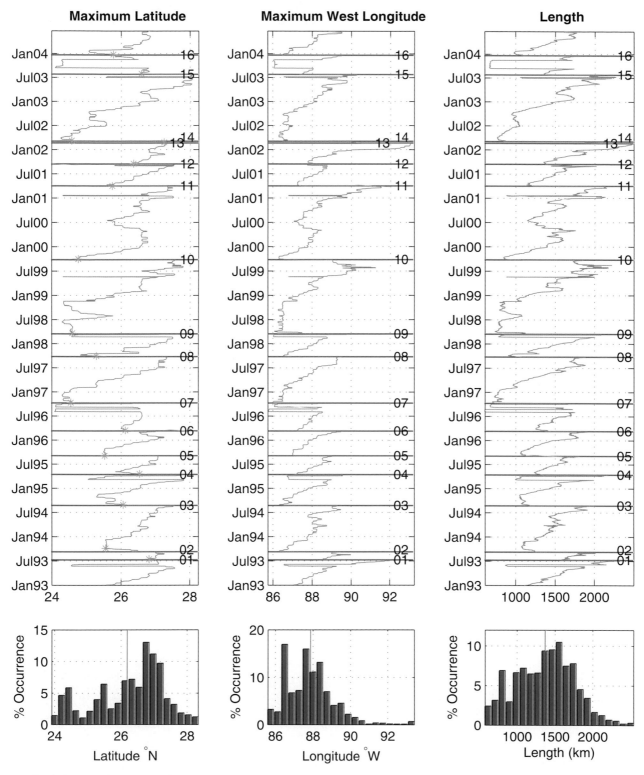

Plate 2. Loop Current maximum northern/western extension and length time series from the altimeter record based on tracking of the 17-cm contour are shown in the upper panels. The horizontal red lines identify the 16 Loop Current eddy separation times. Percent occurrence histograms are shown in the lower panels. Vertical red lines overlaid on the histograms show the mean values (see summary statistics in Table 1 for values).

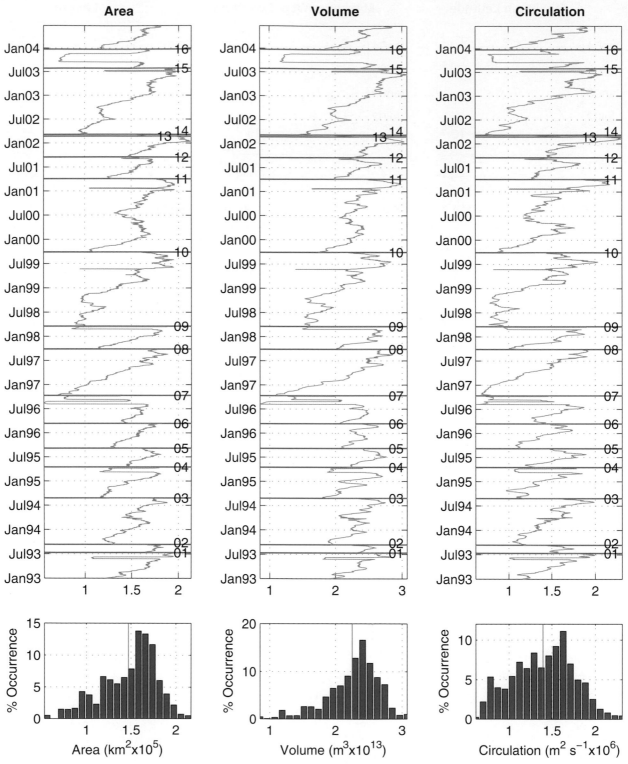

Plate 3. Loop Current area, volume and circulation time series from the altimeter record based on tracking of the 17-cm contour are shown in the upper panels. The horizontal red lines identify the 16 Loop Current eddy separation times. Percent occurrence histograms are shown in the lower panels. Vertical red lines overlaid on the histograms show the mean values (see summary statistics in Table 1 for values).

Table 2. Correlation Matrix for Loop Current Metrics

	Area	Volume	Circulation	Maximum North Latitude	Maximum West Longitude	Length
Area	1.00	0.95	0.93	0.95	0.89	0.95
Volume	0.95	1.00	0.90	0.88	0.80	0.85
Circulation	0.93	0.90	1.00	0.91	0.88	0.94
Max. North Lat.	0.95	0.88	0.91	1.00	0.81	0.94
Max. West Lon.	0.89	0.80	0.88	0.81	1.00	0.94
Length	0.95	0.85	0.94	0.94	0.94	1.00

exhibit larger variations independent of separation events. I prefer to use the Loop Current length. The high correlation found between the altimeter-derived metrics, however, supports the use of surrogate time series as proxies for shedding events. For example, there is very high correlation between area, length and circulation and the northward penetration of the Loop Current, a metric that could easily be computed from model simulations. There has been one estimate of the correlation coefficient between area and northward extent of the Loop Current reported, which is based on frontal analysis derived from the thermal imagery [*SAIC*, 1989]. The value found was 0.76. This is significantly less than the 0.95 determined from the altimetric analysis and may be related to the difficulties associated with identifying the Loop Current boundary in thermal imagery and the smoothing inherent in the processing of the altimeter data. A more detailed comparison of these two remote sensing techniques is described in the next section.

Loop Current Metrics: Comparisons With Satellite Imagery

Direct comparisons of the Loop Current thermal fronts seen in SST imagery with the Loop Current position determined using the synthetic altimetry product (SSH anomaly plus mean) were made to evaluate the Loop Current tracking technique. The Loop Current 17-cm tracking contour was overlaid on SST images sampled during times of good thermal contrast and cloud free conditions. In general, the qualitative agreement is quite good, with the 17-cm contour tracking the Loop Current front within a relatively consistent offset to the inside of the surface thermal front as seen in the imagery (e. g., see Plate 1).

To quantify the offset between the 17-cm contour and the surface thermal front, I made a direct comparison of the location of the 17-cm contour relative to the SST front at the northernmost point of the Loop Current intrusion. Five-day composite SST images from 1993 through 1999 [*Casey and Cornilon*, 1999] coincident with the altimeter maps

were sampled along longitude meridians within +3° and -2° degrees of the point of maximum northern intrusion of the 17-cm contour of SSH. The agreement is good during times of strong thermal contrast; however, the 17-cm contour is consistently south of the SST front. Tracking this offset in the thermal imagery is difficult because of seasonal variations in the thermal signature of the Loop Current and surface waters north of the Loop Current. Nevertheless, during times of good thermal contrast, the maximum temperature gradient is a robust locator of the Loop Current front with the maximum absolute values of the north/south gradient corresponding to the location of the front. Mean values of the north/south temperature gradient as a function of position relative to the 17-cm Loop Current contour are used to quantify the relative offset. The largest negative mean gradient values are located to the north of the thermal front, with little mean gradient to the south over the nearly isothermal Loop Current waters (Figure 1). The extrema value of the mean gradient is offset north of the Loop Current tracking contour by 0.44° of latitude. This is approximately 50 km and in very good agreement with the back-of-the-envelope estimate described in the previous section for the offset between the tracking contour and the surface thermal front. The variability about this bias offset is remarkably small, given the occurrence of warm water filaments and other fine scale frontal variations on the periphery of the Loop Current that are not well sampled by the altimeters. No comparisons can be made, however, during much of the time period, which highlights the difficulty of continuously monitoring the Loop Current with thermal imagery alone.

Loop Current Penetration and Eddy Separation

A primary reason for computing objective metrics is to monitor the time-dependent behavior of the Loop Current system as it intrudes and sheds rings, which are the dominant source of the upper ocean circulation variability in the deep GOM. Clearly a Loop Current intrusion and eddy separation event is associated with each of the peaks in the

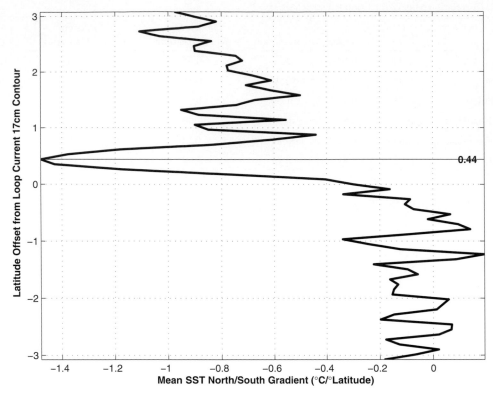

Figure 1. The mean north/south gradient of SST is shown as a function of the latitude offset relative to the northernmost point of 17-cm SSH contour tracking the Loop Current position. The extrema value corresponding to the offset between the surface thermal front and the 17-cm tracking contour is 0.44° of latitude, or approximately 50 km.

metric time series shown Plates 2 & 3. The exact timing of the event, however, is dependent on the criteria selected for the definition of separation, which in turn ultimately defines what counts as a Loop Current eddy. This exercise is further complicated by the ambiguity of associating an exact time with what is clearly a continuous and complicated process. Nevertheless, an objective definition greatly simplifies monitoring of Loop Current eddy separation and comparing observations with dynamical theories and/or model simulations.

We identify timing of Loop Current eddy separation events using the Loop Current length time series since the breaking of 17-cm contour between the Loop Current and a detaching eddy into separate contours causes a discrete change in Loop Current length equal to the circumference of the eddy. The day that this event occurs is identified as the "time" of eddy separation. In cases where the eddy subsequently reattaches to the Loop Current, the final detachment time is used. This requires manual discrimination, which means that identifying separation events cannot be completely automated. The time series of SSH maps must be used to differentiate between cases where an eddy reattaches or two eddies separate within a short time period; nevertheless, the separation

events are identified objectively since the two cases are easily distinguished. The separation events so identified are shown on each of the subplots of the metric time series in Plates 2 and 3. The Loop Current length time series and SSH maps of each of the sixteen eddy separation events at the time of final detachment in the 1993 through 2003 record are shown in Plate 4.

In Table 3, the date of final detachment, separation period, name and eddy area at the time of separation are tabulated for each of the 16 observed events. Horizon Marine, Inc. (HMI) names the eddies in alphabetical order as anticyclones shed from the Loop Current and/or impact offshore operations in the northern GOM. A complete list to date is located on the web at http://horizonmarine.com/namedlces.html. The names appear in the weekly EddyWatch™ reports provided to the GOM offshore oil and gas industry by subscription from HMI. All separation events identified using the SSH 17-cm tracking contour were monitored by the EddyWatch™ program, although a number of smaller anticyclonic eddies (7 total) were also named, causing the breaks in the alphabetical sequence. Only one marginal eddy separation event was identified by the objective tracking procedure (Eddy Odessa/Nansen, Eddy 12), which dissipated so quickly that an area estimate for the eddy

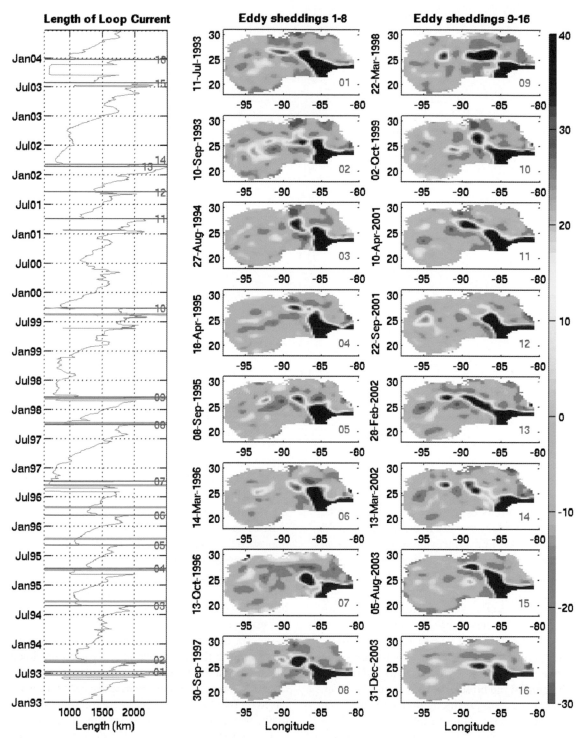

Plate 4. The 16 Loop Current eddy separation events identified in the altimeter record are shown above. Sea surface height maps on the separation dates are shown in the panels to the right (note that the values above 40 cm and below -30 cm have been clipped). Eddy separation times were objectively determined by breaking of the 17-cm tracking contour, which causes a discrete change in the Loop Current length (left panel). The length time series is overlaid with red lines corresponding to the 16 events identified. Gray lines show the 10 separations times determined using the subjective method described in *Sturges and Leben* [2000].

Table 3. Ring separation events from altimetric record 1993-2004. The Eddy Ulysses separation date (*) is a subjective estimate based on imagery and near real-time altimetry.

Eddy Number	Date	Separation Period (months)	Industry Eddy Name	Area (km^2)
1	11 Jul 1993	11.5	Whopper	24,183
2	10 Sep 1993	2.0	Xtra	38,481
3	27 Aug 1994	11.5	Yucatan	43,022
4	18 Apr 1995	7.5	Zapp	21,337
5	8 Sep 1995	4.5	Aggie	24,899
6	14 Mar 1996	6	Biloxi	24,912
7	13 Oct 1996	7	Creole	49,644
8	30 Sep 1997	11.5	El Dorado	49,229
9	22 Mar 1998	5.5	Fourchon	89,143
10	2 Oct 1999	18.5	Juggernaut	40,325
11	10 Apr 2001	18.5	Millennium	45,705
12	22 Sep 2001	5.5	Odessa/Nansen	?
13	28 Feb 2002	5.5	Pelagic	22,119
14	13 Mar 2002	0.5	Quick	49,936
15	5 Aug 2003	17	Sargassum	25,302
16	31 Dec 2003	5	Titanic	33,278
17	*15 Sep 2004	9.5	Ulysses	

could not be made. These smaller eddies are of Loop Current origin, but form on the outer edge of the Loop Current through the interaction of frontal cyclones with the current. This type of small anticyclonic eddy was observed in the northeast GOM during the DeSoto Canyon Eddy Intrusion Study [*Hamilton et al.*, 2000]. Other small named eddies originate as primary Loop Current eddies split and/or shed smaller anticyclonic eddies after separation.

Several notable Loop Current events occurred in the altimeter record. Seven of the sixteen eddies (Eddies 1,4,7,9,11,15, and 16) detached and reattached before final separation. Most of these were short lived; however, a two-month long detachment/reattachment period occurred in the fall of 2003 before the separation of Eddy 16 from the Loop Current. During this detachment event, the interaction of the detached anticyclone with the strong cyclone to the north may have arrested the westward β-induced drift through the eastward velocity induced by the dipole pair formed by the cyclone to the north and the detached anticyclone to the south. The shortest eddy separation period (two weeks) is associated with the first observations of a nearly simultaneous separation of two eddies from the Loop Current. These selected

events and other details can be examined in the 11.5 year-long SSH animation included on the CD-ROM accompanying this monograph.

In Table 4, I compare the subjective ring separation dates published in *Sturges and Leben* [2000] with the objective estimates derived using the 17-cm Loop Current tracking contour for the overlapping events considered. The subjective ring separation dates are also plotted on the Loop Current length time series (see gray lines in left panel of Plate 4). The subjective estimates were found by simply examining each SSH map to determine whether a ring appeared to be separating and continuing to track the separation process until completion. The midpoint of the separation process was identified as the time of separation and an expected uncertainty was assigned based on the duration of the separation process. The average difference in the subjective versus objective separation dates was only three days because of the canceling of differences from event to event. The standard deviation (std) of the differences (29 days) is in good agreement with the root mean square (rms) of the subjective technique uncertainty estimate, which is about 3.5 weeks or 25 days. This is a more robust comparison of the two methods and shows that event tracking

Table 4. Comparison of *Sturges and Leben* [2000] subjective ring separation dates and periods with the objective estimates based on the 17-cm Loop Current tracking contour.

	Sturges and Leben (2000)			17-cm contour results		
Event	Date	Separation Period (months)	Uncertainty (weeks)	Date	Separation Period (months)	Date Difference (days)
1	22 Jun 1993	11	4	11 Jul 1993	11.5	-19
2	19 Sep 1993	3	1	10 Sep 1993	2	9
3	22 Sep 1994	12	4	27 Aug 1994	11.5	26
4	8 Apr 1995	7	4	18 Apr 1995	7.5	-10
5	18 Oct 1995	6	4	8 Sep 1995	4.5	40
6	30 Apr 1996	6	5	14 Mar 1996	6	47
7	Sep 1996	5	4	13 Oct 1996	7	-28
8	11 Oct 1997	13	3	30 Sep 1997	11.5	11
9	14 Mar 1998	5	2.5	22 Mar 1998	5.5	-8
10	22 Aug 1999	17	2	2 Oct 1999	18.5	-41
	Mean	8.5			8.6	3
	Standard Deviation	4.5	3.5 rms		4.8	29

within the uncertainty of an expert/subjective technique can be obtained with the objective technique. There is less scatter in the subjective versus objective separation period (4.5 versus 4.8 month std., respectively), which is presumably caused by subjective smoothing of separation times when evaluating the SSH maps within the separation window.

A histogram of the updated distribution is shown in the upper panel of Figure 2. The 16 events that occurred during the altimetric record are shown in dark gray to highlight the contribution of the more recent events to the histogram of the complete observational record. The histograms of events before and after 1993 are shown in the middle and lower panels, respectively. The mean for the entire compilation is 9.4 months and the mode is 6 months. Using only the pre-1993 record, the mean is 9.96 and the mode is 6 and 9 months, whereas for the altimeter-derived objective tracking periods the mean is 8.6 months and the mode is 6 and 11.5 months. These differences are probably not significant; nevertheless, the change in the distributions between the earlier time period and the altimeter record is striking (see the histograms shown in the lower two panels of Figure 2). This attests to the remarkable variability of the Loop Current eddy separation period.

Loop Current Spectral Analysis

Next I evaluate the subjective versus objective Loop Current tracking technique for estimating the Loop Current spectrum. One method to compute the periodicity of the Loop Current eddy separation is the histogram technique described by *Sturges* [1994]. In this method, a histogram of ring separations is plotted as a function of the inverse separation interval eddy separation cycle to resemble a variance-preserving power spectrum. Hanning passes are used to smooth the distribution while retaining high resolution. The distributions based on the 34 consecutive separation events (1973-1999) as published in *Sturges and Leben* [2000] and an updated version, including the 16 altimeter-derived objective eddy separation intervals through 2003 [a total of 40 consecutive events (see Table 5)], are shown in Figure 3. Both distributions have been smoothed with two Hanning passes.

As expected, the distributions are not much different because of the substantial amount of shared information between the two data sets. Primary peaks in the distributions are found at 6, 9 and 11 months. The one notable difference between the two distributions is the peak appearing near 18 months. More power is seen at this frequency in the smoothed histogram calculated from the extended time series because of the additional events, including two 18.5-month separation periods. The original time series had a maximum separation period of 17 months and only one event of that duration. All of these long-period events occurred after 1998.

The primary limitation of the histogram technique is that, by using only separation periods to estimate the spectrum, no power below the frequency associated with the maximum separation period in the record can be observed. This limits the spectrum to periods less than about 18.5 months for the existing Loop Current observations. By focusing on ring

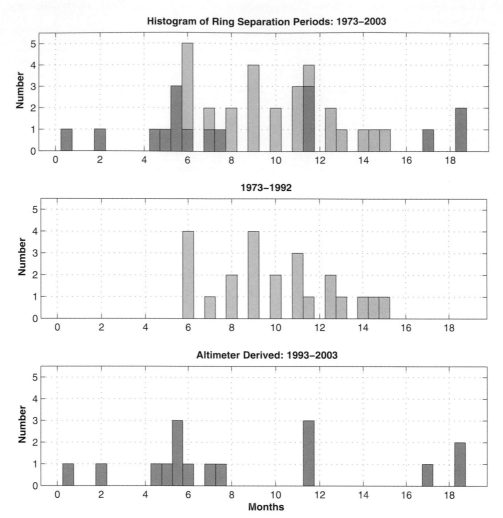

Figure 2. Histogram of the times between ring separations using the data from Table 5 are shown for selected time intervals. The 16 separation events observed during the satellite altimeter record are shown in dark gray, the prior separations in light gray. The range of periods is from 0.5 to 18.5 months, with a mean of 9.4 months and a mode of 6 months.

separations times, one effectively filters the power at all frequencies that are associated with the variability of the Loop Current envelope independent of eddy separation and gives unit power to each individual separation event. This is a reasonable analysis if one is concerned only with the frequency of eddy separation events. More information, however, can be obtained from the spectrum of a continuous time series. For example, the spectrum of the Loop Current length time series contains information on both the frequency and power of the Loop Current length but also the implicit changes in length caused by eddy separation.

The altimeter-derived Loop Current length time series spectrum is shown in Figure 4. Before computing the spectrum, the mean and trend were removed from the 4201 daily values and a full Hanning window taper was applied to the time series to reduce end effects and leakage. Similar to the histogram spectra (Figure 3), there are peaks in the length spectrum at 6, 9 and 11.5 months and little power at exactly 12 months. The dominant peak, however, is at 22 months. This power is associated with both long-period eddy separation events and the far south retreat of Loop Current preceding those separations that cause a very strong low frequency signal in the length time series. This merits further analysis.

A cursory examination of the altimeter-derived metrics finds an apparent relationship between the Loop Current retreat following eddy separation and the subsequent eddy separation period. The far southern retreats are associated with longer separation periods. This is shown in Figure 5. Eddy separation periods are plotted versus the values of

Table 5. A compilation of the 31-year record (July 1973 through June 2004) of LC eddy separation events. A total of 40 consecutive events are listed. Entries through Oct 1986 are from *Vukovich* [1988]; other entries prior to July 1992 are from Table 1 of *Sturges* [1994] using corrections based on *Berger* [1993]. The event in July 1992 is from *Sturges and Leben* [2000]. Data beginning in 1993 are based on objective tracking of the Loop Current using satellite altimeter data, this study.

Date	Separation Period (months)	Date	Separation Period (months)
July 1973		May–June(?) 1989	12.5
April 1974	9	August 1990	14.5
January 1975	9	Aug–Sep 1991	12.5
July 1975	6	19 July 1992	11.5
August 1976	13	11 Jul 1993	11.5
March 1977	7	10 Sep 1993	2
June 1978	15	27 Aug 1994	11.5
April 1979	10	18 Apr 1995	7.5
January 1980	9	8 Sep 1995	4.5
March 1981	14	14 Mar 1996	6
November 1981	8	13 Oct 1996	7
May 1982	6	30 Sep 1997	11.5
March 1983	10	22 Mar 1998	5.5
February 1984	11	2 Oct 1999	18.5
August 1984	6	10 Apr 2001	18.5
July 1985	11	22 Sep 2001	5.5
January 1986	6	28 Feb 2002	5
October 1986	9	13 Mar 2002	0.5
September 1987	11	5 Aug 2003	17
May 1988	8	31 Dec 2003	5

the Loop Current northern extent (Loop Current maximum latitude) immediately following the previous eddy separation event. These northern Loop Current extent values were calculated by finding the minimum value of the Loop Current 17-cm contour maximum latitude coordinate in the five-day window following separation of a Loop Current eddy. The values are also plotted overlaid on the maximum latitude time series (see green stars in left panel of Plate 2). I elected to edit the latitude of one retreat value (number 3 at 26.34°N) that did not reflect the initial state of the Loop Current because in that event a warm filament affected the 17-cm tracking contour just after eddy separation. Instead, I choose a value of 25.6°N that corresponds to the latitude where the Loop Current remained for nearly three months

just after separation. Otherwise, the Loop Current retreat values are relatively stable and can be calculated in a variety of ways by using mean or extrema values within a variety of windows.

There is a distinct break in the distribution of values near the retreat latitude of 25°N. This break is also reflected in the overall distribution which shows a bimodal distribution centered at 25°N in the histogram of the Loop Current maximum latitude (lower left panel of Plate 2). During this time period from 1993 through 2003, the average separation period following retreat is 16.2 months when the entire Loop Current retreats to below 25°N, much longer than the 5.5 month average for the cases where part of the Loop Current remains north of 25°N. Above about 26°N, there is a strong linear relationship between period and retreat with the exception of event 3, which has already been discussed. Below 26°N there is more scatter in the data, most notably events 6, 8, 9, and 12; nevertheless, the overall correlation of retreat to period is –0.88 (-0.83 with unedited point number 3).

Investigation of the incongruent period associated with separation event 8 leads to an interesting relationship. While events 10, 11 and 15 are clustered along the line extrapolating the nearly linear relationship of retreat to period above 26°N, the period associated with event 8 is much shorter. Note, however, that if the periods of events 8 and 9 are combined and plotted with the initial retreat of event number 8, then the point falls within the cluster of the other three southernmost retreat points. In fact, if all of the events exhibiting the incongruent periods (events 6, 8 and 12) are combined with the period of the next eddy separation event (events 7, 9, and 13, respectively), then a nearly perfect linear relationship is found between retreat latitude and separation period (Figure 6). The overall correlation of retreat to period with this combination of separation events is –0.99 (-0.96 with unedited point number 3). Why this may be physically plausible will be explored in the following discussion.

4. DISCUSSION

My overall goal in this study was to show that an objective Loop Current tracking technique could give eddy separation periods comparable to those determined by subjective tracking. This goal has been achieved. Furthermore, the objective technique provides a continuous time series of metrics for study of Loop Current intrusion and eddy separation. The time series presented, their statistics and the identified separation events, which have been validated by comparison with other observations, are the primary results of the successful implementation of this method. Although the Loop Current separation is not a well-defined discrete event, using the breaking of the tracking contour works remarkably well. The

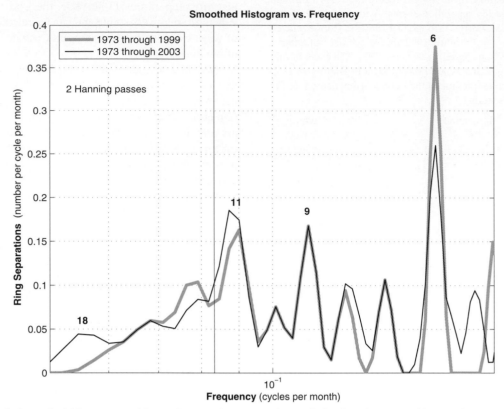

Figure 3. Smoothed histograms of Loop Current ring separation periods plotted to resemble a variance-preserving spectra after Sturges (1994) calculated from 1973-1999 events [*Sturges and Leben*, 2000] and 1973-2003 events including the objective separation estimates from the altimeter record (1993-2003) as described in this paper. The distributions have been smoothed by two Hanning passes. Values at intervals shorter than 5 months and longer than 20 months are off the scale. The vertical solid line is the annual frequency.

objective estimates fall within the subjective error estimates for separation events from the same time period reported in *Sturges and Leben* [2000]. Given that nonsystematic errors in separation time do not cancel when calculating the separation period, more work still needs to be done to estimate the uncertainty in the separation period determined by the objective tracking technique. For example, the 17-cm contour may break at different times or stages within the continuous separation process for the wide variety of events that have been observed. Using several tracking contours and looking at the spread of the estimated separation times may be one way to quantify this error.

The fundamental result of the spectral analyses performed on the objective tracking analyses for this study is that the Loop Current exhibits significant power near 6, 9, and 11 months, and little or no power exactly at the annual frequency in agreement with the results presented in *Sturges and Leben* [2000]. These are robust results that are exhibited by the histogram, smoothed histogram and FFT spectra analyses.

The conspicuous absence of annual power merits further discussion. It has been speculated that the lack of power at 12 months is caused by a beat-frequency effect [*Sturges,* 1992; *Sturges and Leben*, 2000]. The premise is that the power at a lower frequency modulates the power at the annual period, giving rise to the observed spectral peak. The standard relationship between frequencies is:

$$f_1 \pm f_2 = f_3$$

where f_1 is annual, f_2 is the unknown, and f_3 is the peak observed near 11 months. The unknown values vary from 132 to 276 months for the corresponding 11- to 11.5- month peaks seen in the smoothed histogram and FFT spectra (Figures 3 and 4).

Unfortunately, these values are beyond the resolution of the 11.5-year (138-month) time series of altimeter observations used to calculate the FFT spectrum and cannot be studied using the histogram technique from the longer time series of separation periods. The lowest frequency observed in the FFT spectrum (Figure 4) is near 67 months, which is approximately half of the decadal time scale associated with climate signals over the North Atlantic and may be a harmonic of the decadal signal yet to be observed using a longer data record. A much longer record is needed to resolve the power associated with the 11- and 12-month signals. Using a conservative estimate of two times the Raleigh criterion [*Emery and Thompson*, 1998], a 22- to 46-year-long times series would be required to resolve the annual signal from 11- to 11.5-month peaks, respectively.

One caveat concerning the objective Loop Current tracking and annual signals is that the altimetric analysis uses along-track data that has been high-pass filtered. This processing may remove the basin-scale steric signal associated with surface heating and cooling that can be as large as 5 to 8 cm in the Gulf [this estimate is based on unfiltered T/P and Jason over the Gulf] and exhibit power at the annual frequency. This signal, however, would be present in model simulations incorporating realistic heat fluxes. The overall effect on separation periods should be minimal, but some additional power may be present in FFT spectra from model data that is not present in the altimetric analysis.

Another conclusion of this study is that the altimeter-derived estimates of Loop Current metric statistics are in very good agreement with earlier studies based on in situ and radiometry observations. These earlier studies span time periods comparable to the altimetric record with little or no overlap in time and support the conjecture that the fundamental Loop Current behavior is nearly stationary. The agreement of mean and standard deviations and the implied stationarity is not surprising considering the topographic confinement of the current and the fundamental physical control of the dominant eddy-shedding cycle. The distributions of the individual Loop Current metrics shown in the histograms included in Plates 2 and 3, however, are clearly non-Gaussian. Although the mean and standard deviation of the metric time series are stable over different time periods, even in the presence of the irregular eddy separation cycle it is unlikely that the overall distributions are stationary because of the strong influence of the irregular eddy separation cycle and the limited number of total events observed. This effect is seen in the separation period histograms for the time periods of 1973–1992 and 1993–2003 shown in the lower panels of Figure 2. One might argue that the differences in the distributions are an artifact of the techniques and data coverage used to identify eddy separation and estimate the separation period. A counterpoint to this argument is

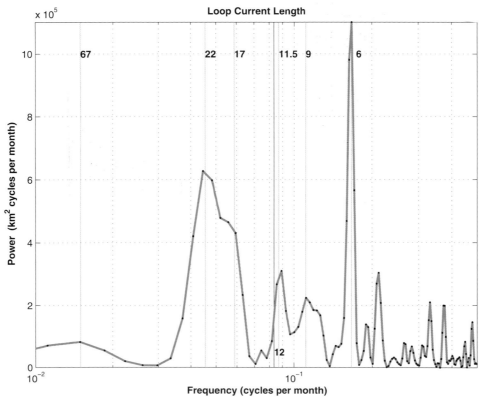

Figure 4. FFT power spectrum of the Loop Current length estimated from the daily 1/1/1993 through 7/1/2004 altimeter-derived time series. The light gray lines highlight periods (in months) associated with selected frequencies.

If the linear relationship holds, then one could conclude that we have observed nearly the full dynamic range of Loop Current separation periods, which I estimate to be about zero (simultaneous separation of two Loop Current eddies) to 22 months. The upper bound is based on the period determined by a port-to-port retreat of the Loop Current, in which the entire current retreats below 24°N immediately following a separation event. This is a realistic latitude for the southernmost retreat possible.

In conclusion, my discussion has been intentionally speculative to motivate future investigations of Loop Current dynamics through the careful analysis of the available observations. I specifically selected the Loop Current length for monitoring Loop Current separation primarily because it is a relatively easy metric to estimate from numerical model simulations. This is to encourage more quantitative comparisons of observations to model experiments. The comprehensive set of Loop Current metrics in this paper and the SSH animation included on the companion CD-ROM are intended to assist in these efforts.

Acknowledgements. I thank Kevin Corcoran for his help calculating the Loop Current metrics and plotting figures, Peter Gimeno for helping with data processing, and Dr. Chad Fox for collaborating with me to develop the MATLAB® Loop Current metric toolbox. Dr. Kenneth S. Casey kindly provided the SST data used in this study. Tony Sturges' insightful suggestions and comments contributed substantially to this work. The U.S. Minerals Management Service (MMS 1435-01-02-CT-31152; MMS 1435-01-04-CA-32645), the National Science Foundation (OCE-324688) and NASA (contract 1221120) provided funding for this study. I greatly appreciate their support.

REFERENCES

Abascal, A. J., J. Sheinbaum, J. Candela, J. Ochoa, and A. Badan (2003), Analysis of flow variability in the Yucatan Channel, *J. Geophys. Res.*, 108, No. 12, 3381, doi:10.1029/2003JC001922.

Berger, T. (1993), Loop Current eddy shedding cycle. *LATEX Program Newsleter*, Vol. 2, No. 24, College of Geosciences, Texas A&M University. [Available from Dept. of Oceanography, Texas A&M University, College Station, TX 77843–3146.]

Berger, T. J., P. Hamilton, J. J. Singer, R. R. Leben, G. H. Born and C. A. Fox (1996), Louisiana/Texas Shelf Physical Oceanography Program Eddy Circulation Study: Final Synthesis Report. Volume I: Technical Report, OCS Study MMS 96-0051, U.S. Dept. of the Interior, Minerals Management Service, Gulf of Mexico OCS Region, New Orleans, LA. 324 pp.

Bunge, L., J. Ochoa, A. Badan, J. Candela, and J. Sheinbaum (2002), Deep flows in the Yucatan Channel and their relation to changes in the Loop Current extension, *J. Geophys. Res.*, 107, 3223, doi:10.1029/2001JC001256.

Casey, K. S, and P. Cornilon (1999), A comparison of satellite and in situ based sea surface temperature climatologies, *J. of Climate*, Vol. 12, No. 6, 1848–1863.

Candela, J., J. Sheinbaum, J. Ochoa, A. Badan and R. Leben (2002), The potential vorticity flux through the Yucatan Channel and the Loop Current in the Gulf of Mexico, *Geophys. Res. Lett.*, 29, No. 22, 2059, doi:10.1029/2002GL015587.

Chérubim, L. M., W. Sturges and E. P. Chassignet (2005), Deep flow variability in the vicinity of the Yucatan Straits from a high-resolution numerical simulation, *J. Geophys. Res.*, 110, C04009, doi:10.1029/2004JC002280.

Cressman, G. P. (1959), An operational objective analysis system, *Mon. Weather Rev.*, 87, 367–374.

Emery, W. J., and R. E. Thompson (1998), *Data Analysis Methods in Physical Oceanography*, 634 pp, Pergamon, Tarrytown, N.Y.

Ezar, T., L. –Y. Oey, H.-C Lee, and W. Sturges (2003), The variability of currents in the Yucatan Channel: Analysis of results from a numerical ocean model, *J. Geophys. Res.*, 108(C1), 3012, doi:1029/2002JC001509.

Fratantoni, P. S., T. N. Lee, G. Podesta, and F. Muller-Karger (1998), The influence of Loop Current perturbations on the formation and evolution of Tortugas eddies in the southern Stratis of Florida, *J. Geophys. Res.*, 103, 24,759–24,779.

Hamilton P., Berger, T.J., J.H. Churchill, R.R. Leben, T.N. Lee, J.J. Singer, W. Sturges, and E. Waddell (2000), Desoto Canyon Eddy Intrusion Study; Final Report, Volume II: Technical Report, OCS Study MMS 2000-080, U.S. Dept. of Interior, Minerals Management Service, Gulf of Mexico OCS Region, New Orleans, LA. 269 pp.

Hurlburt, H. E., and J. D. Thompson (1980), A numerical study of Loop Current intrusions and eddy shedding, *J. Phys. Oceanogr.*, 10, 1611–1651.

Koblinsky et al. (1999), NASA Ocean Altimeter Pathfinder Project, Report 1: Data Processing Handbook, NASA/TM-1998-208605.

Leben, R. R., and G. H. Born (1993), Tracking Loop Current eddies with satellite altimetry, *Adv. Space. Res.*, 13, 325–333.

Leben, R. R., G. H. Born and B. R. Engebreth (2002), Operational altimeter data processing for mesoscale monitoring, *Marine Geodesy*, 25, 3–18.

Lee, T. N., K. Leaman, E. Williams, T. Berger and L. Atkinson (1995), Florida Current meanders and gyre formation in the southern Straits of Florida, *J. Geophys. Res.*, 100, 8607–8620.

Maul, G. A., and F. M. Vukovich (1993), The relationship between variations in the Gulf of Mexico Loop Current and Straits of Florida volume transport, *J. Phys. Oceanogr.* 23, 785–796.

Murphy, S. J., H. E. Hurlburt, J. J. O'Brien (1999), The connectivity of eddy variability in the Caribbean Sea, the Gulf of Mexico and the Atlantic Ocean, *J. Geophys. Res.*, 104, 1431–1453.

Nowlin, W. D., Jr., A. E. Jochens, S. F. DiMarco, R. O. Reid, and M. K. Howard (2001), Deepwater Physical Oceanography Reanalysis and Synthesis of Historical Data: Synthesis Report, OCS Study MMS 2001-064, U.S. Dept. of the Interior, Minerals Management Service, Gulf of Mexico OCS Region, New Orleans, LA. 530 pp.

Oey, L.-Y., H.-C. Lee and W. J. Schmitz (2003), Effects of winds and Caribbean eddies on the frequency of Loop Current eddy shedding: A numerical modeling study, *J. Geophys. Res.*, 108, 3324, doi:10.1029/20002JC001698.

Ohlmann, J. C., P. P. Niiler, C. A. Fox, and R. R. Leben (2001), Eddy energy and shelf interactions in the Gulf of Mexico, *J. Geophys. Res.*, 106, 2605–2620.

Pichevin, T. and D. Nof (1997), The momentum imbalance paradox, *Tellus*, 49A, 298–319.

Tierney, C. C., M. E. Parke, and G. H. Born (1998), An investigation of ocean tides derived from along-track altimetry, *J. Geophys. Res.*, 103, 10,273–10.287.

SAIC (1989), Gulf of Mexico Physical Oceanography Program, *Final Report: Year 5. Volume II: Technical Report*, OCS report/MMS-89-0068, U. S. Department of the interior, Minerals Management Service, Gulf of Mexico OCS Regional Office, New Orleans, LA. 333 pp.

Sturges, W. (1992), The spectrum of Loop Current variability from gappy data, *J. Phys. Oceanogr.*, 22, 1245–1256.

Sturges, W. (1994), The frequency of ring separations from the Loop Current, *J. Geophys. Oceanogr.*, 24, 1647–1651.

Sturges, W. and R. Leben (2000), Frequency of ring separations from the Loop Current in the Gulf of Mexico: A revised estimate, *J. Phys. Oceanogr.*, 30, 1814–1819.

Vukovich, F. M. (1988), Loop Current boundary variations. *J. Geophys. Res.*, 93, 15,585–15,591.

Vukovich, F. M. (1995), An updated evaluation of the Loop Current's eddy shedding frequency, *J. Geophys. Res.*, 100, 8655–8659.

Vukovich, F. M., and G. A. Maul (1985), Cyclonic eddies in the eastern Gulf of Mexico, *J. Phys. Oceanogr.*, 15, 105–117.

Wang, Y. M. (2001), GSFC00 mean sea surface, gravity anomaly and vertical gravity gradient from satellite altimeter data, *J. Geophys. Res.*, 106, 31167–31174.

Welsh S. U., and M. Inoue (2000), Loop Current rings and the deep circulation in the Gulf of Mexico. *J. Geophys. Res.*, 105, 16,951–16,959.

Zavala-Hidalgo, J., S. L. Morey and J. J. O'Brien (2003), Cyclonic eddies northeast of the Campeche Bank from altimetry data, *J. Phys. Oceanog.*, 33, 623–629.

——————————

Robert R. Leben, 431 UCB, Boulder, CO 80309-0431 (leben@colorado.edu)

Burged
of Mexic

Circulation
Geophysic
Copyright
10.1029/16

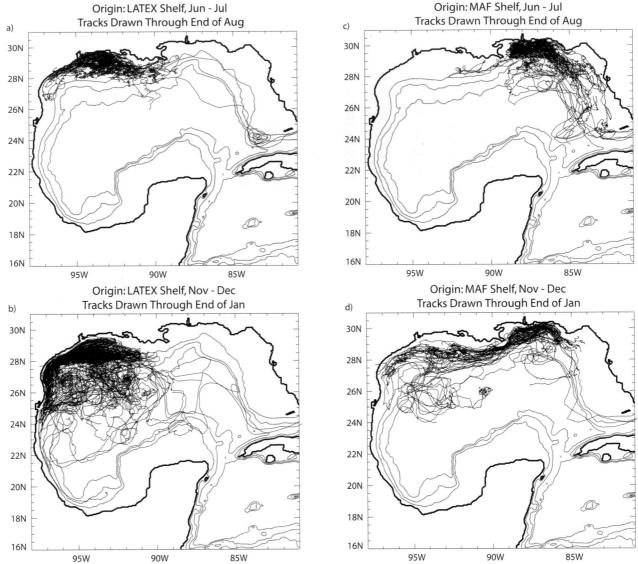

Figure 4. Trajectories for combined LATEX, SCULP I and SCULP II 1 m drogued drifters. (a) Drifters existing within the defined LATEX shelf region in the summer months, (b) Drifters existing within the defined LATEX shelf region in the winter months, (c) Drifters existing within the defined MAFLA shelf region in the summer, and (d) Drifters existing within the defined MAFLA shelf region in the winter. The 200 m, 1000 m, and 2000 m isobaths are drawn shown by the gray contour lines. Adapted from *Morey et al.* [2003a].

exogenous variables, prescribed from data without dependence on the ocean model fields. Winds are derived from the Comprehensive Ocean–Atmosphere Data Set (COADS) [*DaSilva*, 1994]. The sea surface temperature is relaxed to monthly climatology with a relaxation coefficient of 1 m/d (giving a relaxation time scale in days equivalent to the mixed layer depth in meters), which primarily has the effect of preventing temperature overshoots in the shallowest quiescent locations. There is no relaxation of the surface salinity to climatology. A surface salinity flux has the effect

of uniformly evaporating a net of 46 cm/yr of water from the Gulf, and is chosen to balance the volume flux by rivers averaged over the year. The model is run with monthly climatology river discharge derived primarily from the United States Geological Survey data for the U.S. rivers, and time constant values for the Mexican rivers derived from multiple sources. The simulation is initialized from rest with 1994 World Ocean Atlas [*National Oceanic and Atmospheric Administration*, 1994] January climatology temperature and salinity and run for a total of 13 years with model data saved

suggests se
transported

Preferre
low-salinit
simulated
Confluence
Bay of Can
vation requ
confluence
ity water fr
current was
where the a
direction. I
across the
the summer
low-salinity
across the s

To deterr
salinity wa
export func
et al. [2003
defined as

where

is the comp
water. Here,
two dimens
a reference
ter export f
low-salinity
purpose is c
along the 20
around the C
tion calculat
the location
exported in

In the sou
export is acti
ilar seasonal
the duration
north. This c
of the down
salinity wate
this current
The most vi
during the sp

for analysis at 48-hour intervals beginning in the fourth year of model integration. Numerical details of the simulation are as described in *Morey et al.* [2003a].

The modeled transport through the Yucatan Channel has no seasonal component and is approximately 27 Sv (10^6 m^3 s^{-1}). This value is comparable to previous estimates of 28 Sv [*Gordon,* 1967; *Roemmich,* 1981] and somewhat larger than the 23.8 Sv average measured by *Sheinbaum* [2002]. The average LC eddy shedding period from the model is 9.3 months, compared to 9.5 months reported by *Sturges and Leben* [2000], with periods ranging from 3.5 months to nearly two years. Aperiodicity in the eddy shedding is consistent with the previous literature. The model mean velocity field shows the northern boundary of the LC at 26.5°N and anticyclonic features to the west of the LC, similar to the mean circulation inferred from contours of mean dynamic topography relative to 1000 m.

SHELF CIRCULATION CLIMATOLOGY

The monthly climatology of simulated upper ocean currents on the continental shelves of the northern and western GOM is consistent with the circulation patterns inferred above from the climatological wind forcing, but not consistent with a buoyancy driven flow. A downcoast current is well defined from west of the Mississippi River Delta through the southern Bay of Campeche during the fall and winter months (Figure 5). This current is consistent with the direction of the along-coast component of the wind stress (primarily southward) during these months (Plate 1). *Zavala-Hidalgo et al.* [2003b] showed that the monthly mean current direction near the coast computed from their model correlates highly with the along-coast wind stress component, particularly over the East Mexico Shelf. This result agrees with *Cochrane and Kelly* [1986], *Cho et al.* [1998], and *Nowlin et al.* [1998]. Due to the geometry of the coastline, the sign of the along-coast wind stress component changes for the eastern Bay of Campeche, over the Campeche Bank, where an upcoast current is apparent. These opposing flows result in a confluence in the southern Bay of Campeche during the fall and winter months.

During the spring months the average wind direction is oriented more zonally toward the west (Figure 2). Associated with this shift in the wind direction is a northward migration of the point at which the sign of the along-coast wind stress component changes. In April, downcoast flow over the LATEX shelf is opposed by upcoast flow along the East Mexico Shelf. This along-coast migration of the convergence in the along-coast wind component and the modeled along-coast flow can be seen from the time versus along-coast distance plots of these quantities (Plate 1). By careful inspec-

tion, a slope of the zero contour (the point at which the sign of the along-coast component changes) from about 1000 km to 2000 km along the coast in the figure is evident from about mid-March to mid-April. The bending of the coastline (and isobaths) at the bight where the LATEX Shelf meets the East Mexico Shelf (about 1800 km in Plate 1) causes an irregular pattern in the along-coast wind stress component, resulting in a strong convergence that will be discussed later in connection with cross-shelf transport.

A complete flow reversal over the LATEX and East Mexico shelves from downcoast during the winter to upcoast is seen in the July monthly climatology of the model velocity field. Weak eastward flow is evident over the MAFLA shelf toward De Soto Canyon at this time. September begins the development of the downcoast current along the LATEX Shelf, again migrating along the coast to the Bay of Campeche. As for the spring, this fall migration of the flow reversal can be seen in the time versus along-coast distance plots of the along-coast wind stress and velocity climatology (Plate 1).

The most striking feature in the NCOM GOM simulation climatology is perhaps the complete reversal of the coastally attached current along the East Mexico and LATEX shelves. If the along-coast wind stress is the primary driving mechanism of this current, one would also expect that coastal upwelling occurs along the narrow East Mexico Shelf. Zonal temperature sections averaged over the fall–winter and spring–summer periods illustrate the subsurface response to the shifting winds (Figure 6). During the spring and summer months, this region is strongly stratified. Isotherms slope upward toward the coast consistent with the upwelling favorable winds. The temperature at 50m over the shelf is actually 1 to 2°C cooler during the summer months than during the winter as a consequence of the upwelling. *Zavala-Hidalgo et al.* [2003b] present subsurface historical hydrographic data which support the existence of summertime upwelling of cold water over the East Mexico Shelf. In their analysis of the data, an area of cold subsurface water can be seen along the East Mexico Shelf, with warmer water offshore, suggesting isotherms tilting upward toward the coast.

During the winter, a temperature inversion can be seen near the surface, which is possible due to the presence of low-salinity water along the coast (Plate 2). Outside of this coastal low-salinity water, the upper ocean is only weakly stratified. This wintertime cold, fresh coastally-trapped current can be seen from satellite thermal imagery, as shown in *Zavala-Hidalgo et al.* [2003b].

SALINITY CLIMATOLOGY

The large volume of fresh water flowing into the Gulf of Mexico from the major rivers, particularly the Mississippi

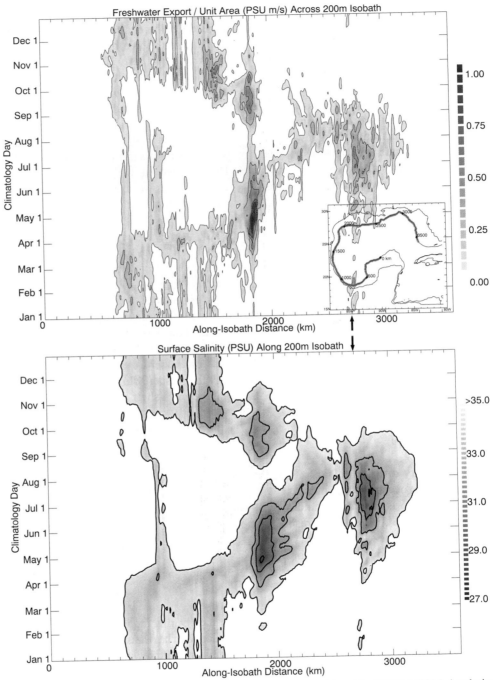

Figure 8. Top: Climatology fresh water export per unit area, as computed from the NCOM GOM simulation data with equation (1) across the 200 m isobath as a function of along-isobath distance (shown at the bottom) and time. Inset: The 200-m isobath used in the calculation is highlighted with along-isobath distance indicated. Bottom: NCOM GOM simulation salinity climatology along the same isobath as in the top figure. Arrows indicate the approximate location of the Mississippi River Delta.

The entrainment of low-salinity water from the outer shelf into the currents of the deep Gulf happens not in a predictable pattern at the same time every year, but rather happens as short lived episodes (several weeks to months) dependent on the state of the chaotic eddy field and the LC. However, due to the climatological wind-driven transport of this buoyant water over the shelf, there is a strong tendency for these entrainment events to occur at certain locations and certain months of the year, which results in the seasonal patterns seen in the salinity field.

It is common that river-discharged water contains large amounts of nutrients that enhance biological productivity [*Rabalais, et al.*, 2002]. The result is that the low-salinity water formed near rivers is often rich in chlorophyll. The episodic entrainment events of this riverine-influenced water are therefore also commonly evident in images of ocean color observed from space [*Gilbes et al.*, 1996; *Del Castillo et al.*, 2001; *Toner et al.*, 2003] and recently from in-situ observations [*Biggs et al.*, 2005, This Volume). Maps of chlorophyll *a* concentration monthly climatology created from SeaWiFS satellite ocean color data show a pattern similar to the model surface salinity (Plate 2). Both the low-salinity tongues in the summer off the WFS and in the winter in the Bay of Campeche appear as high chlorophyll tongues from the satellite data. Further, peaks in the monthly climatology chlorophyll concentration are concurrent with the surface salinity minima in locations influenced by the transport of river-discharged fresh water (Figure 7), with the signal dissipating progressively away from the fresh water source (as seen in panels a through d of Figure 7). Monthly climatology chlorophyll *a* time series compare well with salinity annual cycles derived from the model and the WOA98 data set near the hearts of these low-salinity tongues (Figure 7). In the De Soto Canyon region, the chlorophyll and WOA98 salinity time series are (negatively) correlated with correlation coefficient magnitudes of 0.88 and 0.82 in boxes A and B of Plate 2, respectively, and of 0.69 in the Bay of Campeche (Plate 2 Box F). This strongly suggests a link between the transport of low-salinity water formed by riverine discharge and the biology of the open GOM.

CONCLUSIONS

A high resolution model and observational data have been used to examine the seasonal variability of circulation on the continental shelves of the northern and western GOM with consequences for the upper ocean salinity and biology. Despite large gradients in the buoyancy flux, the circulation on the shelves is primarily wind-driven. Northerly winds during the winter months force a downcoast flow along the LATEX and East Mexico Shelves toward the Bay of Campeche, transporting low-salinity water along this path. This current meets an opposing upcoast current and this confluence zone retreats northward with the shifting winds during the spring. By summer, upcoast flow is evident along the East Mexico Shelf and there is eastward transport over the MAFLA Shelf toward De Soto Canyon. This results in a pool

of low-salinity water over the LATEX and MAFLA Shelves, spreading across the shelf break at the western edge of this canyon. The downcoast current along the LATEX and East Mexico Shelves again develops during the fall months.

When buoyant low-salinity water is found over the outer shelves near the shelf break, it can be entrained in currents associated with the offshore mesoscale eddy field and transported over deeper waters away from the shelf. The interaction of the shelf water with the eddy field offshore forms a mechanism for effectively redistributing the fresh water input by rivers throughout the GOM. The annual circulation pattern results in seasonally varying distribution of low-salinity water over the shelves, and subsequently governs the variability of cross-shelf export of fresh water. Low-salinity water is exported offshore in the Bay of Campeche primarily during the late fall and winter months. The time window during which this cross-shelf exchange is seen in the model climatology grows progressively to the north along the East Mexico Shelf, due to the migration of the low-salinity water along the shelf by the seasonally shifting winds. An active area of cross-shelf fresh water export is seen from fall through spring at the northern end of the East Mexico Shelf, where it meets the wide LATEX Shelf. During the summer months, vigorous entrainment of low-salinity water into the offshore eddy field occurs over De Soto Canyon, influenced by the eastward transport of Mississippi River water across the MAFLA Shelf.

This annual cycle in the cross-shelf fresh water export pattern results in striking features in the salinity climatology throughout the GOM. A low-salinity tongue is evident in the climatology in the Bay of Campeche during the winter, as a downcoast current transporting low-salinity water meets an opposing current forcing the buoyant water further offshore where it can be entrained and transported northward. A similar summertime feature can be seen emanating from the MAFLA Shelf across the western edge of De Soto Canyon and then southeastward along the western edge of the WFS. Here the low-salinity tongue in the climatology closely follows the edge of the mean position of the LC. Similar patterns can be found in the chlorophyll *a* concentration climatology produced from SeaWiFS satellite ocean color data. This suggests a connection between the seasonal circulation on the continental shelves, the redistribution of water discharged by the major rivers in the Gulf, and the hydrographic and biological characteristics of the upper ocean throughout the GOM.

Acknowledgments. This project was sponsored by the ONR Secretary of the Navy grant awarded to J. J. O'Brien, the NASA Office of Earth Science, and the U.S. Dept. of the Interior Minerals Management Service grant for this monograph. Simulations were performed on the IBM SPs at FSU and the Naval Oceanographic Office with time provided by the DoD HPCMO. The authors thank Paul Martin and Alan Wallcraft for development of the NCOM, and the Goddard Earth Sciences Data and Information Services Center/DAAC for the production and distribution of the SeaWiFS data. These activities are sponsored by NASA's Earth Science Enterprise.

REFERENCES

Biggs, D. C., and F. E. Müller-Karger (1994), Ship and satellite observations of chlorophyll stocks in interacting cyclone-anticyclone eddy pairs in the western Gulf of Mexico, *J. Geophys. Res., 99*, 7371–7384.

Cho, K., R. O. Reid, and W. D. Nowlin, Jr. (1998), Objectively mapped stream function fields on the Texas–Louisiana shelf based on 32 months of moored current meter data. *J. Geophys. Res., 103*, 10377–10,390.

Cochrane, J. D., and F. J. Kelly (1986), Low-frequency circulation on the Texas–Louisiana continental shelf. *J. Geophys. Res., 91*, 10,645–10,659.

Comision Nacional del Agua (2002), Compendio basico del agua en Mexico, 96pp.

Conkright, M., S. Levitus, T. O'Brien, T. Boyer, J. Antonov, and C. Stephens (1998), "World Ocean Atlas 1998 CD-ROM Data Set Documentation". Tech. Report 15, NODC Internal Report, Silver Spring, MD.

DaSilva, A., A. C. Young, and S. Levitus (1994), "Atlas of Surface Marine Data 1994, Volume 1: Algorithms and Procedures," *NOAA Atlas NESDIS 6*, U. S. Department of Commerce, Washington.

Del Castillo, C. E., P. G. Coble, R. N., Conmy, F. E. Müller-Karger, L. Vanderbloemen, and G. A. Vargo (2001), Multispectral in situ measurements of organic matter and chlorophyll fluorescence in seawater: Documenting the intrusion of the Mississippi River plume in the West Florida Shelf, *Limnol. Oceanogr.*, 46, 1836–1843.

Fratantoni, P. S., T. N. Lee, G. Podesta, and F. Muller-Karger (1998), The influence of the Loop Current perturbations on the formation and evolution of Tortugas eddies in the southern Straits of Florida. *J. Geophys. Res., 103*, 24,759–779.

Gilbes, F., C. Thomas, J. J. Walsh, and F. E. Müller-Karger (1996), An episodic chlorophyll plume on the West Florida Shelf, *Cont. Shelf Res., 16*, 1201–1224.

Gill (1982), A. E., Atmosphere – Ocean Dynamics, Academic Press, San Diego, 662pp.

Gordon, A. (1967), Circulation of the Caribbean Sea. *J. Geophys. Res, 72*, 6207–6223.

Holland, W.R., J.C. Chow, and F.O. Bryan (1998), Application of a third-order upwind scheme in the NCAR Ocean Model. *J. Climate, 11*, 1487–1493.

Martin, P.J. (2000), "A description of the Navy Coastal Ocean Model Version 1.0". NRL Report: NRL/FR/7322-009962, Naval Research Laboratory, Stennis Space Center, MS, 39 pp.

Mellor, G.L., and T. Yamada (1974), A hierarchy of turbulence closure models for planetary boundary layers. *J. Atmos. Sci., 31*, 1791–1806.

Mooers, C.N.K., and G. Maul (1998), Intra-Americas sea circulation, *The Sea*, A. Robinson and K.H. Brink, Eds., The Global Coastal Ocean, Regional Studies and Syntheses, Vol. 11, Wiley and Sons, 183–208.

Morey, S. L., and J. J. O'Brien (2002), The spring transition from horizontal to vertical thermal stratification on a mid-latitude continental shelf, *J. Geophys. Res., 107*(C8), 3097, doi: 10.1029/2001JC000826.

Morey, S. L., P. J. Martin, J. J. O'Brien, A. A. Wallcraft, and J. Zavala-Hidalgo (2003a), Export pathways for river discharged fresh water in the northern Gulf of Mexico, *J. Geophys. Res., 108*(C10), 3303, doi:10.1029/2002JC001674.

Morey, S. L., W. W. Schroeder, J. J. O'Brien, and J. Zavala-Hidalgo (2003b), The annual cycle of riverine influence in the eastern Gulf of Mexico basin, *Geophys. Res. Lett., 30*(16), doi:10.1029/2003GL017348.

Müller-Karger, F. E., J. J. Walsh, R. H. Evans, and M. B. Meyers (1991), On the seasonal phytoplankton concentration and sea surface temperature cycles on the Gulf of Mexico as determined by satellites, *J. Geophys. Res.*, 96, 12,645–12,665.

National Oceanic and Atmospheric Administration (1994), *World Ocean Atlas 1994* [CD-ROM NODC-43], National Oceanographic Data Center, Washington, DC.

Nowlin, W. D., Jr., A. E. Jochens, S. F. DiMarco, R. O. Reid, and M. K. Howard (2001), Deepwater physical oceanography reanalysis and synthesis of historical data: Synthesis report, *OCS Stud. MMS 2001-064*, 528 pp., U. S. Dep. of the Inter., Mineral. Manage. Serv., Gulf of Mex. OCS Reg., New Orleans, LA.

Ohlmann, J. C., P. P. Niiler, C. A. Fox, and R. R. Leben (2001), Eddy energy and shelf interactions in the Gulf of Mexico. *J. Geophys. Res., 106*, 2605–2620.

Rabalais, N. N., R. E. Turner, Q. Dortch, D. Justic, V. J. Bierman, and W. J. Wiseman (2002), Nutrient-enhanced productivity in the northern Gulf of Mexico: past, present and future. *Hydrobiologia, 475*, 39–63.

Roemmich, D. (1981), Circulation of the Caribbean Sea: A well-resolved inverse problem, *J. Geophys. Res., 86*, 7993–8005.

Sheinbaum, J., J. Candela, A. Badan, and J. Ochoa (2002), Flow structure and transport in the Yucatan Channel, *Geophys. Res. Let., 29*, doi: 10.1029/2001GL013990.

Sturges, W., and R. Leben (2000), Frequency of ring separations from the Loop Current in the Gulf of Mexico: A revised estimate. *J. Phys. Ocean., 30*, 1814–1819.

Toner, M., A. D. Kirwan Jr., A. C. Poje, L. H. Kantha, F. E. Müller-Karger, and C. K. R. T. Jones (2003), Chlorophyll dispersal by eddy-eddy interactions in the Gulf of Mexico, *J. Geophys. Res., 108*(C4), 3105, doi:10.1029/2002JC001499.

Vidal, V. M. V., F. V. Vidal, and J. M Perezmolero (1992), Collision of a Loop Current anticyclonic ring against the continental shelf slope of the western Gulf of Mexico, *J. Geophys., Res., 97*, 2155–2172.

Vukovich, F. M., G. A. Maul (1985), Cyclonic eddies in the eastern Gulf of Mexico. *J. Phys. Oceanogr., 15*, 105–117.

Weisberg, R. H., Z. Li, and F. Muller-Karger (2001), West Florida shelf response to local wind forcing: April 1998. *J. Geophys. Res., 106*, 31,239–262.

Yang, H., R. H. Weisberg, P. P. Niiler, W. Sturges, and W. Johnson (1999), Lagrangian circulation and forbidden zone on the West Florida Shelf. *Cont. Shelf Res, 19*, 1221–1245.

Zavala-Hidalgo, J., A. Pares-Sierra, and J. Ochoa (2002), Seasonal variability of the temperature and heat fluxes in the Gulf of Mexico, *Atmosfera, 15*, 81–104.

Zavala-Hidalgo, J., S. L. Morey, and J. J. O'Brien (2003a), Cyclonic eddies northeast of the Campeche Bank from altimetry data, *J. Phys. Ocean., 33*, 623–629.

Zavala-Hidalgo, J., S. L. Morey, and J. J. O'Brien (2003b), Seasonal circulation on the western shelf of the Gulf of Mexico using a high-resolution numerical model, *J. Geophys. Res., 108*(C12), 3389, doi:10.1029/2003JC001879.

Steven L. Morey and James J. O'Brien, Center for Ocean – Atmospheric Prediction Studies, Florida State University, Tallahassee, FL 32306-2840 (morey@coaps.fsu.edu)

Jorge Zavala-Hidalgo, Centro de Ciencias de la Atmosfera, Universidad Nacional Autonoma de Mexico Cd. Universitaria, México DF, 04510 México

Low-Frequency Circulation Over the Texas–Louisiana Continental Shelf

Worth D. Nowlin Jr., Ann E. Jochens, Steven F. DiMarco,
Robert O. Reid, and Matthew K. Howard

Department of Oceanography, Texas A&M University, College Station, Texas

Low-frequency circulation over the Texas-Louisiana continental shelf is examined. Currents over the inner shelf are upcoast (Rio Grande to Mississippi River) in summer and downcoast in nonsummer and are driven by an annual cycle of winds. This results in an annual signal for salinity, with lowest salinity waters occurring (a) in late spring along the inner portion of the western shelf when downcoast flows carry the high discharges from the Mississippi-Atchafalya and other rivers to the Mexican border, and (b) in summer over the inner and outer eastern shelf when the upcoast flow causes a pooling of the discharges from the Mississippi-Atchafalya rivers over that shelf. Upcoast winds during summer also result in high salinities over the western shelf due to advection from off Mexico and upwelling. Currents over the outer shelf are variable, but predominantly upcoast throughout the year, probably a result of the integrated effects of anticyclonic eddies impinging on the shelf edge. Comparison of currents in the weather band (2–10 d) with the mesoscale band (10-100 d) suggests the shelf is divided at approximately the 50-m isobath. The weather band predominates over the inner shelf, reflecting frequent passage of fronts over the region. The mesoscale band predominates over the outer shelf, indicating the presence of offshelf eddies that frequent this region.

1. INTRODUCTION

1.1. Objectives

The relevant physical processes that drive the circulation of the Texas-Louisiana shelf have widely varying temporal and spatial scales. The basic description of the general circulation of this shelf has been the subject of several investigations [e.g., *Cochrane and Kelly*, 1986; *Cho et al.*, 1998; *Nowlin et al.*, 1998a, 1998b]. Our objective is to characterize the general circulation over the Texas–Louisiana

continental shelf on time scales from annual and seasonal to mesoscale and weather band periods (greater than two days) and quantify the relative strengths of the processes responsible for these scales. In addition, we also will characterize the spatial scales, particularly cross-shelf scales, of property variability associated with spatially distributed processes on the shelf, e.g., mesoscale eddy interaction with the shelf circulation.

1.2. Review of Key Previous Studies

Cochrane and Kelly [1986] suggested a quasi-annual pattern of low-frequency circulation over the Texas–Louisiana shelf. They considered mainly hydrographic data taken monthly for three years (January 1963 to December 1965)

Circulation in the Gulf of Mexico: Observations and Models
Geophysical Monograph Series 161
Copyright 2005 by the American Geophysical Union.
10.1029/161GM17

aboard the *M/V GUS III* inshore of the 110-m isobath [*Temple et al.*, 1977] supplemented by current measurements [e.g., *McGrail and Carnes,* 1983; *McGrail*, 1985; *Halper et al.*, 1988; *Brooks*, 1984]. They used monthly fields of geopotential anomaly of the sea surface relative to 70 db as indicators for circulation. Over the inner shelf, the annual cycle consisted of downcoast flow (defined here to be in the direction that a shelf wave would propagate, directed from the Mississippi River toward the Rio Grande River) except in July and August when upcoast flow dominated. This cycle was found to be coherent with the annual cycle of low-frequency, alongshelf winds, which are generally upcoast in summer and downcoast otherwise. At the shelf edge near the 200-m isobath, they pictured the flow generally to be toward the east or northeast (upcoast), but available data were sparse. That flow at the shelf edge was envisioned as the return flow required by continuity of a closed circulation over the shelf.

Wang [1996] confirmed the annual cycle of winds and its effects on inner shelf currents using objectively analyzed winds and observed currents from April 1992 through November 1994. For the same period, *Cho et al.* [1998] constructed monthly stream function fields based on current meter data and examined the quasi-annual circulation pattern using empirical orthogonal functions (EOFs) of the stream function fields.

Li et al. [1997] considered the mean and interannual variability of fields of temperature, salinity, and geopotential anomaly over the shelf using hydrographic data from 1961 through 1994. They examined the average conditions, variance, and selected anomalies for spring (May), summer (July/August), and fall (November). From patterns of geopotential anomaly of 0 db relative to 70 db, they inferred downcoast flow on the inner shelf except during the summer months when there was upcoast flow. They showed these flows to be associated with the alongshelf wind component, with enhancement due to buoyancy forcing by the Mississippi-Atchafalaya river discharge during the spring. The evidence for upcoast flow at the shelf edge was not so clear.

Oey [1995] conducted a modeling study of how the cyclonic gyre shelf circulation, proposed by *Cochrane and Kelly* [1986], was forced by wind, buoyancy from the Mississippi-Atchafalaya river discharge, and Loop Current eddies. He concluded the westward flow over the inner shelf was due primarily to wind, but with some contribution by river discharge and Loop Current eddies, and the flows elsewhere over the shelf were driven primarily by Loop Current eddies and modulated by the wind curl. His peak fall transports over the inner shelf were 0.21 to 0.25 Sv, with 40–48% being due to the wind, 28–33% due to river discharge buoyancy effects, and 19–32% due to Loop Current eddies.

1.3. Overview of the LATEX Field Program

The Texas–Louisiana Shelf Circulation and Transport Processes Study (LATEX A) was the largest of the three components of the Louisiana–Texas Shelf Physical Oceanography Program (LATEX) sponsored by the Minerals Management Service of the Department of the Interior. It covered the Texas–Louisiana continental shelf from the 10-m isobath to the continental slope between the Mississippi River and the Rio Grande River and consisted of five field components: moored ocean measurements, drifting buoys, hydrography including a conductivity-temperature-depth (CTD) instrument and expendable bathythermograph probes (XBTs), shipboard acoustic Doppler current profiling (ADCP), and meteorological measurements. A synthesis of the results of LATEX A is given in *Nowlin et al.* [1998a, 1998b].

The LATEX A field program was conducted from April 1992 to December 1994. Of the two major components to the field program, moored measurements and hydrography, only the moored measurements are considered here; the hydrography will be discussed in another paper. Thirty-six moorings with current meters, wave gauges, meteorological buoys, and inverted echo sounders provided a shelf wide network of current, temperature, and salinity time series (Figure 1). Moorings were deployed, maintained, or recovered on eighteen cruises during the field program. Two instruments, near top and bottom, were placed on all moorings; an additional mid-depth instrument was placed on moorings in water depths of ~50 m or more. Details regarding instrumentation, data returns, data processing, basic statistics and plots are provided in *Nowlin et al.* [1998a, 1998b] and the LATEX data and technical reports referenced therein.

The inner (outer) shelf is defined as that part of the LATEX shelf that is inshore (offshore) of about the 50-m isobath; the shelf edge is at ~200-m isobath. Summer is defined as June through August and nonsummer is September through May.

1.4. Contents of This Paper

The approach taken in this paper is to begin with long-term and seasonal patterns of circulation and then proceed to describe the characteristics of the circulation in the mesoscale and weather bands. We begin with a description of the mean circulation based on the LATEX results (Section 2). Then in Section 3 we describe the seasonal, quasi-annual pattern of circulation forced mainly by the wind regime. Mesoscale and weather band variability are presented in Section 4 along

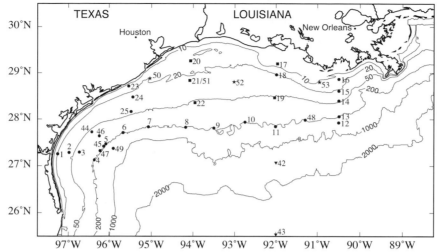

Figure 1. Locations for LATEX A current meter moorings and meteorological buoys (solid circle denotes a current mooring, star a meteorological buoy, square a current mooring with meteorological buoy attached, and triangle an inverted echo sounder).

with estimates of spatial and temporal scales of variability. Finally, we give a brief summary in Section 5.

2. MEAN CIRCULATION FROM LATEX OBSERVATIONS

2.1. Record-length 10-m Velocity Components

Horizontal currents, salinity, and temperature were observed from April 1992 through November 1994 at 31 mooring locations (Figure 1) on the Texas–Louisiana shelf as part of LATEX A. Records from 27 of the moorings had at least one year's worth of data points and were used in this study. Each time series record was processed first with a 3-hour low-pass, cosine-Lanczos filter (3-hr LP) to produce hourly data sets for all instrument types. A 40-hr LP filter was applied to each 3-hr LP record without gaps being filled. Record-length means were computed using the 40-hr LP records. Due to gaps in the records, the number of samples varies by location and by data type; the number of 6-hourly data points ranged from 461 to 3452 for different locations and depths.

Table 1 gives the record-length mean and standard deviation of alongshelf and cross-shelf velocity components. The standard deviations of current are high relative to the mean. To determine if the signs of the mean currents are meaningful, we determined the standard error of the estimate of the mean at the 95% confidence level by normalizing twice the standard deviation of the mean by the square root of the effective degrees of freedom of each time series. The effective degrees of freedom were determined using the

autocorrelation scales of the broadband signal. The mean current values in bold in Table 1 indicate that the standard error of the estimate of the mean is small enough that, at 95% confidence, the sign of the mean will not change.

The pattern of mean 10-m, alongshelf velocity component exhibits a two regime structure (Figure 2). The 0 cm·s⁻¹ isotach generally parallels the 50-m isobath. Inshore is a regime where the mean flow is downcoast; offshore the mean flow is upcoast. In spite of the high variability, the standard errors of the estimates of the means are such that qualitative direction is not in question for most locations (Table 1). On the inner shelf, the strongest downcoast flows of ~8–11 cm·s⁻¹ are located at the bend in the Texas coast. The mean alongshelf currents throughout the water column over the inner shelf generally follow the basic pattern of downcoast flow (see values along the 10-m, 20-m, and 50-m isobaths in Table 1).

These patterns of record-length flow are consistent with the findings of *Cochrane and Kelly* [1986]. The period of downcoast alongshelf winds, which force downcoast flow over the inner shelf, is much longer than the period of upcoast alongshelf winds (see Section 3.1). The former is approximately of nine month duration (nonsummer), while the latter is of three month duration (summer). Thus the temporal mean field shows downcoast flow over the inner shelf.

At the shelf edge, the mean flow is strongly upcoast at ~8–12 cm·s⁻¹ over the western half of the shelf (moorings 4–8 and 49) as compared to the eastern half at ~1–5 cm·s⁻¹ (moorings 9–13 and 48). The standard deviations of the upper alongshelf currents at shelf edge moorings for the western half are approximately twice those for the

Table 1. Record-length mean and standard deviation of along- and cross-shelf velocity components from 6-hourly, 40-hour low-pass filtered current meter data. None means there was no middle instrument. The bold indicates those values for which, at the 95% confidence level, the mean direction will not change. S = short record (< 1 year of data). Alongshelf (cross-shelf) components are positive upcoast (onshore). See Figure 1 for mooring locations and associated isobaths.

Mooring Isobath	No	(a) Top Current Meters Alongshelf velocity Mean (cm·s⁻¹)	Std. Dev. (cm·s⁻¹)	Cross-shelf velocity Mean (cm·s⁻¹)	Std. Dev. (cm·s⁻¹)	(b) Middle Current Meters Alongshelf velocity Mean (cm·s⁻¹)	Std. Dev. (cm·s⁻¹)	Cross-shelf velocity Mean (cm·s⁻¹)	Std. Dev. (cm·s⁻¹)	(c) Bottom Current Meters Alongshelf velocity Mean (cm·s⁻¹)	Std. Dev. (cm·s⁻¹)	Cross-shelf velocity Mean (cm·s⁻¹)	Std. Dev. (cm·s⁻¹)
10 m	16	**-3.3**	9.5	0.9	4.5	none	none	none	none	**-0.9**	4.8	0.2	5.1
	17	**-7.0**	15.3	-0.8	6.2	none	none	none	none	**-2.7**	7.3	**2.4**	6.7
	20	**-11.5**	21.0	1.7	10.6	none	none	none	none	-1.0	8.7	-0.1	5.5
	23	**-8.6**	21.4	0.0	3.6	none	none	none	none	**-4.5**	13.6	-0.5	6.9
	1	**-4.8**	17.9	0.2	3.7	none	none	none	none	-2.6 S	10.0 S	-1.4 S	4.1 S
20 m	15	-1.5	13.6	1.7	7.5	none	none	none	none	**-1.7**	6.0	0.3	3.1
	18	-0.9	14.9	0.1	5.2	none	none	none	none	0.0	6.7	0.3	4.9
	21	**-3.4**	11.0	-0.1	5.1	none	none	none	none	-0.6	7.2	**-0.7**	3.9
	24	**-9.4**	20.0	**-1.7**	6.5	none	none	none	none	**-2.8**	10.4	**-1.0**	5.5
	2	**-7.5**	19.8	-1.0	6.6	none	none	none	none	**-2.1**	12.5	**-1.7**	5.8
50 m	14	1.0	11.3	-0.4	8.2	**2.5**	10.2	**-1.8**	6.4	0.5	6.8	**-0.7**	3.7
	19	**-4.6**	15.8	**3.4**	11.0	-0.2 S	7.5 S	0.3 S	4.6 S	0.0	4.2	**-0.2**	2.1
	22	-2.6	15.0	**5.0**	9.9	0.7	8.1	**1.4**	5.3	**-1.3**	5.9	-0.4	3.2
	25	**-4.9**	12.8	-1.0	6.0	**-3.3**	13.0	-0.9	5.4	-1.9 S	12.8 S	-1.3 S	5.2 S
	3	-0.2	14.4	-0.2	7.9	-1.5 S	11.7 S	-0.3 S	5.1 S	**-2.3**	8.4	**-1.4**	3.8
200 m	13	3.1	15.3	0.3	8.3	0.3	12.8	0.5	3.2	**-3.4**	8.4	-0.2	2.6
	48	2.0	11.1	0.8	7.5	**-1.3**	7.4	0.2	3.3	0.5 S	3.7 S	0.4 S	2.2 S
	11	1.5	9.9	-0.4	6.9	0.7	9.9	**0.8**	4.1	**-2.0**	5.2	**-1.2**	2.1
	10	**4.3**	10.6	**1.8**	8.4	1.0	8.3	**1.1**	3.3	0.1	2.2	**0.1**	1.2
	9	**4.5**	12.1	-0.9	8.9	-0.1	7.1	**1.3**	3.5	**-1.0**	3.1	**0.1**	1.8
	8	**7.6**	14.6	-0.2	8.6	-1.0	12.4	-0.6	4.3	**-0.6**	2.7	**0.5**	1.3
	7	**12.6**	20.6	0.6	8.2	0.5	12.3	0.5	3.7	-0.3	7.7	0.0	1.4
	6	8.6	18.0	0.2	9.6	1.9	15.2	**0.9**	3.9	**-2.0**	6.6	**0.7**	1.4
	5	**8.2**	18.1	**2.5**	10.3	1.6	14.3	1.0	5.8	-0.9	6.3	0.3	1.6
	4	1.9	18.4	2.2	9.1	-1.7	13.4	0.7	5.7	-2.6	10.1	-0.1	2.7
500 m	12	1.5	18.7	-0.6	10.9	-2.4 S	13.8 S	2.0 S	5.0 S	-1.4 S	2.2 S	-0.5 S	0.7 S
	49	**9.3**	23.9	2.5	12.2	5.5	17.7	1.0	8.1	**-2.2**	3.3	-1.7	3.4

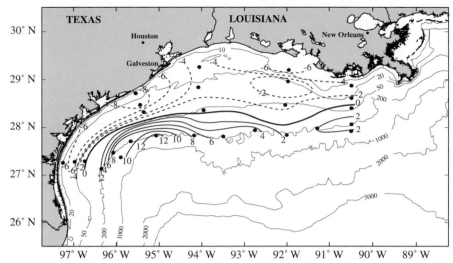

Figure 2. Record-length means of alongshelf 40-hour, low-pass, 10-m velocity components (cm·s^{-1}). Solid circles show instrument locations. Depth contours are in meters. Alongshelf currents are positive upcoast (Rio Grande to Mississippi River).

eastern half, suggesting that the greatest variability also is over the western half. Over the outer shelf (along the 200-m isobath), upcoast flow dominates in the upper 100 m. Dominant eastward directions for the average alongshelf currents at 100 m, however, are not as clear as those at 10 m. Mean flows at mid-depth are greatest at the western shelf edge. Below 100 m the mean current is generally downcoast and the variability is much smaller than at 10 m.

We attribute the larger mean currents at the western shelf edge to Loop Current eddies, with their significant current speeds, shown by our examination of the sea surface height anomaly maps to be adjacent to the western shelf edge for a significant portion of the measurement program [*Nowlin et al.*, 1998a, 1998b; discussed in Section 3.3 of this paper].

The variability of cross-shelf velocity component is seen to be high relative to the mean (Table 1). Fewer than half of the records gave significant directions for the cross-shelf component; no clear patterns are evident in these data.

The mean salinity and temperature fields (not shown) support the organization of the shelf properties into inner and outer shelf regimes. The mean isohalines and isothermals at 10-m depth approximately parallel the isobaths, with salinity increasing offshore and the thermal gradient displaying warming from inshore to offshore. In particular, the 34 isohaline approximately follows the 50-m isobath. It effectively provides a dividing line between a strong salinity gradient structure over the shelf inshore of the 50-m isobath and a weak gradient structure over the outer shelf. Thus this isohaline at 10-m depth also marks the line between the two regimes, divided at approximately the

50-m isobath, seen in the mean 10-m, alongshelf velocity component.

Over the inner shelf, the mean salinity increases with depth. This basically reflects the presence at the surface of low salinity water derived from the downcoast flow over the inner shelf of waters mixed with the discharge from the Mississippi and Atchafalaya rivers. At the shelf edge, however, a salinity maximum occurs at the 100-m instruments, as compared to the 10-m or near bottom instruments. The source of this maximum is from the Subtropical Underwater, which is brought into the Gulf by the Loop Current and into the western Gulf by the Loop Current eddies that propagate into the northwest Gulf.

2.2. Mean 10-m Velocity Stream Function

Cho et al. [1998] deduced the near-surface (~10 m), low-frequency circulation and its variability on the Texas–Louisiana continental shelf based on the discretely sampled LATEX A current data. They chose a stream function representation, thus constraining the estimated flow to be horizontally nondivergent. Stream function fields were prepared for the upper current measurement levels using monthly averages for each of the 32 months during which data were collected.

The shelf-wide field of mean 10-m velocity stream function is shown in Figure 3. They showed that the ratio of the rms residual error to the rms speed was 0.38 (or 0.14 in relative error variance), indicating that the record length mean flow field is represented reasonably well by the stream function equation used. Moreover, this mean field cor-

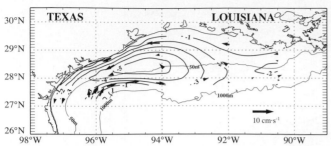

Figure 3. Shelf-wide mean 10-m velocity stream function field using record-length mean of LATEX observations. Arrows represent corresponding mean current vectors from upper meters. The unit of stream function is 10^7 cm^2·s^{-1} [after *Cho et al.*, 1998].

responds well in terms of implied geopotential anomaly range to the mean geopotential anomaly range obtained by *Li et al.* [1997] using station data from all of the LATEX A hydrographic cruises.

This mean stream function field is consistent with the record-length, 10-m, alongshelf flow representation of Figure 2. Both representations as well as the mean geopotential anomaly pattern of *Li et al.* [1997] show downcoast flow over the inner shelf and upcoast flow over the outer shelf with minimum speeds near the 50-m isobath. Maximum speeds over the inner shelf are seen near the bend in the Texas coast. Also in Figure 3 we see a pattern for the cross-shelf flow that is consistent with the upper meters at moorings 19 and 22 (Table 1).

In Section 3 we investigate the low-frequency variability of this quasi-annual pattern of currents.

3. QUASI-ANNUAL, SEASONAL VARIABILITY

3.1. Quasi-Periodic, Near-Annual Circulation Forced by Winds

A dense network of wind velocity measurement sites was in place over the northwestern Gulf of Mexico during the LATEX A measurement field period. For this period *Wang et al.* [1998] derived hourly fields of 10-m level wind velocity components using optimal statistical interpolation. Monthly averages of alongshelf components of wind stress from six locations nearly equally spaced along the 20-m isobath between 90.5°W and 26°N are shown in Figure 4. This time sequence of alongshelf wind forcing shows downcoast forcing during the nonsummer regimes and upcoast forcing during the summer regimes. The strong quasi-periodic, near-annual cycle is quite clear and characteristic of wind forcing over this inner shelf. The pattern observed during LATEX is consistent with long-term means shown by *Cochrane and Kelly* [1986] and *Nowlin et al.* [1998a]. This sequence is very well related to the along-coast component of low-frequency flow over the inner shelf as we shall see.

Cho et al. [1998] produced fields of the mean 10-m velocity stream functions for nonsummer and for summer. These are shown as Figure 5 upper and lower, respectively. Clearly the currents are stronger in nonsummer when wind-forced flow over the inner shelf is downcoast. For nonsummer there is some upcoast flow over the outer shelf, though it is much weaker than the coastal current. For summer the flow is upcoast across the shelf. Combining these patterns

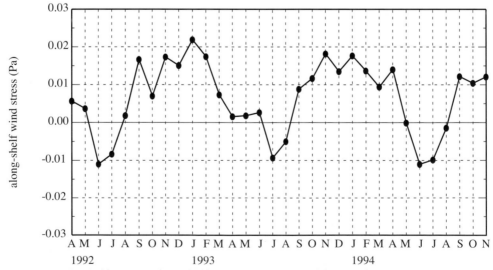

Figure 4. Sequence of monthly mean, along-shelf wind stress characterizing the Texas–Louisiana inner shelf for the period April 1992 through November 1994 (downcoast is positive). Values were calculated from hourly analyzed wind components produced by *Wang* [1996] at six, nearly equally spaced, points along the 20-m isobath from 90.5°W to 26°N, averaged for the six points, then averaged over each month [after *Cho et al.*, 1998].

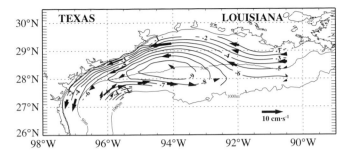

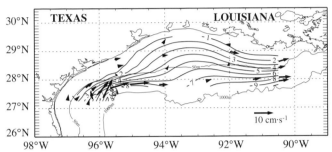

Figure 5. Shelf-wide mean 10-m velocity stream function field using (top) nonsummer mean and (bottom) summer mean of LATEX observations. Arrows represent corresponding mean current vectors from upper meters. The unit of stream function is 10^7 cm^2·s^{-1} [after *Cho et al.*, 1998].

results in the record-length mean stream line pattern shown in Figure 3. Of course the mean more closely resembles the nonsummer pattern because then currents are much stronger than in summer.

To make the relationships of these patterns of current and the time series of alongshelf wind more quantitative, *Cho et al.* [1998] examined the first two EOF modes calculated from the 32 monthly mean records of LATEX A currents at 10 m. The EOF patterns and amplitude time series are shown in Figures 6 and 7, respectively. The amplitude time series of the first EOF mode compares very favorably with the time series of nearshore alongshelf wind shown in Figure 4. *Cho et al.* [1998] calculated the squared correlation between the two series as 0.83 and the variance of the first EOF mode from the long-term mean currents as 0.89. Thus the total variance of the monthly circulation that is wind driven is nearly 74%. This is clearly much larger than the estimate (40–48%) by *Oey* [1995] from his model study (Section 1.2).

EOF mode 2 accounts for only 3% of the variance. Its amplitude time series bears no resemblance to the alongshelf wind forcing. The pattern for this second mode looks very much as though it represents circulation forced by eddies over the continental slope as discussed in section 3.3. For more detail of the low-frequency variability of currents during the LATEX period, velocity stream function patterns based on the monthly mean currents are given in *Cho et al.* [1998].

3.2. Annual Variability Based on Fitting Time Series to Annual Signal

For record-length means and autocorrelations, the methodology applied to the LATEX A time series was described in section 2.1. For other analyses, gaps of two weeks or less in the 3-hr LP time series were filled using the maximum entropy auto-regressive method [*Anderson*, 1974; *Smylie et al.*, 1973; *Ulrych et al.*, 1973] and then filtered with the 40-hr LP filter. The 40-hr LP filter was applied to remove tidal and inertial signals. [For details see *DiMarco et al.*, 1997.]

For analysis of the annual signals, a 15-day low-pass filter was applied to the 40-hr LP records to reduce their high frequency variability. The amplitudes of the annual signals of alongshelf current, cross-shelf current, salinity, and temperature then were determined for each time series. We define the annual signal as the fundamental plus its first two harmonics: 365.25 d, 182.625 d, and 121.75 d. Only two harmonics were used so that annual and mesoscale band signals (see section 4.1) would be better separated. We used the method of cyclic descent to remove the mean and fit the specified frequencies iteratively by least squares [*Bloomfield*, 1976; *DiMarco*, 1998].

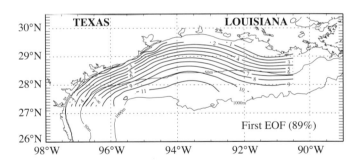

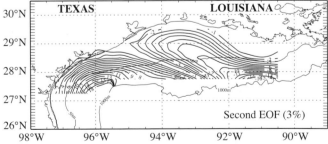

Figure 6. Structure of the (top) first and (bottom) second empirical orthogonal functions (EOFs) calculated from the 32 monthly mean, 10-m velocity stream function fields. The percent of variance accounted for by each mode is indicated. Contours for the first EOF and the feature over the eastern shelf in the second EOF have negative values; contours for the feature over the western shelf in the second EOF have positive values [after *Cho et al.*, 1998].

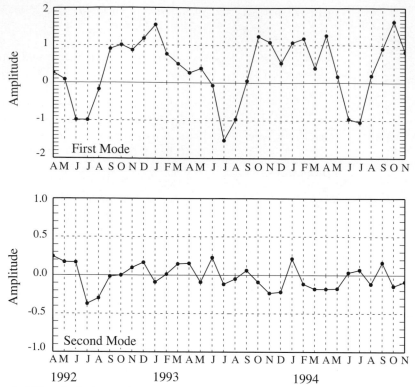

Figure 7. Amplitude (10^7 cm$^2 \cdot$s^{-1}) time series over 32 months for the (top) first and (bottom) second EOFs based on the monthly mean, near-surface velocity stream functions [after *Cho et al.*, 1998].

The percent of variance for six EOF modes for the annual signals of cross-shelf and alongshelf 10-m currents, salinity and temperature are given in Table 2. The major part (> 60%) of the variance for the current components and temperature is given by the first mode with substantially less variance (< 20%) in higher modes, while for salinity the first two modes both have substantial variance (57% and 31%) and together account for 88% of the variance.

Table 2. Percent of variance of the six EOF modes for the annual signal of along- and cross-shelf velocity components, salinity, and temperature. The major part of the variance for each component is within the modes shown above the line in the table.

Mode	Alongshelf component (%)	Cross-shelf component (%)	Salinity (%)	Temperature (%)
1	66.6	61.8	56.6	93.4
2	13.3	17.0	31.2	5.5
3	8.2	11.3	1.6	0.6
4	6.9	4.4	2.9	0.4
5	3.5	3.5	3.6	0.1
6	1.5	2.0	4.1	0.1

The amplitude time series for the first two modes of annual signal of salinity are shown in Figure 8, and the pattern for the salinity anomaly field constructed from combined EOF modes 1 and 2 is shown in Figure 9.

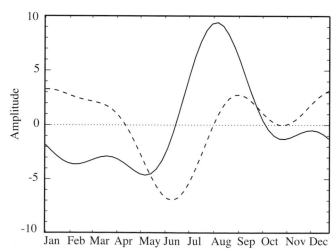

Figure 8. Time series of amplitudes of the EOF modes 1 (solid) and 2 (dashed) for the annual signal of salinity. Mode 1 contains 56.6% of the variance and mode 2 contains 31.2%.

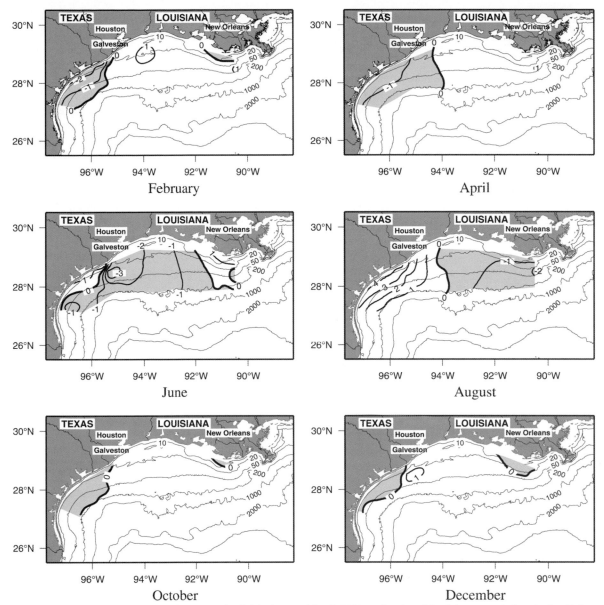

Figure 9. Monthly average salinity anomaly fields for combined EOF modes 1 and 2 of the annual signal of salinity. Positive values denote saltier water, relative to the mean; negative values denote fresher water. Light shading denotes low salinity water.

The first modal pattern (not shown) divides the shelf into two regimes at approximately 94°W, with the eastern pattern being increasingly negative to the east and the western pattern being increasingly positive to the west. Together with the amplitude time series (Figure 8, solid line), the temporal phasing of the simultaneous freshening over the eastern shelf and increasing salinity over the western shelf in summer leads us to associate the first mode of salinity variability with the seasonal cycle of alongshelf currents driven by the annual cycle of the winds.

The second EOF mode for the annual signal of salinity is closely associated with the annual variability of river discharge and the movement of low salinity water across the shelf to the shelf edge. The spatial pattern of mode 2 (not shown) shows positive EOF values over the shelf, except inshore of the 20-m isobath, with highest absolute values generally occurring west of ~92°W and seaward of the 20-m isobath. This pattern suggests that the variability of salinity is smallest in those regions immediately adjacent to rivers and that the peak river discharge in spring does not significantly alter the already low salinity water close to the coast. The temporal amplitudes (Figure 8, dashed line) of this mode decrease starting in April and are lowest in mid-June. The temporal and spatial patterns together suggest that a freshening occurs over most of the western shelf approximately one month after the April–May peak discharge from the Mississippi-Atchafalaya River system. The maxima in the amplitude of mode 2 from August to April are about half as large as, and of opposite sign to, the minimum of the spring–summer freshening trend. Therefore we associate the variability in salinity of this mode over the shelf with the annual cycle of river discharge.

The effects may be seen in the pattern for the combined modes (Figure 9). The western shelf freshens, relative to the mean, from October through May, when the alongshelf flows are downcoast, carrying low salinity water from the Mississippi-Atchafalaya rivers to the west. This freshening increases from fall to spring, with the lowest salinity water occurring during April–May when the river discharge peaks. The western shelf then experiences an increase in salinity, which starts in May and peaks during summer, when the upcoast currents move saltier water into the region from the southwest. In contrast, the eastern shelf experiences the freshening during summer when the wind-driven upcoast currents cause the river discharge to pool over the eastern shelf. From October to May, the salinity on the eastern shelf increases relative to the summer, which is consistent with the transport of the river discharge out of the region by downcoast, alongshelf currents in nonsummer.

Whitaker [1971] found that two highs in coastal water level at Galveston and at many other locations around the Gulf occurred during the year. He only could account for the high in September that was related to the thermal effect. It is evident from the LATEX studies that the secondary high he observed in May at Galveston is due to the effect of low salinity water. Results of the *GUS III* cruises support this deduction. Clearly the annual river discharge variability with the help of wind-driven coastal currents are the drivers of this spring minimum in salinity and maximum in sea level.

3.3. Major Circulation Patterns Forced by Off-Shelf Current Eddies

The spatial pattern for EOF mode 2 of the low frequency circulation over the LATEX shelf from *Cho et al.* [1998] (Figure 8) looks very much like circulation forced by eddies located over the continental slope and impinging on the shelf. Such eddies occur frequently. The depths of the isothermal surfaces at or near the 20°C isotherm have been shown to be excellent indicators of the presence, configuration, and strength of eddies in the Gulf of Mexico by *Leipper* [1970]

and others. Based on temperature data from LATEX A, the LATEX Eddy Circulation Study (LATEX C), and the Gulf of Mexico Cetacean Study (GulfCet), we constructed distributions of the depth of the 20°C isotherm for twelve periods during 1992–1994. LATEX A data used were CTD temperature; LATEX C data were AXBT profiles of temperature versus depth; and GulfCet data used were a combination of CTD and XBT temperature profiles. Vectors representing 10-m current velocities at LATEX A moorings averaged over the sampling periods plus the two preceding weeks were added to these fields. Examples for three periods (August 1992, July–August 1993, and July–August 1994) are shown in Figure 10 [see *Nowlin et al.* 1998b for all 12 periods].

In all but three of the 12 periods an anticyclone was present off the western region of the shelf, and the measured currents give evidence that at least the flow over the outer shelf was influenced by their presence. The pattern of circulation off the eastern shelf region was not so uniform, although evidence for the presence of an anticyclone is seen in the 20°C isotherm depth distributions and the currents at the shelf edge during 7 of the 12 periods. In addition to large anticyclones derived from the Loop Current, numerous smaller cyclonic and anticyclonic features are seen over the continental slope which may influence the outer shelf circulation [e.g., see *Hamilton*, 1992; *Hamilton et al.*, 2002]. Seldom are the currents seen to have the same alongshelf direction along the entire LATEX shelf edge. It is clear that the currents at the shelf edge are dominated by the offshore eddy field.

As a quantitative measure of the effects of off-shelf eddies on currents over the outer shelf, we computed correlations between currents and gradients of sea surface height (SSH) based on their geostrophic relationship. We correlated the satellite-derived approximately cross-isobath components of SSH gradients [*Leben et al.*, 2002] with approximately along-isobath components of measured velocity taken from the middle instrument at nominal depths of 100 m on each mooring, positioned between the surface and bottom boundary layers. Average temperatures at that depth are near 20°C. We correlated east-west SSH gradients with north-south velocity components at moorings 4 and 5 and north-south SSH gradients with east-west velocity components at moorings 6 through 12. The current records were smoothed with a 5-day running mean and then sampled daily. The SSH gradient was derived from daily fields of SSH, provided on a 0.25 degree grid, as the average gradient of the four SSH values points surrounding each mooring. The correlations for moorings 4, 5, 6, 9, 10 and 11 are statistically significant at the 95% level; that for mooring 11 is significant with 99% confidence. Results are given in Table 3.

The SSH gradients (velocity components) are positive when the SSH (component) is increasing (directed) to the north or east. Thus the positive correlations at moorings 4 and 5, along the generally north-south isobaths, indicate that northward (southward) velocity components are significantly

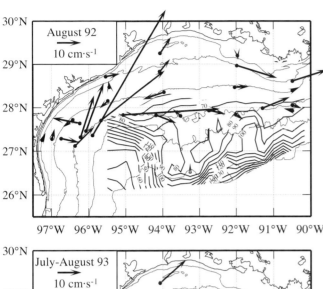

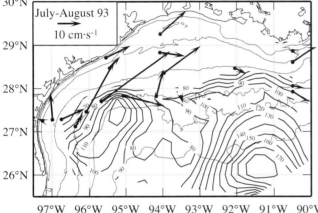

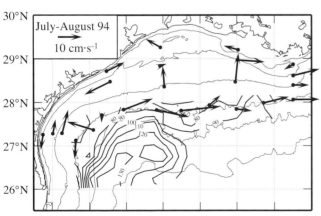

Figure 10. Depth of 20°C isotherm (m) for three summers in 1992–1994 based on temperature data from LATEX A, LATEX C, and GulfCet. Vectors represent 10-m current velocities at LATEX A moorings averaged over the sampling periods plus the preceding 14 days.

Table 3. Correlations between north-south (east-west) components of SSH gradient and east-west (north-south) velocity components as measured at the shelf edge during LATEX A.

Mooring	C	N	Edof	95%	99%
	East-west SSH gradient versus north-south velocity component				
04	0.343	601	41	.301	.389
05	0.437	492	19	.433	.549
	North-south SSH gradient versus east-west velocity component				
06	-0.411	565	29	.355	.456
07	-0.404	458	15	.482	.606
08	-0.129	678	74	.226	.294
09	-0.447	635	21	.413	.526
10	-0.332	678	43	.294	.380
11	-0.350	510	52	.268	.348
12	-0.443	153	6	.707	.834

C is the correlation coefficient
N is the number of samples
Edof is the effective degrees of freedom [*Emery and Thomson*, 1997]
95%: Correlation is statistically different from 0 at the 95% confidence level when C exceeds this number.
99%: Correlation is statistically different from 0 at the 99% confidence level when C exceeds this number.

correlated with westward limbs of anticyclones (cyclones), as expected. Likewise, the negative correlations along the generally east-west isobaths, where moorings 6-12 were located, are of the expected sign. For example, eastward velocity components should be negatively correlated with the northern limbs of cyclones. Clearly there is good quantitative evidence for forcing of currents over the outer LATEX shelf by eddies located at the shelf edge.

4. MESOSCALE AND WEATHER-BAND VARIABILITY; KINETIC ENERGY INVENTORY

4.1. Mesoscale and Weather-Band Energy Levels

In this section, we compare and discuss distributions of kinetic energy over the Texas–Louisiana shelf for weather band (2- to 10-d) and mesoscale (10- to 100-day) periods. We selected 0.5 cpd (two days) as the high frequency limit on the weather band because this is approximately the half amplitude frequency of the 40-hr LP filter used to process the data. The low frequency limit was set by determining the frequency of passage of cold fronts over the shelf during LATEX A [*Nowlin et al.*, 1998a]. Ignoring the summer situ-

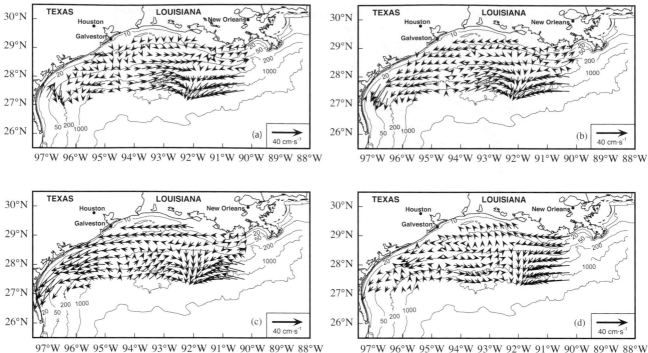

Figure 13. Gridded currents for 4 November 1992 at (a) 0000 UTC, (b) 0600 UTC, and (c) 1800 UTC and for 5 November 1993 at (d) 1200 UTC.

[*Wang et al.*, 1998]. During frontal passage, currents over the inner shelf turned downcoast with speeds initially of ~3–10 cm·s⁻¹ and increasing to ~10–30 cm·s⁻¹ (Figure 13 b and c). Currents over the outer shelf to the east of 94°W continued to be dominated by the off-shelf eddies, while currents to the west intensified and became directed downcoast. By 0600 UTC on 5 November 1992, the front had passed over the shelf and was east of the LATEX region over deeper Gulf waters. Winds were directed southward over the LATEX shelf but with diminishing speeds throughout 5 November. Currents were directed downcoast and also had diminishing speeds (Figure 13d). *Wang* [1996] estimated the response time of the currents to the winds of this and similar atmospheric events to be 4–6 hr over the shelf off the south Texas coast and ~10 hr over the LATEX inner shelf region off the Mississippi River Delta.

The two cold air outbreak examples discussed here reflect typical patterns of the currents, air-sea exchanges, and property changes associated with cold front passages. The frontal passage often begins in the northwest part of the LATEX shelf and moves southeastward. Thus the waters in the eastern LATEX area may experience the effects of the event later than waters in the western LATEX area. During these events, the water column becomes well mixed. Currents over the inner shelf intensify in the downcoast direction

in response to the downcoast along-shelf component of the wind. The front even can turn upcoast flow into downcoast flow within hours. On-shelf flow can occur over the shelf break or at mid-shelf to provide replacement water for the waters carried downcoast over the inner shelf. Note for example in Figure 13c the on-shelf component of flow at the 50-m isobath over the central part of the LATEX shelf (92°W to 94°W). The outer shelf circulation generally remains under the influence of the off-shelf mesoscale features and currents of the open ocean, although the currents off the south Texas outer shelf do respond to the winds of these events. The greatest effects in water property changes occur over the inner shelf in water depths < ~50 m. Sea temperatures drop due to the air-sea exchanges induced by the event. Salinity increases over the inner shelf are probably due to the on-shelf flow of more saline waters from the open Gulf and the outer shelf. This flow occurs to replace waters that are moved rapidly downcoast.

Winter cyclogenesis can result in intense storms. Approximately 10 of these storms per year occur in the northern Gulf [*Saucier*, 1949; *Johnson et al.*, 1984]. One occurred during the period 12–13 March 1993, when a combination of upper-level dynamics and an existing baroclinic zone in the western Gulf of Mexico produced an intense extratropical cyclone. This system was comparable in strength to a category 1 hurricane, and was dubbed by many as the

"Storm of the Century." However, over a 30-year period from winter 1965/66 to winter 1995/96 there were 7 cyclogenesis events classed as intensity category 4 storms, including this one, under the classification system of *Hsu* [1993], so the March 1993 superstorm was not a "100-year storm" [see *Nowlin et al.*, 1998a]. Along the storm's path (Figure 14), winds reached 40 m·s⁻¹, generating seas more than 7 m high and producing storm surges of more than 3 m along parts of the northeastern Gulf coast [see *Schumann et al.*, 1995]. Expanded descriptions of the meteorological extrema and accounts of the devastation appear in *Lott* [1993] and *Kocin et al.* [1995]. Meteorological descriptions and the effectiveness of weather forecasting of this storm are discussed in *Uccellini et al.* [1995] and *Caplan* [1995]. *Wang et al.* [1998] presented meteorological and oceanographic fields for this superstorm.

In the early hours of 12 March 1993, the soon-to-be "Storm of the Century" was a disorganized area of low pressure located over northern Mexico just south of the Texas border (Figure 14). Within 12 hours, this low-pressure feature moved into the coastal waters off south Texas, where it consolidated and deepened. Wind speeds at NDBC buoy 42020 increased from about 8 to 18 m·s⁻¹ as the wind direction changed from east to northwest. Wind speeds there exceeded 10 m·s⁻¹ for the next 36 hours. Over the second 12-hr period, central pressures in the cyclone dropped dramatically (17 mb in 12 hours) as the cyclone moved east-northeast. By 0000 UTC 13 March wind speeds at the Bullwinkle oil platform near Southwest Pass, Louisiana (BURL1), had exceeded 25

m·s⁻¹, and would continue to do so for the next 12 hours. The storm made landfall near Pensacola, Florida, around 0900 UTC 13 March and continued to deepen for the next 15 hours or so, although at a much slower rate, as it continued northward along the eastern U.S. seaboard.

Figure 15 comprises four snapshots, taken before, during, and after passage of the superstorm, which show gridded current vectors over the Texas–Louisiana shelf on a 0.25° by 0.25° grid. The current fields are based on instantaneous current meter observations at the indicated times and were gridded using the adjustable tension continuous curvature surface gridding algorithm of *Smith and Wessel* [1990]. Twenty-four hours before the storm entered the Gulf, winds were relatively light from the east or southeast. The current patterns do not appear to be responding directly to the wind field (Figure 15a). The strongest currents were in the southwest region of the shelf and were associated with Eddy Vázquez, which was located just off the shelf.

Twenty-four hours later at 1200 UTC 12 March, the currents had responded to the strong downcoast wind component that had developed along the entire coast from the Mississippi delta to south Texas [see *Wang et al.*, 1998, for all wind fields]. Organized currents over the inner shelf region were flowing west and south along the coast (Figure 15b). East of 96°W, easterly winds seaward of the shelf break (~200-m isobath) and northeasterly winds over the shelf, combined with the requirement to replace water transported downcoast in the highly divergent nearshore regime from 96°W to the Mississippi delta, produced strong onshore

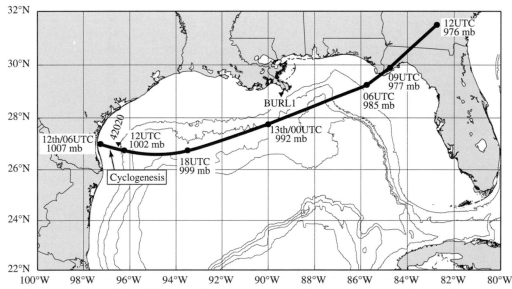

Figure 14. Track across the Gulf of Mexico and central pressures of the surface low associated with the superstorm of 12-14 March 1993 [track and pressure data from *Schumann et al.*, 1995]. Bathymetric contours shown are 200, 1000, 2000, and 3000 m. Locations of meteorological buoy stations, 42020 and BURL1, are shown (inverted triangles).

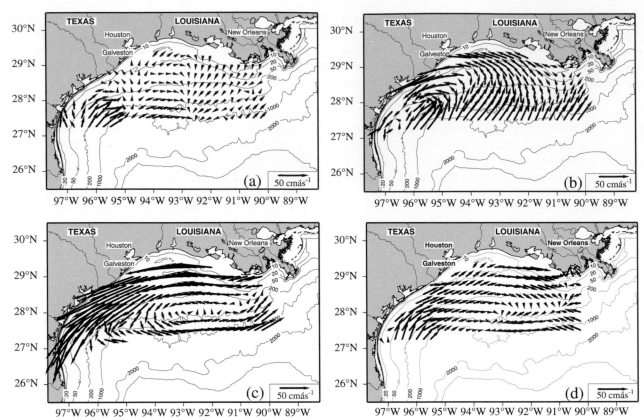

Figure 15. Gridded currents for (a) 11 March 1993 1200 UTC, (b) 12 March 1993 1200 UTC, (c) 12 March 1993 1800 UTC, and (d) 14 March 1993 1000 UTC, showing conditions during the passage of the "Storm of the Century".

flow over the eastern half of the Texas–Louisiana shelf. The influence of Eddy Vázquez on the circulation diminished as the coastal jet began to overwhelm the northern arm of the eddy. The water flow over the shelf was cyclonic with the center of the cyclone occurring near the 50-m isobath in the southwest region of the shelf just north and west of the storm center.

During 1400–2200 UTC 12 March (e.g., Figure 15c) central pressures in the cyclone deepened rapidly (approximately 1.5 mb·h⁻¹), and wind speeds continued to increase. The center of the oceanic cyclone remained between the 50- and 200-m isobaths and followed the storm as it tracked eastward. Although the coastal flow continued to intensify and transport water south, a strong northward transport continued across a broad region (96°W to 91°W) at the shelf break. This net transport of water onto the shelf compensated for the downcoast export of water along the inner shelf.

As the storm center moved east of New Orleans, wind directions over the Texas–Louisiana shelf shifted from northeast to northwest. Downcoast currents over the inner shelf intensified in the eastern part of the region. The eastern limit of the strong southward arm of the shelf circulation migrated

eastward with the storm and by 2200 UTC surface water was being exported from the shelf everywhere along the shelf break except in a relatively narrow region just west of the Mississippi delta. Currents measured along the south Texas coast reached their peak values of 80–120 cm·s⁻¹ between 1600–1730 UTC.

An hour after the storm center made landfall, strong winds still persisted over the Texas–Louisiana shelf and were nearly uniform from the northwest. Downcoast of Galveston, coastal winds had a strong downcoast component and currents there were also directed strongly downcoast. Currents over the inner shelf from 91° to 93°W were upcoast (eastward) in response to the upcoast wind component. East of 91°W the currents remained downcoast, presumably not having yet had time to respond to the wind shift. Between 93° and 95°W the flow across the shelf was toward the coast, probably in response to the large scale, alongshelf divergence of flow over the inner shelf. In the southwest near the shelf break, Eddy Vázquez appears to have been reasserting its influence on the flow.

By 1000 UTC 14 March (Figure 15d), when the storm was centered near Maine, wind speeds over the Texas–Louisiana shelf were greatly diminished, with variable direction. Rather than a weak cyclone-anticyclone pair with a zone of shoreward flow between them, the current pattern shown is a broad anticyclonic flow similar to the mean summer pattern. This remarkably different flow pattern is consonant with upcoast flow regularly seen in summer in response to upcoast wind components.

Examination of measured currents and output from a shelf circulation model [*Nowlin et al.*, 1998b] revealed that the passage of the superstorm through the region triggered a shelf wave with a dominant mode traveling about 5 m·s⁻¹ downcoast. Clearly, the forcing and oceanic response to this superstorm were within the weather-band.

Wang et al. [1998] determined the sensible heat flux associated with this superstorm. These fluxes were 0–80 W·m⁻², with lowest values farthest offshore, at 1200 UTC on 12 March 1993 when the storm center was located at the coast near 17°N (Figure 14). Fluxes increased as the storm passed over the shelf. They intensified to over 240 W·m⁻² on the western shelf by 0000 UTC on 13 March 1993 after the storm had passed off the LATEX shelf. *Wang et al.* [1998] noted that the sensible heat fluxes associated with the superstorm were considerably larger than those of the cold air outbreak, as is consistent with the differences in intensity of the causal storms.

Figure 16 shows the water level (mean sea level removed) for stations at Sabine Pass and Galveston Pleasure Pier for the times associated with both the November 1992 cold air outbreak and the March 1993 superstorm. For the March 1993 superstorm (Figure 16c and 16d), these records show an increase in water level which peaked about the time the storm entered the Gulf. The increase appears to be part of the normal tidal cycle except the magnitude appears larger than normal. This enhanced amplitude is probably associated with the initially increasing strength of the downcoast coastal current (e.g., Figure 15b). From 1200 UTC 12 March to about the time of the storm's landfall, water levels at Galveston and Sabine Pass drop to extremely low levels. It appears water was moved away from the coast under the action of winds after the storm center passed to the east of these locations. Following landfall of the storm, wind speeds in the Gulf began to decrease and simultaneously, water levels began to rise, returning to pre-storm pattern in approximately 3–4 d.

In contrast, the water levels did not respond as dramatically to the cold air outbreak in November 1992 (Figure 16a and b). The pattern of response, however, was similar. First there was a drop in water level associated with the time of frontal passage. Note, however, that the tidal signal is still prevalent in the record during the outbreak, whereas it is not during the March 1993 superstorm. Second, after frontal

passage, the water levels began to rise and relaxed back to pre-storm levels after approximately 3 d.

It is important to note from these water level and current time series that the oceanic response over the shelf to these types of energetic atmospheric storms is rapid and shelf-wide. The circulation patterns established by the storms do not persist after the storms leave the Gulf, and the water level and circulation patterns quickly return to normal patterns within a few days [for a discussion of the similar response of the shelf to passage of hurricanes see *Nowlin et al.*, 1998b, and *Keen and Glenn,* 1999].

Mesoscale-Band Illustration: Loop Current Eddy Vázquez (Eddy V): The influence of mesoscale band phenomena on currents and water properties of the Texas–Louisiana shelf is well illustrated by the case of one Loop Current Eddy (LCE), Eddy Vázquez (Eddy V). This eddy had the greatest impact on Texas–Louisiana shelf circulation of the five LCEs observed during the LATEX field program [*Nowlin et al.,* 1998b]. Hydrographic, current meter, and drifter data, sea surface height fields from satellite altimeter data, and sea surface temperature fields from satellite AVHRR data were used to study the evolution of this eddy.

Eddy V was an identifiable anticyclonic feature from September 1992 through September 1993. The eddy moved into the northwest corner of the open Gulf in November 1992 and remained there through March 1993 (e.g., Figure 17). In April, the eddy split into north (Eddy Vₙ) and south (Eddy Vₛ) parts. During April and May, Eddy Vₙ translated westward and encroached on the shelf edge. A cyclonic eddy developed to the west of Eddy Vₙ. This cyclone-anticyclone pair had a significant influence on the shelf circulation and property distributions as Eddy Vₙ was dissipating throughout spring and summer 1993. Eddy V and its northern remnant persisted for approximately 10 months in the northwest Gulf.

The geopotential anomaly field of 3 db relative to 400 db, based on data collected from 26 April–18 May 1993, shows both Eddy Vₙ and the associated cyclone to its west (Figure 18). The 10-day average current vectors for the period 29 April through 9 May 1993 are superimposed on the geopotential anomaly field. They show eastward currents up to 50 cm·s⁻¹ associated with the northern periphery of Eddy Vₙ and southward alongshelf flow of about 15 cm·s⁻¹ associated with the western cyclone. An onshelf current jet was between the anticyclone and cyclone.

The influence of the cyclone-anticyclone pair on water properties is evident in temperature and salinity fields (not shown) from April/May 1993. Warm waters of Eddy Vₙ were moved onto the shelf at about 95.5°W by the northwestward flowing jet between Eddy Vₙ and the western cyclone. These warm waters then bifurcated to flow to the east along the

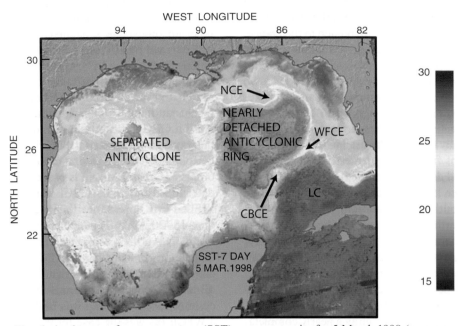

Plate 3. A satellite-derived sea surface temperature (SST) map composite for 5 March 1998 (an average over 7 days, nominal). SST values (in ° C) are color-coded as indicated by the bar to the right.

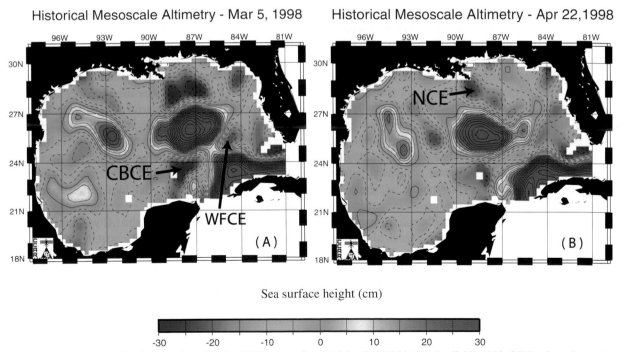

Plate 4. Satellite-derived sea surface height (SSH) maps for: (A) March 5, 1998, (B) April 22, 1998. SSH values (in cm) are color-coded as indicated by the bar below the maps.

the entire LC boundary, forming a dense cyclonic shield in February 1999. The tip of the LC was then held near 27°, east of 89°W, by the blocking cyclones in this shield until May 1999. During this interval small anticyclones were shed from both sides of the LC. The tip of the LC began trying to push to the west between holes in the cyclonic shield around the periphery of the LC and finally broke through in a substantial way in June and July 1999. According to *Leben* [this volume] there was a separation on 2 October 1999. So, in the time interval between the ring separation in March 1998 and the following ring separation in October 1999, penetration blocking twice delayed the progress of the penetration of the LC into the GOM. One major blocking sequence occurred in 1998 when the tip of the LC was in a (comparatively) southerly location; other penetration blocking events of shorter duration took place at ~ 27°N in 1999. Together these situations, along with neck-filling and weak reattachments, added ~ 14 months to the relevant separation interval published by *Leben* [this volume] of ~ 18.5 months. A more detailed consideration of penetration blocking and delays in eddy shedding is contained in section 5 of the appendix.

Plates 3 and 4 are clearly at variance with the claim by *VM85*, as noted in the Introduction, of the dominance of the large eddy detachment process by WFCEs only (relative to CBCEs). Both cyclone types, perhaps in non-amplified or LCFE form for one of them, are typically present at some time during the eddy detachment configurations in Plates 2, 3, and 4. This is also the case in most of the existing maps involving the detachment process in the southeasternmost GOM that this author has viewed (and where adequate data are available to make this decision). Plate A-19 in the appendix presents the earliest well documented example of a LC configuration known to me which involves at detachment a cyclone pair of roughly equal strength at the throat of an embryonic large warm-core eddy, similar to that shown in Plates 3 and 4A. Plate A-19 is based on data from 1967 that was acquired by *Nowlin et al.* [1968] and used to describe the first clearly detached anticyclone; please also see *Nowlin and Hubertz* [1972]. The temperature contours on the map shown in Plate A-19 were adapted from a figure on page A6-2 in a report by *Molinari and Festa* [1988]. The historical map section of the appendix associated with the present article contains more detail on this event.

In the Introduction, we initially considered LC configurations that were based on hydrographic data (Plates 1 and 2). In this section on amplified frontal eddies as examined with modern SST and SSH maps of the GOM we have discussed in detail a large anticyclonic eddy detachment and separation in the southeasternmost GOM that occurred in the spring of 1998. One recent LC configuration was based on SST data (Plate 3), and a pair of configurations (Plate 4) were based

on SSH data. The presentation of the recent maps in Plates 3 and 4 in combination with Plates 1 and 2 is meant to give the reader a feeling for the characteristics of maps which use different types of data acquired at much different times to define the configuration of the LC and its eddy field. The details tend to be different for nearly each case of eddy shedding. But in general terms, Plates 3 and 4 resemble Plates 1 and 2 as examples of the view of first detachment as initially developed by *CO72* using hydrographic data.

However, the LC configurations and eddy-like features shown in Plates 3 and 4 are but a preliminary and special glimpse at those present in the composite SSH/SST database [*Leben's* Movie]. We now continue this investigation of the detachment of large and medium-sized eddies from the LC, together with a brief consideration of some other more or less concurrent processes, by considering a different set of SSH/SST, maps from October 2000 through June 2001. In qualitative terms, one observes diverse situations with respect to configurations of the LC that are associated with medium-sized and large eddy detachment. At the start of this section we defined two types (or modes or mechanisms) of detachment which are named with respect to the principal physical process responsible for each detachment mode, cyclonic penetration (CyDet) and westward propagation (WpDet). Plates 3 and 4 depicted an example of one type (in CyDet mode) of situation where the LC was originally oriented roughly northwest/southeast, from the Yucatan Straits up to ~ 27°N.

The other general or prototypic descriptive mechanism for detachment of large and medium-sized anticyclones from the LC, earlier called a WpDet mode, occurs when the tip of the LC is oriented more east/west (not perfectly of course) than northwest-southeast. WpDet mode tends to occur after the LC either recaptures its previously detached eddy (a strong reattachment), or has undergone pronounced neck-filling after a strong necking-down or near-detachment. In WpDet mode, westward propagation of the nearly-detached eddy at the tip of the LC plays the dominant initial role in the detachment process, with cyclonic features tending to play a subsidiary finishing role. An illustrative example of WpDet mode is now described using SST and SSH maps from the time frame late 2000 to mid-2001.

An example of partially necking-down the LC that was dominated by one cyclone from the east (a WFCE) was observed [*Leben's* Movie] to begin in October, 2000. This necking-down continued for about six weeks, with neck-filling (but without a substantial change in orientation of the tip of the LC). However, westward bending of the LC tip became noticeable in the latter part of November and was more pronounced in December, especially the latter part of that month when the LC also began to penetrate further

(~ northwestward) into the GOM. Necking-down the LC was clearly underway again on roughly 9 January 2001. Plate 5A, a SSH map for 23 January 2001, contains a strong necking-down of the LC that is more pronounced than the necking-down noted above for the October–December 2000 time frame. The necking-down in Plate 5A was also dominated at this time by a WFCE protruding out into the eastern GOM from the WFS (a common occurrence). A CBLCFE is also present on 23 January. Both features are indicated on Plate 5A. The smaller or medium-sized anticyclonic feature in the northwestern GOM in Plate 5A may be traced by consulting the SSH database for late 2000 and early 2001. The relevant maps show that previously detached (mostly small) anticyclonic features had coalesced in the northwestern Gulf near 26.5°N and 92.5°W. All of the comments in this and other paragraphs that are not explicitly exhibited on Plate 5A (or other plates that are based on SSH maps) may be verified either by visiting the SSH historical database of individual maps [Leben, 2004] or by slowly playing Leben's Movie.

A neck-filling began almost immediately in early February, 2001, and there was a well developed thick-necked LC by 15 February (Plate 5B). A comparison of Plates 5A and 5B may be used as a definition a strong necing-down followed by neck-filling. At this time the leading edge of the LC had extended westward to ~ 90°W, the axis of the LC having become bent into a mostly east-west orientation at its tip. The CBLCFE on Plate 5A had, by the time of Plate 5B, amplified along the boundary of the LC to become a CBCE. In the later part of February and throughout March 2001 small anticyclones were shed by cyclones from both the eastern and western tips of the LC. Penetration of the LC into the GOM after the situation indicated on 15 February 2001 (Plate 5B) continued throughout March 2001, during which time the LC extended quite a bit farther west and somewhat north. This was indicated, for example, by SST as well as SSH maps throughout March. The example given here in Plate 6 is a SST map for 26 March 2001. Plate 6 clearly exhibits two prominent cyclones on the eastern side of the LC (WFCEs, one at ~24°N, the other at ~26°N), which however coalesced later to form one larger WFCE (please see Plate 7A, the SSH map for 10 April 2001). In Plate 6 the LC is strongly extended to the northwest. The SSH map for 26 March also indicates an extensive penetration of the LC, out to ~ 91–92°W.

Plate 6 (and Plate 7A as well) also show a small anticyclone in the process of detachment from a nearly detached much larger ring to the northeast of the LC (a process noted, perhaps for the first time, by Leben and Born [1993]). The evolution of this feature may be followed in Plate 7B. Leben [this volume] registered a ring separation on 10 April 2001 (the SSH map in Plate 7A is for the same date). The large embryonic anticyclonic ring at the western tip of the LC

on Plate 7A is just beginning to pull away from the rest of the LC due to westward propagation. The LC configuration in Plate 7A is somewhat visually similar to the separation scenario in some of the results by Hurlburt and Thompson [1980, 1982]. This pulling away becomes more pronounced in the SSH map frames for 15–25 April, at which time cyclones begin to slip in behind the westward propagating ring and the LC, and initiate final visual detachment. The cyclones involved in final detachment were a NCE and a CBCE. Further development of the cyclonic structures between the LC and the ring, as depicted on Plate 7B (the SSH map for 15 June 2001, Plate 7B), may be viewed by referring to Leben's Movie during May and early June. Also, the incipient small anticyclone that was present to the northeast of the LC in Plates 6 and 7A had detached and propagated from 87°W to about 90°W by the time of Plate 7B, perhaps trapped along the continental slope in the northern Gulf.

In Plate 7A, the LC was in an approximately east-west orientation at its tip; after separation and visual detachment it did not return to a southerly-located port-to-port mode, remaining extended to ~26–27°N (in Plate 7B) in contrast to the situations shown in Plate 4B. That is, after a WpDet separation/detachment, the LC does not typically retract sharply into a southerly-located port-to-port mode (Plate 7, and Leben's Movie), as it usually does after a CyDet (Plate 4). From the time of Plate 7A, the detached ring seen in Plate 7B had propagated into the western GOM to reside totally west of 90°W. As illustrated by Plates 7A and 7B, WpDet anticyclones may tend after separation to become medium-sized (as opposed to large) rather quickly (~ 2 months), though again this is not always the case. By the end of July and early August 2001[Leben's Movie] the detached ring in Plate 7B had captured an older anticyclonic feature in the westernmost Gulf (centered near 95°W and 26°N on 15 June 2001 in Plate 7B). On the SSH map for 31 July 2001, for example, this comparatively large coalesced anticyclonic feature was occupying the area (nominal) between 24°–27°N, 91–96°W.

Leben's Movie has also been used to identify several examples (~ 15) of delays in further penetration of the LC into the GOM (penetration blocking), occurring during 12 separation intervals. Two blocking sequences were involved in some of the longer separation intervals. Section 5 of the appendix contains a detailed discussion of all of these cases, covering ~ 10 years of data. The processes involved tend to take place most readily when the tip of the LC is either in a southerly port-to-port mode and/or moves northwesterly and becomes delayed in the vicinity of ~ 26.5–27°N. An example of a long stay for the tip of the LC (during 1998) in the southerly region (~ 24.5–25.5°N) was presented earlier in this section of the present work. This separation interval was also further

Historical Mesoscale Altimetry - Jan 23, 2001

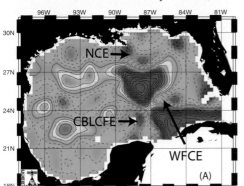

Historical Mesoscale Altimetry - Feb 15, 2001

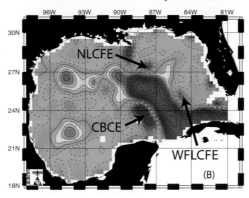

Plate 5. Satellite-derived sea surface height (SSH) maps for: (A) January 23, 2001, (B) February 15, 2001. SSH values (in cm) are color-coded as indicated by the bar below the maps.

Sea surface height (cm)

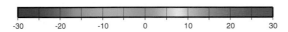

WEST LONGITUDE

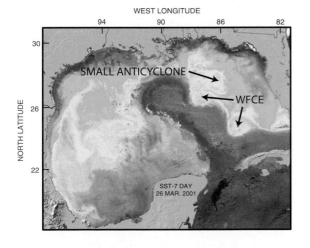

Plate 6. A satellite-derived sea surface temperataure (SST) map composite for 26 March 2001 (an average over 7 days, nominal). SST values (in ° C) are color-coded as indicated by the bar to the right of the map.

Historical Mesoscale Altimetry - Apr 10, 2001

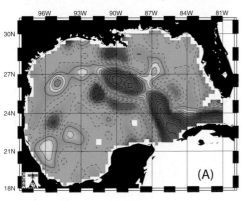

Historical Mesoscale Altimetry - Jun 15, 2001

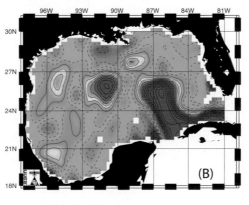

Plate 7. Satellite-derived sea surface height (SSH) maps for: (A) April 10 2001, (B) June 15, 2001. SSH values (in cm) are color-coded as indicated by the bar below the maps.

Sea surface height (cm)

extended later by two shorter duration cyclonic shield events near ~ 27°N as well as a few reattachment sequences. An example of a long duration northerly-located blocking configuration by a cyclonic shield is mentioned now. According to *Leben* [this volume], an eddy separated from the LC in early October 1999. An eddy separation did not occur in the year 2000. *Leben's* Movie shows that throughout most of this year the tip of the LC was in extended mode, in the latitude range 26–27°N, without much change in position or orientation. During most of this time a peripheral cyclonic shield of variable packing density was in place, appearing to block further development (eddy separation and final detachment) of the LC system for ~ 12 months. As usual, smaller anticyclones were shed through gaps in the cyclonic shield during that interval of blocking. Weak attachment played a minor role in delaying this separation interval in late 1999. Neck-filling played a role in January and February 2001.

DISCUSSION

Both descriptively- and dynamically-based results and ideas related to the diverse components of the composite process of the shedding of anticyclonic rings from the LC are the subject of a large number of publications. Different methods of data acquisition and descriptive interpretation have been used, along with several considerations involving theoretical studies, as well as analyses of diverse numerical model results (for example, *Behringer et al.* [1977]; *Brooks et al.* [1984a, 1984b]; *Bunge et al.* [2002]; *Candela et al.* [2002, 2003]; *Chassingnet et al.* [this volume]; *Chérubin et al.* [2005a, 2005b]; *CO72*; *Dietrich et al.* [1997]; *Elliot* [1979, 1982]; *Ezer et al.* [2003]; *Fratantoni* [1998]; *FEAL98*; *Hurlburt and Thompson* [1980, 1982]; *Hurlburt* [1986]; *Ichiye* [1962]; *Leben* [this volume]; *Maul* [1977, 1978]; *Nof and Pinchevin* (2001); *Nowlin et al.* [1968]; *Oey* [1996]; *Oey et al.* [2003]; *Oey et al.* [this volume]; Reid [1972]; *Sturges et al.* [1993]; *Sturges and Leben* [2000]; *Vukovich et al.* [1979]; *VM85, Vukovich* [1986, 1988a, 1988b, 1995]; *Wallcraft* [1986]; *Welsh and Inoue* [2000]; *ZEAL02;* and *ZEAL03*). One may get a variety of impressions of events occurring in the Loop Current System in the eastern GOM and their descriptive mechanisms from different data sets, as well as from interpretations by various authors, as noted throughout this article. Theoretical and numerical studies have been helpful, but are not yet clearly realistic.

Descriptive interest in the past, based on more or less all methods of observation, has focused on northwestward penetration of the LC into the GOM and on the dates or date ranges and time scales associated with the subsequent separation of large anticyclones from the LC. One component of the previously prevailing table of separations, as published

by *Sturges and Leben* [2000], involved results based on SSH maps from 1992 to 1999. *Sturges and Leben* [2000] state (in the footnote to their Table I) that in their table of ring separations entries that started in July 1973 and continued to October 1986 were taken from *Vukovich* [1988a; his Table 1]. Please also see *Vukovich* [1995] for additional discussion of some aspects of the separation process, including the definition of separation and variations in warm-core eddy sizes. The intermediate period from 1987 through 1991 was discussed by *Sturges* [1994]. The earliest published table of separations is probably by *Behringer et al.* [1977], based on hydrographic data. The most recent update is by *Leben* [this volume].

The existing database on eddy characteristics does not involve a large number of realizations, and is perhaps deceptively inhomogeneous if results from all methodologies and analyses and interpretations are included. *Sturges* [1993] examined the question of whether the monthly distribution of ring separations could reliably be distinguished from random, and concluded on the basis of the database in hand at that time that it could not. Figure 7 by *Leben* [this volume] contains a distribution of ring separation intervals for the results obtained in the 1973–1992 time frame (involving ~ 22 eddy separations) which appears to differ from recent results from the 1993–2003 time frame (involving ~ 16 eddy separations). *Leben* [this volume] refers to this difference as striking and notes that the histograms in his Plates 2 and 3 are non-Gaussian and that it is doubtful that the overall distribution is stationary. Heterogeneity is also found in the published collage of numerical model results on the circulation in the GOM, which have to a large extent concentrated on eddy separation.

The present article emphasizes prototypical warm-core eddy detachment phases, in the context of a complex composite eddy-shedding process. New results are introduced, based primarily on the modern SSH data in *Leben's* Movie. The study of blocking cyclones and their arrangement as a cyclonic shield around the periphery of the LC is a secondary focus (new results are also described and discussed). Penetration blocking appears to influence the time intervals between eddy separations as well (a detailed investigation may be found in appendix section 5). *Sturges et al.* [1993] was the first publication to emphasize the pervasive influence of reattachment processes, although *Brooks et al.* [1984a, 1984b] and *Vukovich* [1988a] had earlier discussed reattachment briefly. Reattachments of diverse types also affect the onset of separation. Both *Sturges et al.* [1993] and *Sturges and Leben* [2000] are notable for their past recognition and consideration of the complexity of the composite separation process. According to *Sturges and Leben* [2000], anticyclonic ring formation and separation events involve several stages and are not quick and simple, but long and drawn out. This is clearly an accurate synopsis of the overall situation as observed using visualizations of modern SSH data. Whereas *Sturges and Leben* [2000] is one of the few previous publications that has recognized the diversity and complexity of the composite separation process for large and medium sized anticyclonic rings from the LC, *Oey et al.* [this volume] is one of the first theoretical and numerical model oriented publications to begin to do so, also see *ZEAL02, Oey et al.* [2003], and *Chérubin et al.* [2005a, 2005b].

Leben [this volume, his Plate 4] shows a composite SSH map at his definition of time of separation for all of the ~ 16 eddy separation events that he observed in this position using the modern SSH database. The new quantitative definition of separation by *Leben* [this volume] involves the tracking and determination of the onset of pulling away by an embryonic eddy from a specific SSH contour (17 cm) associated with the LC. It is a prerequisite of this definition that there be no further reattachment of this contour to the LC. A clear-cut visualization of final detachment/separation, which is much more difficult to determine/specify, requires that cyclones slip in behind the detached eddy without further reattachment. This latter process can be very drawn out if the rather prolific very weak reattachments (touching attachments at SSH contours less than the 17 cm level) are included.

The separation intervals published by *Vukovich* [1988a] were based on a combination of AVHRR-based infrared observations and in situ data. SST maps based on infrared data are not useful in identifying LC configurations in the warmest weather months; please see *Leben and Born* [1993] in this regard as well as, for example, *Vukovich et al.* [1979]. In addition, cloud cover inhibits AVHRR data acquisition. *Vukovich* [1988a, his page 10,585], for example, refers to frontal positions plotted for each clear sky day as being typically available 5–9 days per month when the necessary temperature contrasts were observable. According to *Sturges and Welsh* [1990], SST maps are based on data acquired from a variably mixed Ekman Layer. Perhaps it is an accident (this data set is actually very small), but the separation dates published by *Leben* [this volume] that are based on modern SSH data occur mostly in the warm weather months. It is also apparent that SST observations have been used in a variety of interpretive modes, as described and discussed in more detail in the appendix for this article. Please also note that observations of the penetration of the LC into the GOM and subsequent detachment of eddies in the mid- and late 1960s and 1970s were based on hydrographic cruises that mostly took place in the fair weather months (~late spring, summer, early fall).

It turns out that frontal analysis maps based on AVHRR data can be misleading with respect to the identification of cyclones on the western or Campeche Bank side of the LC, to a large extent because the surface expression for CBCEs can be relatively weak (see appendix section 2 for more discussion). *CO72* was the first to observe that cyclonic features in the southeasternmost GOM located on the WFS side of the LC exhibited significantly colder surface temperatures than similar features on the CB side. Another key ingredient of the difficulty in observing CBCEs was anticipated by *CO72* (his pp. 105–106), in the context of outlining the problems (the observation of short time and space scales) associated with using standard hydrographic surveys (neither very synoptic nor frequently taken) to determine the mechanism of anticyclone detachment in the southeasternmost GOM. Some of the factors mentioned by *CO72* were the rapidity and irregularity with which cyclonic features cut across the LC from its CB side. These considerations with respect to the determination of warm-core eddy detachment mechanisms might also apply to other methods of data acquisition.

Cyclone pairs, with one cyclone located respectively on each of the east and west sides of the LC boundary in the southeasternmost GOM, are typically observed in amplified form on SSH maps to cut across the LC to trigger a detachment or near-detachment of large anticyclonic rings. This kind of result in which cyclones play the dominant role, here called a CyDet mode of detachment, typically yields a comparatively large, sometimes elliptical warm-core eddy. See, for example, Plates 3, 4 and 5A in this article. Illustrations of a few other configurations of the LC in CyDet mode may be found in the appendix, and many other examples are present in *Leben's* Movie. The label CyDet is used since cyclones play the dominant role, with westward propagation also involved. CyDet mode is almost always the initial type of detachment or a near-detachment but rarely a final detachment. In a CyDet mode configuration, the LC axis orientation is often more or less northwest-southeast at the onset of first detachment, depending on the degree of previous westward bending. The amplified form of the LCFEs involved are called CBCEs when they originate from the CB area west of the LC, and WFCEs when they originate from the direction of the WFS, and these features have substantial vertical penetration scale (~800–1000 m). The specific trigger for penetration across the LC, and/or the joining of the cyclone pairs in question in CyDet mode has not yet been sorted out to the best of my knowledge, although instabilities are probably involved.

Another type of detachment, called WpDet mode, is observed to occur when the LC is bent over toward the west as well. (An example of a prototypic WpDet mode circa spring 2001 is presented in Plates 5B and 7 in this article.) This mode of detachment is called WpDet because the prin-

SUMMARY AND CONCLUSIONS

Leben [this volume] contains new and important results on the characteristics of eddies at the defined time of separation. This study by *Leben* contains an update of eddy separation dates and timescales, and includes a description of the methodology and its limitations. It includes a new quantitative definition of time of separation; earlier criteria used by *Sturges and Leben* [2000] were qualitative. Modern satellite altimetry observations [*Leben's* Movie] are also the baseline data that are used in the present study of descriptive detachment mechanisms and the processes that contribute to penetration blocking (as well as brief discussions of other component phenomena). As an aid in interpreting past investigations involving other data types, the present article also contains a joint consideration of readily available results based on hydrographic and SST data. Material in this vein is presented in detail in the appendix. A couple of concurrent SSH/SST intercomparison datasets were found (Plates A-1 through A-4, along with A-13 and A-14).

The principal focus of this article has been on updating and extending the observational description of the process of detachment of large and medium-sized anticyclones from the LC in the context of a composite eddy-shedding sequence. Detachments were found to occur typically (though not invariably) in two general configurations (as described in the next paragraph). A secondary focus of this article is the role of cyclones (sometimes in the form of a shield) on its periphery in blocking (or inhibiting or impeding) the northwestward penetration of the tip of the LC into the GOM (see appendix section 5 for details). The shedding or leakage of small anticyclones through gaps in this cyclonic shield may be a significant component of the penetration blocking process, as also noted on the basis of numerical model results by ZEAL02. Reattachments influence the duration of separation intervals. Other processes that are connected with the observed composite sequence of events associated with eddy shedding by the LC have been noted in passing.

One of the descriptive mechanisms of anticyclonic eddy detachment from the LC is called in this article CyDet mode. This is because cyclonic intrusions into the LC operating approximately as pairs, one on each side, are the primary factors, with westward propagation playing a subsidiary role. CyDet mode is typically characteristic of large anticyclonic eddy detachments from the LC in the southeasternmost GOM just north of the Yucatan Straits. This occurs after sufficient penetration of the LC into the GOM, and starts when the axis of the LC is oriented more or less northwest-southeast. The normal outcome of a CyDet is to allow the tip of the LC to reorient from northwest-southeast to a more east-west direction. The LC may retreat to a southerly-located port-to-port mode after a CyDet.

The details of the two detachment modes identified in this study are visually different, and in contrast to the cyclonic-pair-related pinch-off mode (a CyDet) so noticeable in the southeasternmost GOM in most data sets, WpDet mode initially looks like a direct pulling apart of a ring from the LC by westward propagation (hence WpDet). This mode usually occurs when the tip of the LC is reoriented more or less east/west. WpDet mode typically involves the reattachment to the LC of an anticyclonic eddy that was previously detached or strongly necked-down in CyDet mode. Cyclones play a role, but more of a subsidiary one, in finishing off a final visual detachment in WpDet mode. For WpDet, a cyclone on the northern side of the LC (a NCE) may participate in finishing a detachment, typically associated with a CBCE as well. In a WpDet sequence, the tip of the LC typically (but not always) does not retract much south afterward, and medium-sized rings are often involved.

Mixtures of CyDet and WpDet modes may also occur, involving both cyclonic intrusions and westward propagation in roughly equal proportion. WpDet mode is connected to *Leben's* [this volume] definition of a separation, which focuses on the initial stages of a final detachment. This choice is probably a good idea, because final detachment is often difficult to pinpoint pictorially. The modern view of eddy detachment and separation events mentioned throughout this article and its appendix may be checked out by viewing the animation of SSH (*Leben's* Movie) on the CD at the end of this book. The appendix also contains a few maps of historical interest in this regard.

Cyclonic features that in an embryonic form are similar to Loop Current Frontal Eddies (so called and also abbreviated LCFEs by *FEAL98*), and in particular amplified versions of these LCFEs, are distinctly involved in CyDet mode. Large or amplified cyclonic features on both sides of the LC in the southeastern GOM (called CBCEs when on the west, and WFCEs on the east) are in general terms more or less equally important (in CyDet mode) with respect to detachment. However, one particular cyclone type is often dominant or more active at any given time in any given detachment sequence. The cyclones associated with CyDet mode also move around the boundary of the LC. *ZEAL03* (their p. 625) pointed out that the CBCEs present in their database were always near the CB at the time that LC anticyclones were being shed in the southeasternmost GOM. The CBCE as found in Figure 2 by *Maul* [1977], as well as a similar feature observed by *CO72* (Plate A-22), were based on subsurface temperature data. But *VM85,* in the abstract of their study that was essentially dependent on sea surface temperature data as well as some in-situ observations, over-emphasized the role played by one cyclone type, located on the eastern side of the boundary of the LC (called a WFCE in this article). The possibility that cyclones of various sizes, which routinely show up on all sides of the LC in modern SSH maps, might not always have a readily measurable temperature (or temperature gradient) signal was not considered. The significance of the results by *CO72* [*Schmitz*, 2003] were almost ignored in the literature for ~ 30 years (until *ZEAL02* and *ZEAL03*), with the notable exception of *Elliot* [1979, 1982].

The vertical structure for CBCEs is in general very energetic at depth, as first observed by *CO72*. *ZEAL03* (their Figure 5) plotted a temperature section through a CBCE (refer to Plate A-7) and observed that the surface temperature signal was very weak at the time of observation. This kind of finding illustrates one of the problems encountered in the detection of CBCEs with the techniques used by (for example) *VM85* and *Vukovich et al.* [1979]. Another example of a temperature section across a CBCE (Figure A-1), with results similar to the *ZEAL03* example just noted, was also given in appendix section 2. Temperature sections across WFCEs, initially published by *VM85* (their Plate 3), clearly showed that WFCEs are also usually associated with deeply penetrating isotherm depth displacements, as well as a strong surface signal. *VM85* (their p.111) stated that the surface signature due to outcropping for the WFCEs that they had observed (a prime example is shown in Plate A-5) was a result of upwelled subsurface water, as opposed to entrainment of shelf water. However, *Chassingnet et al.* [this volume, see their section 6], in their SeaWiFS maps, show possible examples of both of these mechanisms. CO72 had also noticed early on that the cyclones on the WFS side of the LC tended to have a colder, fresher surface signature than those on the west. The amplitude of the outcropping at the sea surface for a WFCE is a key reason why this kind of cyclonic feature is more readily visible in AVHRR data than a CBCE.

Detachment processes and penetration blocking, which have been the focus of this article, although important contributions to the complicated composite mechanism of the shedding of large and medium sized anticyclonic eddies from the LC in the GOM, are only two of the several processes involved. Other component processes have been mentioned in passing. For example, prolific examples of various forms of reattachment between the LC and the anticyclonic eddies that have been detached or nearly detached from it show up readily in sequences of modern SSH maps [in *Leben's* Movie]. Types of reattachment include eddy recapture by the LC, along with neck-filling associated with a strong, nearly complete necking-down, and a weak or touching attachment. The space-time averaging associated with SSH data processing might on occasion lead to a detachment looking like a near-detach-

ment, or vice-versa. Observations of the touching mode of behavior of anticyclonic features in the GOM using SSH data could also be affected by temporal/spatial averaging. Frontal eddies and their amplified versions are also responsible for the detachment of comparatively small anticyclones from the LC or from its larger current rings in the eastern GOM. The detachment of small anticyclones from the LC as suggested by the modern SSH dataset is perhaps much more common than previously discussed in any publication. It is also the case that amplified frontal eddies may split rings as they propagate through the western GOM (please see, for example, *Biggs et al.* [1996]; splitting may have been first suggested in print by *Elliot* [1982], his p. 1295). In the western GOM, anticyclonic features also merge and get larger (temporarily).

It turns out that penetration blocking by cyclones around the periphery of the LC is much more common than currently thought. The penetration-blocking phenomena (an example of which was first mentioned by *ZEAL02*) is considered in detail in the appendix. The options are more complex than envisioned by *ZEAL02*. ~ 15 examples are analyzed in appendix section 5, utilizing *Leben's* Movie as the primary data source. Inspection of specific cases of penetration impediment, as described in detail in this appendix section, shows qualitatively that longer- duration penetration blocking leads to longer separation intervals, but reattachment processes are also involved in delaying separation. For five cases in Table 3 by *Leben* [this volume] involving separation intervals of 11.5 months and longer, blocking cyclones were in place for 6–14 months, occuring in three cases in two separate general locations (~ 24.5–25.5°N, ~26–27°N). Leben [this volume] mentions a bimodal distribution of the retreat latitude of the LC after separation. For the set of 7 separation intervals in *Leben's* Table 3 in the vicinity of 5.5 months long (4–7.5 month range, centered about 3 separation intervals of 5.5 months), delays in separation appear to occur for ~ a few months.

Frontal eddies in the FC and LC, and perhaps in the YC, develop in a relatively thin frontal zone (~15–35 km wide, say) along their left-hand boundary looking downstream, where comparatively large upper-layer horizontal (cyclonic) shears are usually observed to occur. This type of uppermost layer cyclonic horizontal shear zone, typically accompanied by high vertical shear, is a prominent characteristic [*Schmitz*, 1996] of more or less all portions of the named western boundary current segments that are part of the Gulf Stream System, with the LC being one example. See section 6 of the appendix for this article for a discussion of possible frontal eddies in the vicinity of the Yucatan Straits. Upstream conditions are discussed in more detail in appendix section 6.

Possible mechanisms for the amplification of frontal eddies might include baroclinic [*Hurlburt*, 1986, his pp. 2380–2383] or mixed baroclinic-barotropic instability of the LC at, for example, locations near the northern edge of the Campeche Banks where the stabilizing effects of bottom topography (slope) abruptly cease. Please also see *Chérubin et al.* [2005a] and *Oey et al.* [this volume]. *Chérubin et al.* [2005b] demonstrate in their numerical model results that the LC regime in the GOM meets the criteria for both barotropic and baroclinic instability. *Chérubin et al.* [2005a, 2005b] also found a deep-reaching cyclonic shield around the periphery of the LC in their model results. Figure 4 by *Oey et al.* [this volume] contains results from a numerical experiment that exhibits model analogues of the deep cyclones originally observed by CO72. *Oey et al.* [2003] have recently considered the reattachment issue in the context of a numerical experiment; please also see Dietrich et al. (1997). Studies of numerical prediction for the GOM are now available [*Chassingnet et al.*, this volume; *Kantha et al.*, this volume]. The intercomparison of descriptive results with inferences from various numerical experiments involving the GOM circulation, as presented briefly in this paragraph, is expanded upon in the last section of the appendix.

Four principal conclusions from this work, based primarily on a descriptive examination of modern SSH data [mostly in *Leben's* Movie], follow. First, there are two prototypic modes of detachment of anticyclonic rings from the LC, depending on the relative influence of peripheral cyclones and westward propagation. The case where cyclones dominate normally occurs first, at a location in the southeasternmost GOM just north of the Straits of Yucatan. The mode where westward propagation dominates is typically the "final detachment," and usually occurs further north after one (or more) kinds of reattachment(s) between the LC and its recently detached ring. These modes (mode mixing is also possible) are described in detail throughout this article and its appendix.

Second, penetration of the LC into the GOM may be blocked by cyclones, which are often arranged in a peripheral cyclonic shield of variable packing density around the LC when it is extended into the Gulf. Blocking tends to cluster in two general areas. One is located in the vicinity of 24.5–25.5°N when the LC is in a southerly port-to-port mode. The other occurs when the tip of the LC has advanced to ~ 26–27°N. Blocking by cyclones for a few to several months along with delays by a variety of reattachments are associated with increases in the time between eddy separations (yielding larger or longer separation intervals). These combined episodes can occur at each of the two areas noted above for one of the longer separation intervals. Reattachment sequences usually last for a week to ~ a few months, but the latter time interval occurs rarely and only when multiple reattachments are involved.

Third, SST data alone are not adequate to clearly determine the descriptive mechanisms of these phenomena, but can be an important supplement to hydrographic or especially SSH data. More hydrographic-type data are needed in the LC, on small spatial and temporal scales across the cyclones around its rim. Moored instrument deployments directly in the LC are nearly nonexistent.

Finally, realistic numerical model results are starting to partially emerge, but are still not adequate. Many sufficiently long new numerical experiments which are thoughtfully intercompared with the data base will be needed to guide the development of clearly realistic determinations of the dynamics of the Loop Current and its Eddy Field.

Acknowledgements. I would like to thank Frank Kelly, Heather Riddles, and Dixie Smith for significant help on this project, as well as Jim Bonner for his support. I also wish to point out that the animation of SSH maps by Bob Leben on the CD at the back of this book is a fabulous asset. I have experienced a fruitful exchange recently with Jorge Zavala-Hidalgo, and his work and collegiality has had a strong influence on this article. Over the years, I have learned a great deal about the Gulf of Mexico from Tony Sturges. Thanks to Mia Shargel for much editorial guidance. This work is dedicated to the memory of John Cochrane, whom I never had the privilege to meet, and to my deceased colleague J. Dana Thompson, physical oceanographer of the Gulf of Mexico extraordinare. Support from the Minerals Management Service is gratefully acknowledged.

REFERENCES

Bane, J. M., Jr., D. A. Brooks and K. R. Lorenson, (1981). Synoptic observations of the three-dimensional structure and propagation of Gulf Stream meanders along the Carolina continental margin. *J. Geophys. Res.*, *86*, 6411–6425.

Behringer, D. W., R. L. Molinari and J. F. Festa, (1977). The variability of anticyclonic flow patterns in the Gulf of Mexico. *J. Geophys. Res.*, *82*, 5469–5476.

Biggs, D. C., G. S. Fargion, P. Hamilton and R. R. Leben, (1996). Cleavage of a Gulf of Mexico Loop Current eddy by a deep water cyclone. *J. Geophys. Res.*, *101*, 20,629–20,641.

Brooks, D. A., S. Nakamoto and F. J. Kelly, (1984a). Investigation of low frequency currents in the Gulf of Mexico, Volume I. Texas A&M Research Foundation, Contract 110483-HHH-1AH, 1984a.

Brooks, D. A., S. Nakamoto and F. J. Kelly, (1984b). Investigation of low frequency currents in the Gulf of Mexico, Volume II. Texas A&M Research Foundation, Contract 110483-HHH-1AH, 1984b.

Bunge, L., Ochoa., J., Badan, A., Candela, J., and J. Scheinbaum, (2002). Deep flows in the Yucatan Channel and their relation to the Loop Current extension. *J. Geophys. Res.*, *107(C12)*, 3233, doi:10.10292001JC001256.

Candela., J., J. Scheinbaum, J. Ochoa, and A. Badan, (2002). The potential vorticity flux through the Yucatan Channel and the Loop Current in the Gulf of Mexico. *Geophys. Res. Lett.*, *29(22)*, 2059,doi:10.1029/2002GL015587.

Candela., J., S. Tabahara, M. Crepon, B. Barnier, and J. Scheinbaum, (2003). Yucatan Channel flow: Observations versus CLIPPERATL6 and MERCATORPAM models. *J. Geophys Res.*, *108(C12)*, 3385, doi:10.1029/2003JC001961.

Chassignet, E. P., H. E. Hurlburt, O. M. Smedstad, C. N. Barron, D. S. Ko, R. C. Rhodes, J. F. Shriver, A. J. Wallcraft, and R. S. Arnone, (this volume). Assessment of ocean prediction systems in the Gulf of Mexico using ocean color.

Chérubin, L. M., W. Sturges, and E. P. Chassignet, (2005a). Deep flow variability in the vicinity of the Yucatan Straits from a high-resolution numerical simulation. *Journal Geophys. Res.*, in press.

Chérubin, L. M., Y. Morel, and E. P. Chassignet, (2005b). The role of cyclones and topography in the Loop Current ring shedding. Submitted to *J. Phys. Ocy.*

Cochrane, J. D., (1969). Currents and waters of the eastern Gulf of Mexico and western Cabibbean, of the western tropical Atlantic, and of the eastern tropical Pacific Ocean. Tech. Report, Texas A& M Univ., Ref. 69-9T, 29–31.

Cochrane, J. D., (1972). Separation of an anticyclone and subsequent developments in the Loop Current (1969), in *Contributions on the Physical Oceanography of the Gulf of Mexico*, edited by L. R. A. Capurro and J. L. Reid, *Texas A & M University Oceanographic Studies, 2*, 91–106. Gulf Publishing Co., Houston, Tex.

Dietrich, D. E., C. A. Lin, A. Mestas-Nunez and D. D. Ko, (1977). A high resolution numerical study of Gulf of Mexico fronts and eddies. *Meterol. Atmos. Phys., 64*, 187–201.

Elliot, B. A., (1979). Anticyclonic rings and the energetics of the circulation of the Gulf of Mexico. PhD Dissertation, 188 pp., Texas A & M Univ., College Station, Tex.

Elliot, B. A., (1982). Anticyclonic rings in the Gulf of Mexico. *J. Phys. Oceanogr., 12*, 1291–1309.

Ezer, T., L-Y. Oey, W. Sturges, and H-C. Lee, (2003). The variability of currents in the Yucatan Channel: Analysis of results from a numerical ocean model. *J. Geophys. Res., 108 (C1)*, 3012, doi:10.1029/2002JC001509.

Fratantoni, P. S., (1998). The formation and evolution of Tortugas Eddies in the southern Straits of Florida and Gulf of Mexico. PhD Dissertation, 189 pp., Univ. Miami, Coral Gables, Florida.

Fratantoni, P. S., T. N. Lee, G. P. Podesta and F. Muller-Karger, (1998). The influence of loop current perturbations on the formation and evolution of tortugas eddies in the southern Straits of Florida. *J. Geophys. Res., 103*, 24,759–24,779.

Hurlburt, H. E., (1986). Dynamic transfer of simulated altimeter data into subsurface information by a numerical ocean model. *J. Geophys. Res. 91*, 2372–2400.

Hurlburt, H. E. and J. D. Thompson, (1980).A numerical study of loop current intrusions and eddy shedding. *J. Phys. Oceanogr., 10*, 1611–1651.

Hurlburt, H. E. and J. D. Thompson, (1982). The dynamics of the loop current and shed eddies in a numerical model of the Gulf of Mexico, in *Hydrodynamics of Semi-enclosed Seas*, edited by J. C. J. Nihoul, Elsevier Science, New York, NY, pp. 243–297.

Ichiye, T., Circulation and water-mass distribution in the Gulf of Mexico. *Geofisica Internl., 2*, 47–76, 1962.

Kantha, L. H., J-K. Choi, K. J. Schaudt, and C. K. Cooper, (this volume). Regional data- assimilative model for operational use in the Gulf of Mexico.

Leben, R. R., (2004). Real-Time Altimetry Project: http://www.ccar.colorado.edu/.

Leben, R. R., (this volume). Satellite observations of Gulf of Mexico mesoscale circulation and variability.

Leben, R. R. and G. H. Born, (1993). Tracking loop current eddies with satellite altimetry. *Adv. Space Res., 13*, 325–333.

Lee, T. N., Florida Current spin-off eddies, (1975). *Deep-Sea Res., 22*, 753–765.

Maul, G. A., (1977). The annual cycle of the Gulf Loop Current, 1: Observations during a one-year time series. *J. Mar. Res., 35*, 29–47.

Maul, G. A., (1978). The 1972-1973 cycle of the Gulf Loop Current, 2: Mass and salt balances of the basin. CICAR-II Symposium on Progress in Marine Research in the Caribbean and Adjacent Regions. *FAO Fisheries Rep., 200 (Suppl.)*, 597–619.

Maul, G. A., and F. M. Vukovich, (1993). The relationship between variations in the Gulf of Mexico Loop Current and Straits of Florida Volume Transport. *J. Phy. Oceanog., 23*, 785–796.

Molinari, R. L. and J. F. Festa, (1978). Ocean thermal and velocity characteristics of the Gulf of Mexico relative to the placement of a moored OTEC plant, final report to the U. S. Department of Energy, DOE/NOAA Interagency Agreement no. BY-76-A-29-1041, 106 pp., NOAA/AOML, Miami, Fla,.

Morrison, J. M. and W. D. Nowlin, Jr., (1977). Repeated nutrient, oxygen, and density sections through the Loop Current. *J. Mar. Res., 35*, 105–126.

Nof, D. and T. Pichevin, (2001). The ballooning of outflows. *J. Phys. Oceanogr., 31*, 3045–3058.

Nowlin, W. D. Jr., and J. M. Hubertz, (1972). Contrasting summer circulation patterns for the eastern Gulf, in *Contributions on the Physical Oceanography of the Gulf of Mexico*, edited by L. R. A. Capurro and J. L. Reid, *Texas A & M University Oceanographic Studies, 2*, 119–137. Gulf Publishing Co., Houston, Tex.

Nowlin, W. D., Jr., D. F. Paskausky, and R. O. Reid, (1968). A detached eddy in the Gulf of Mexico. *J. Mar. Res., 26*, 185–186.

Oey, L. Y., (1996). Simulation of mesoscale variability in the Gulf of Mexico: sensitivity studies, comparison with observations, and trapped wave propagation. *J. Phys. Oceanogr., 26*, 145–175.

Oey, L. Y., H. C. Lee, and W. J. Schmitz Jr., (2003). Effects of winds and Caribbean Eddies on the frequency of Loop Current eddy shedding: A numerical study. *J. Geophys.Res., 108 (C10)*, 3324, doi:10.1029/2002JC001698.

Oey, L-Y., T. Ezer, and H.-C. Lee, (this volume). Loop Current, rings and related circulation in the Gulf of Mexico: A review of numerical models and future challenges.

Reid, R. O., A simple dynamical model of the Loop Current, (1972), in *Contributions on the Physical Oceanography of the Gulf of Mexico*, edited by L. R. A. Capurro and J. L. Reid, *Texas A & M University Oceanographic Studies, 2*, 157–160. Gulf Publishing Co., Houston, Tex.

Schmitz, W. J. Jr., (2003). On the circulation in and around the Gulf of Mexico, Vol. I, A review of the deep water circulation. Available online at the Conrad Blucher Inst., Texas A&M Univ., Corpus Christi, TX: http://www.serf.tamus.edu/gomcirculation/.

Sturges, W., (1994). The frequency of ring separations from the Loop Current. *J. Phys. Oceanogr., 24*, 1647–1651.

Sturges, W. and R. Leben, (2000). Frequency of ring separations from the loop current in the Gulf of Mexico: a revised estimate. *J. Phys. Oceanogr., 30*, 1814 – 1819.

Sturges, W., and S. Welsh, (1990). Wind-driven response of surface infrared signals. *J. Phys. Oceanogr., 20*, 1842 – 1848.

Sturges, W., J. C. Evans, S. Welsh and W. Holland, (1993). Separation of warm-core rings in the Gulf of Mexico. *J. Phys. Oceanogr., 23*, 250 – 268.

Von Arx, W.S., D. F. Bumpus and W. S. Richardson, (1955). On the fine-structure of the Gulf Stream front. *Deep-Sea Res., 3*, 46–65.

Vukovich, F. M., (1986). Aspects of the behavior of cold perturbations in the eastern Gulf of Mexico: a case study. *J. Phys. Oceanogr., 16*, 175–188.

Vukovich, F. M., (1988a). Loop current boundary variations. *J. Geophys. Res., 93*, 15,585–15, 591, 1988a.

Vukovich, F. M., (1988b). On the formation of elongated cold perturbations off the Dry Tortugas. *J. Phys. Oceanogr., 18*, 1054–1059.

Vukovich, F. M., (1995). An updated evaluation of the Loop Current eddy-shedding frequency. *J. Geophys. Res., 100*, 8655–8659.

Vukovich, F. M. and B. W. Crissman, (1986). Aspects of warm rings in the Gulf of Mexico. *J. Geophys. Res., 91*, 2645–2660.

Vukovich, F. M. and G. A. Maul, (1985). Cyclonic eddies in the eastern Gulf of Mexico. *J. Phys. Oceanogr., 15*, 105–117.

Vukovich, F. M., B. W. Crissman, M. Bushnell and W. J. King, (1979). Some aspects of the Gulf of Mexico using satellite and in-situ data. *J. Geophys. Res., 84*, 7749–7768.

Wallcraft, A., (1986). Gulf of Mexico circulation modeling study, Year 2. Progress Report by JAYCOR, submitted to the Minerals Management Service, Metairie, LA. Contract No. 14-12-0001-30073.

Welsh, S. E. and M. Inoue, (2000). Loop current rings and deep circulation in the Gulf of Mexico. *J. Geophys. Res., 105*, 16,951–16, 959.

Zavala-Hidalgo, J., S. L. Morey, and J. J. O'Brien, (2002). On the formation and interaction of cyclonic eddies with the Loop Current using NCOM and a suite of observations. *Oceans 2002 IEEE/MTS Conference Proceedings*, 1463-1466.

Zavala-Hidalgo, J., S. L., Morey, and J. J. O'Brien, (2003). Cyclonic eddies northeast of the Campeche Bank from altimetry data. *J. Phys. Oceanogr., 33*, 623–629.

William Schmitz Jr., WHOI, 1915 Hidden Way, Corpus Christi, TX 78412 (wschmitzjr@stx.rr.com)

Deep-Water Exchange Between the Atlantic, Caribbean, and Gulf of Mexico

Wilton Sturges

Department of Oceanography, Florida State University

The deep waters of the Caribbean Sea are renewed by sporadic flow over the two deep sills connecting with the open Atlantic. In turn, the deep waters of the Gulf of Mexico are renewed by flow from the northwest Caribbean over the sill in Yucatan Channel, between Mexico and Cuba. So we examine the obvious question: if all this renewal water is coming in, what is going out? The view that emerges is a three-layer mean deep flow in the Caribbean and the Gulf. There is strong vertical shear in the water above 800 m associated with the primary upper-layer flow. Surprisingly, this shear continues to at least 1400 m. In the three-layer scenario described here, the upper-most "deep layer" flows into the Gulf at depths of ~800-1100 m as an extension of the upper-layer circulation. The deepest inflow is sporadic, just above sill depth. The third layer, containing the required mean return flow to achieve mass balance, is found between these two mean flows at depths of ~1100 to ~1900 m. There is little information about the horizontal pattern of this deep return flow. The transport in these deep flows into the Gulf and in the return flow is slightly less than 1 Sv. These conclusions are inferred from the computed mean vertical shear, conservation of potential vorticity in the deep layers, the observed salinity gradients, and the assumption that there can be no net flow between 800 and 2000 m. The deep return flow of water from the Gulf of Mexico, having salinity ~34.96 psu, appears to flow from Yucatan Channel all the way back to Windward and Jungfern Passages, leaving the Caribbean above the entering deep renewal water.

1. INTRODUCTION AND BACKGROUND

The essential ideas of renewal of deep water in the Caribbean Sea and Gulf of Mexico have been known for decades. The waters of the Atlantic spill over into the Caribbean through two passages that are ~2000 m deep. Plate 1a shows the positions of the major deep basins and the sills that connect them. The deep opening farthest east is the Anegada-Jungfern Passage Sill. From the work of Worthington in the '60s to the newer studies of *Joyce et al.* [1999, 2001] and *MacCready et al.* [1999], it has usually been assumed that there is renewal over Windward Passage as well. Recently *Johns et al.* [2004, personal communication] have made measurements in Windward Passage. There are, of course, many more observations in the near-surface layers than in the deeper waters flowing in over the sills. See, e.g., *Johns et al.* [2002], *Wilson and Johns* [1997], *Richardson* [2004] and *Mooers and Maul* [1998]. The openings at the far eastern end are only ~1000 m deep.

The density surface of Plate 1a, chosen to match one shown by *Lozier et al.* [1995], lies near the middle of

Circulation in the Gulf of Mexico: Observations and Models
Geophysical Monograph Series 161

the upper-layer flow that becomes the Loop Current-Florida Current-Gulf Stream system. At these depths there is strong vertical shear in the flow, as indicated by the slope of density surfaces in Plate 1a, from ~400 m in the western Caribbean down to ~ 650 m in the Atlantic. At these depths the water properties are continuous across the island arc at the east end of the Caribbean. The mean position of the Loop Current is evident. Plate 1b shows a density surface (σ_t 27.4) near 700–850m. At these depths the flow contains the low-salinity Antarctic Intermediate Water (AAIW) from the far south. (Because of space limitations, a number of figures have been placed in an appendix, on a CD in the back of this book.) Figure A 7, in the appendix, shows salinity on that surface, which can be as low as ~34.76 psu in the southeast, increasing to ~ 34.9 psu in the Gulf. *Schmitz and Richardson* [1991] point out that the Caribbean salinity minimum depends on whether the source has been directly from the south or indirectly from the North Atlantic.

There is a rich literature concerning the mixing of water pouring over sills as it entrains resident water and sinks: see, e.g., *Baringer and Price* [1997], *Bryden and Nurser* [2003], *Killworth* [2001], *or Price and Baringer* [1994]. The main ideas about large-scale renewal as used here, however, are straightforward. Cold water from the Atlantic pours in some sporadic way over the sill at Jungfern into the far southeastern Caribbean as well as over Windward into the western basin. The renewal of the deep waters of the Gulf of Mexico, coming in over the sill in Yucatan, must therefore come first through the Caribbean Sea, not directly from the Atlantic. The other opening between the Gulf and Atlantic, through the Straits of Florida, is only ~800m deep and is the limiting depth for the outflow of Gulf waters that form the Florida Current.

Scientists from the Ensenada group have recently achieved the decades-old dream of maintaining long moorings in the passage between Mexico and Cuba. In a remarkable series of papers they have made huge strides in our knowledge of that flow field, both in the upper layers and the deep [*Bunge et al.* 2002, *Candela et al.* 2002, *Ochoa et al.* 2001, *Rivas et al.*, 2005, *Badan et al.* (this volume)].

We briefly review here the deep-water renewal over the sills. We then ask several questions that follow naturally from our understanding of the inflow. If all this deep flow is coming in, what is going out? Why is the water in the Caribbean, inside the sill at Jungfern, so different from the temperature and salinity of the Atlantic waters just outside? Can we understand the deep temperature budget? That is, if cold deep water keeps flowing in, how can the deep water remain at constant temperature?

2. METHODS

Most of our classical ideas about what the ocean looks like are based on the methods of oceanographers represented by F. Fuglister, J. L. Reid Jr., or L. V. Worthington. They based their analyses on a carefully-selected set of high-quality data, taken by themselves or by trusted colleagues. In recent years we have come to appreciate the enormous effects of variability over time, particularly on the western sides of the ocean. While accuracy is surely an essential feature of a good data set, a measurement good to three decimal places can be misleading if the quantity varies in time in the first decimal place.

And so in this paper I have taken an approach that emphasizes averaging in time. The small "data points" in Plate 1a are not individual hydro stations but averages of 5–10 samples from the available NODC data in a small region. The sampling density results in a "smoothing scale " that is small and irregular. The number of stations ranges widely; in some figures shown later, error bars are given. The data are spread over many decades; considering the large variability at long time scales, however, whether the averaging is adequate remains an open question.

The primary averaged quantity is potential density. In each cluster of hydro data, potential density was computed for each observed sample. This avoids the known issues [e.g., *Lozier et al.* 1995] of forming averages from the mean temperature and salinity and thereby achieving nonexistent densities. Because there are unresolved issues of salinity calibrations in the world-wide data base, I have restricted the data set to data from U. S. ships. The data set was first pruned to remove data points that lie outside three standard deviations in salinity (on a t, S diagram) in broad local regions. Within each data grouping we have chosen the median value rather than the simple mean, because the median is less affected by outliers.

There were several reasons why we decided to form this database rather than using others that might have been easily available. First, many observers leave the Gulf of Mexico out of their domains. Second, I was interested in being able to examine the data in small, narrow regions rather than using averages over perhaps one degree or more; I had hoped to trace the flow from the sills by forming data groups that lie as closely as possible to the deep isobaths. And finally, I was interested in the idea of a cyclonic deep flow near the boundaries dominated by topographic rectification. These reasons forced the use of a specialized data set.

It may seem a straightforward statistical question to determine how many samples are required to suppress the effects of eddy motion, but at least two difficulties arise. When the sampling is nearly random with respect to the eddies, such as lines of hydro stations taken along widely spaced lines

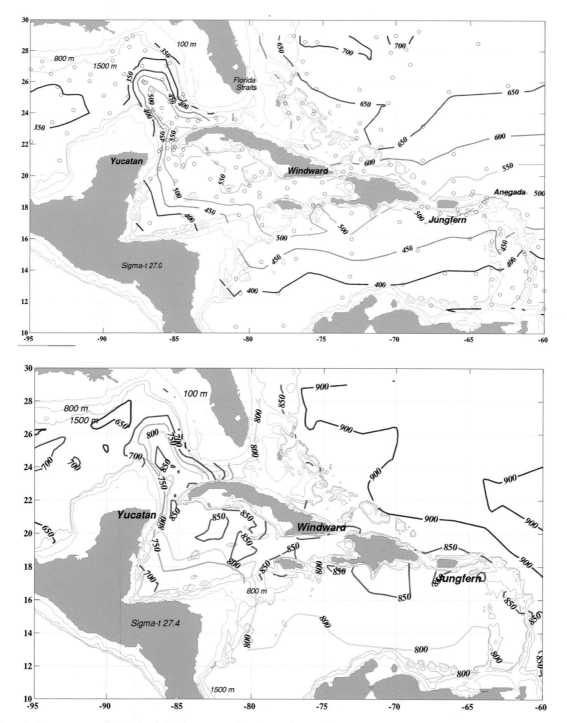

Plate 1. (a, upper panel) Depth of the sigma-t surface 27.0. The small 'data points' are averages of 5–10 hydro stations from the NODC data archives. The region near the Florida Straits is not resolved in this data set. 1b (lower panel), Depth of sigma-t 27.4. (See text). The isobaths 100, 800, 1500 m are labeled.

of longitude, we usually have no idea if any given sample, taken years ago, is in an eddy or not. And we will rarely know whether the sampling was done nearer the edge or the center of an eddy. Second, the field is usually horribly under-sampled, so that we are not certain that we have a decent measure of the standard deviation. If we examine an E-W section in Fuglister's Atlas, for example, we easily find large-amplitude fluctuations in the depth of a deep isotherm. These eddy motions are now known to be common, but because the eddy field is not horizontally homogeneous we continue to have poor measures of it, especially near boundaries. And the bulk of the work here is concerned with properties near the boundaries. The values of standard deviations computed here must therefore be considered as best-guess estimates.

Because (for these maps) we are leaning exclusively on the available data rather than on theoretical ideas, we have used the simplest contouring routine available (within Matlab), using linear interpolation between data groups with no speci-fied horizontal scales of smoothing or interpolation. The data are first interpolated onto a uniform x-y grid at 0.25 degree resolution.

Finally, it must be acknowledged that the use of averaged data has to be examined as critically as the use of data from a snapshot. The mean Gulf Stream looks very different from the instantaneous stream, although some properties are conserved. Swift, narrow flows will appear as slow, broad flows. We hope to get the directions right and the transports approximately right.

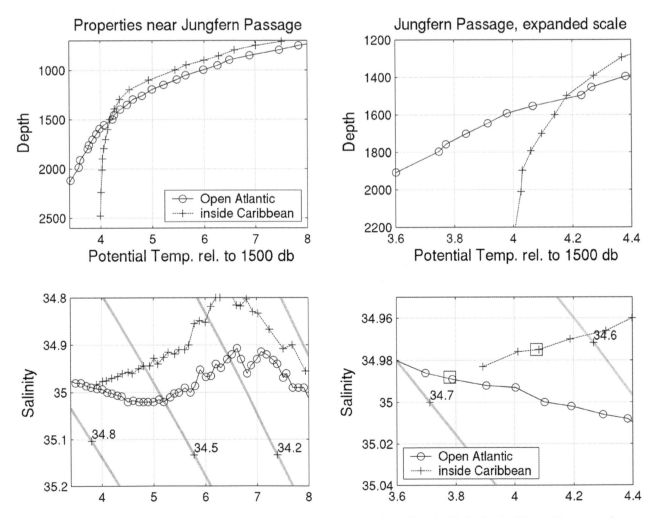

Figure 1. Properties near Jungfern Passage; just outside in the Atlantic and just inside in the Caribbean. Upper panels show potential temperature versus depth; lower panels show potential temperature versus salinity; right-hand panels show an expanded view of the left-hand panels. Sloping lines show potential density relative to 1500 db.

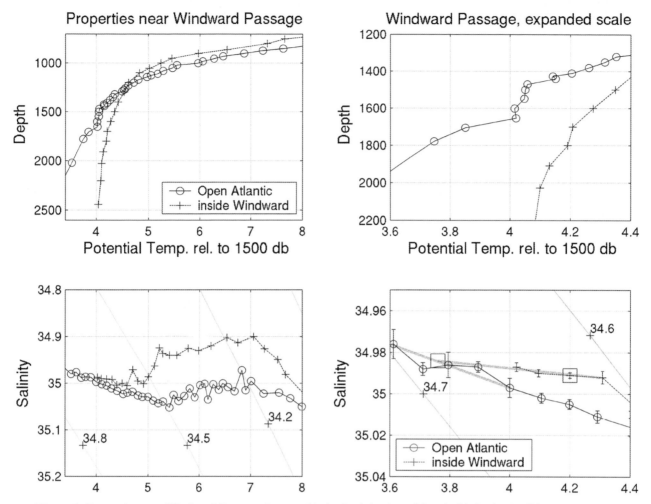

Figure 2. Properties near Windward Passage; just outside in the Atlantic and just inside in the Caribbean. Details as in Figure 1.

3. DEEP INFLOWS

3.1. Inflow at Jungfern Passage

The observations near Jungfern Passage give an excellent introduction to ideas about deep-water renewal over the sills of the Caribbean. The left-hand panels of Figure 1 show temperature versus depth in the upper panel, and temperature versus salinity (t, S) in the lower panel. The right-hand panels show expanded-scale versions of the left-hand panels. These properties are from a collection of data just outside the sill in a region ~ 200 km E-W by ~ 100 km N-S, for the Atlantic data, and similarly (but to the west) just inside the sill for the Caribbean data. Potential temperature inside the Caribbean basins decreases little below ~1500 m, whereas it continues to decrease in the open Atlantic. The salinity minimum inside the Caribbean goes to ~ 34.8 psu here, whereas in the Atlantic to the north it reaches only ~ 34.9. The salinity maximum near 5°C associated with Atlantic water is found on the open Atlantic t, S curve; below that, salinity decreases with depth as a result of Antarctic Bottom Water.

In the lower right panel of Figure 1 we see the remarkable contrast between the water properties inside and out. At temperatures below ~ 5°C salinity decreases with depth in the open Atlantic, whereas salinity increases with depth inside the sills.

If we imagine extending the t, S curve of the Caribbean data, it would intersect the t, S curve of the Atlantic data at the position indicated by the small square symbol near 3.8°C. A second small square, on the Caribbean properties, shows the depth at which the water enters over the sill. That is, if Atlantic water of the characteristics of ~ 3.79°C, 34.988 psu were mixed with the upper Caribbean waters along the inner t, S curve anywhere along this diagram, the resulting mix-

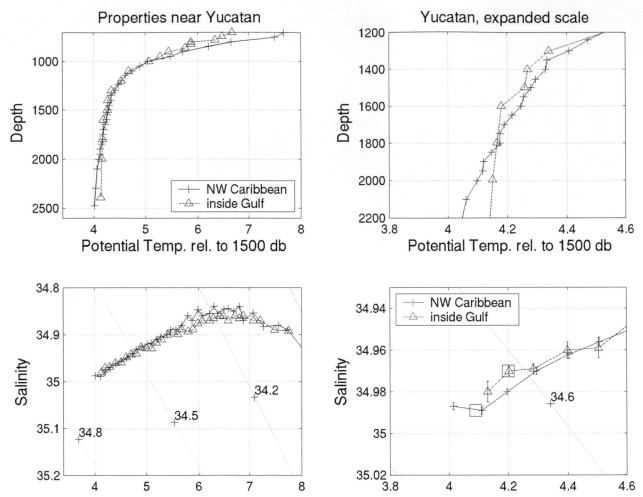

Figure 3. Deep water properties near Yucatan Passage; on the eastern edge of the NW Caribbean and along the eastern boundary inside the Gulf of Mexico. Details as in Figure 1.

ture would have the properties shown in this figure between the two small squares.

The entering water (3.79°C) lies at a depth of ~ 1750 m in the open Atlantic. At those depths, the water on the inside is shown by the small square near 4.09°C. The entering water from the Atlantic mixes by entrainment as it cascades along the bottom and, symbolically on this diagram, all the way along the t, S curve. Small parcels of entering water mix rather quickly, while larger parcels take longer to mix. These entering parcels do not merely plunge to great depth with their original density, of course, but turn to follow the bottom topography, as they mix, and flow approximately along an isobath in geostrophic flow. (The deep outflow leaving the Mediterranean Sea follows a different set of dynamics, making "Meddies," as discussed later; see, e.g., *Nof et al.* 2004.) The process of mixing and entrainment is discussed extensively elsewhere, as mentioned earlier. These ideas are

hardly new. Figure 1 differs only in detail from a similar one based entirely on IGY data [*Sturges* 1970, Figure 1].

A more recent idea is that the deep flow in such basins will preferentially be cyclonic (Northern hemisphere) as a result of the rectification of topographic waves near the bottom (unless the flow is dominated by other effects). The direction of inflowing water here adds to the tendency for cyclonic deep flow, as has been found in other locations: in the Gulf of Mexico [*DeHaan and Sturges* 2005], in the southeastern Caribbean [*Joyce et al.* 2001; *MacCready et al.* 1999], and in the Cayman Basin inside Windward Passage [*W. Johns*, 2004, personal communication], and elsewhere. Cyclonic deep flow is being found in high-resolution numerical models [see the review by *Oey*, 2005, this volume].

It may be appropriate here to anticipate a question to be explored later. Why are the water properties so strikingly different between the open Atlantic and just inside the sills in the Caribbean Sea, when we might have assumed that the water on the inside originates out in the open Atlantic? In other words, what is the source of this odd Caribbean water in the lower right-hand portion of Figure 1?

3.2. Inflow at Windward Passage

Deep water properties near Windward Passage, Figure 2, show one striking difference; the waters inside and outside do not show the conspicuous contrast that is so obvious in the area near Jungfern. The similarities between the water inside and outside Windward Passage suggest an easier exchange with the open ocean above sill depth than is the case at Jungfern-Anegada. Whether the weak temperature gradient inside the sill near 1500–1600m is a consequence of this exchange seems a reasonable (but unanswered) question. In the lower-right panel of Figure 2, I have smoothed the irregularities in the t, S curves for the eye by adding dotted lines. The standard error bars (not standard deviation) are based on small sample sizes and so may not be well determined.

Comparing Figures 1 and 2 we see that the Atlantic water flowing in over the sill at Jungfern is very slightly saltier

than over Windward. Once inside the basins, however, the water inside Jungfern (averaged over several degrees of longitude) is fresher by almost 0.02 psu as a result of the mixing and flow.

3.3. Inflow at Yucatan Channel

Figure 3 shows properties near Yucatan Channel, inside the Gulf near the Florida Escarpment, as compared with those outside Yucatan Channel in the eastern N-W Caribbean. The water-property differences between the two basins are less here than near the Caribbean sills, so the simple idea (from Figure 1) of extrapolating the t, S curve in the Gulf to intersect that from the Caribbean is fraught with uncertainty. The square symbol on the Caribbean t, S curve is placed at the properties found at ~2000 m, the depth of the sill in Yucatan, as being consistent with mixing along the t, S curve in the Gulf up to ~ 4.2°C, relative to 1500db.

Although the differences across the sill are smaller here than on earlier figures, note that the waters of the Caribbean are 0.1°C colder than inside the Gulf. At 2000 m, of course, this is a significant difference.

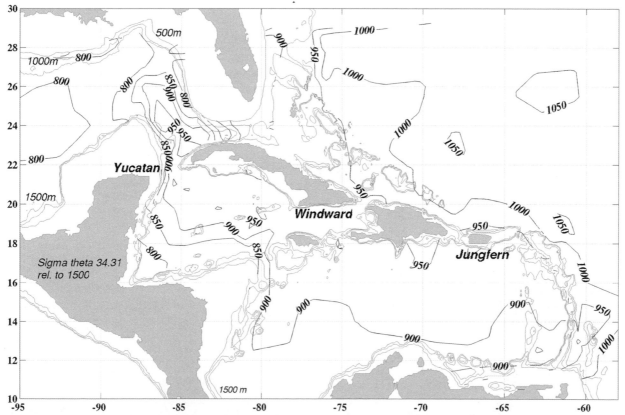

Figure 4. Depth of the potential density surface 34.31 relative to 1500 db. The data are as in Plate 1a. The area near the Florida Straits, or between Florida and Cuba, is not resolved in these data.

4. PROPERTIES ON DEEPER SURFACES

Before studying the deeper flow, it is instructive to examine the lower portion of the upper-layer flow. Plate 1b shows a density surface at the lower depths of the primary upper-layer flow. The salinity on that surface (Figure A 7) shows contours of 34.75–34.85 psu in the far eastern end; these relatively low values are mixed gradually as the flow proceeds to the west. Previous generations of oceanographers were confident that there was an inflow of AAIW at these depths, long before we had good current meter measurements to quantify the transport. If G. Wüst had drawn Figure A 7, he would surely have put several big arrows near 15°N indicating flow to the west. The gradual decrease of salinity from east to west is not a guarantee of flow to the west, but it has often been used as a suggestive indicator.

4.1. Evidence of Deep Shear

The main flow into the Gulf from the Caribbean leaves through the Florida Straits, where the controlling depth is ~800 m. Therefore we might expect the vertical

shear to vanish below that depth. So it comes as a surprise to examine the depth not only of the surface shown in Plate 1b, but also of the potential density surface $\sigma_{1.5} = 34.31$, Figure 4, clearly at the bottom of the main upper-layer flow. The depth of this surface increases from 800 m in the NW to ~950 m in the SE Caribbean. The slope is more pronounced in the western corner of the northwestern basin beneath the flow that forms to become the Yucatan Current; the slope has the same sign as in the upper flow.

So we are faced with a riddle. The limiting depth in the Florida Straits is ~800 m. Yet Figure 4 shows that this density surface obviously has not become level; the vertical shear of the upper flow remains.

It is unlikely that there is substantial net flow to the north into the Gulf below the controlling depth of the Florida Straits. There could be a mean inflow in a limited region, but only if there is a mean compensating return flow to the south. Because the vertical shear continues at this depth and quite a bit deeper, the obvious question becomes At what depth does the mean velocity go to zero and begin to flow to the south? By this we clearly refer to the horizontally averaged velocity represented in the geostrophic shear. We know now, from the results of the recent mooring program (see refer-

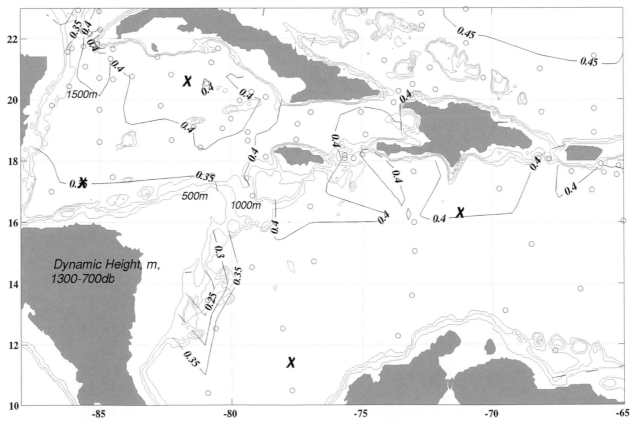

Figure 5. Mean dynamic height between 700 and 1300 db, in dyn. m.

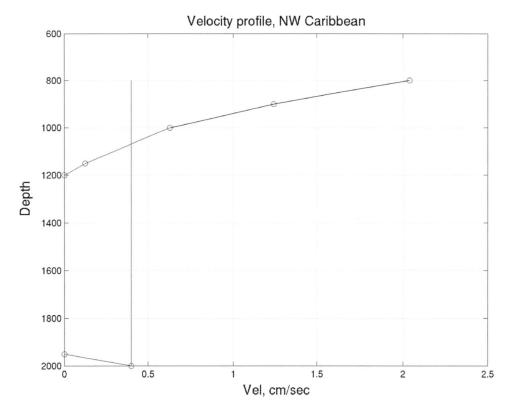

Figure 6. Mean vertical shear in the NW Caribbean basin between the Xs of Figure 5; absolute velocity is determined by mass conservation between 800 and 2000m (see text). The vertical line at 0.4 cm/sec shows the location of the required zero velocity cross-over.

ences earlier), that there is considerable horizontal variability in the mean flow in Yucatan.

And so I have examined a sequence of maps like Figure 4, going deeper and deeper. The surprising result is that the density surfaces continue to slope, with the same sign, down to almost 1500 m, where the signal disappears into the noise.

4.2. The Mean Shear in Deep Water

Because these deep density surfaces slope so consistently, the signal of mean dynamic height in the deep water becomes useful even though it is small. Figure 5 shows the dynamic height between 1300 and 700 db in the Caribbean; Figures 6 and 7 show the vertical shape of the profile (see also Figure A1 in the appendix). There is only a ~4 cm height difference across the region but it is well above the noise level. One purpose of Figure 5, however, is to show the selected locations (marked by the four Xs) where the dynamic height topography is relatively flat, away from regions of strong shear, and where the sampling is adequate. I compiled the

dynamic heights carefully at those 4 locations to determine the mean geostrophic shear in deep water.

Figure 6 shows the computed mean vertical shear and estimated absolute velocity between the two "X" locations in the NW Caribbean. The difference signal is not reliably different from zero below ~1300 m, as is plotted in Figure 6. As a working hypothesis, I assumed that there is no net horizontal flow below ~800 m. If there is flow into the Gulf in a layer below 800 m, as there appears to be, then this flow has to return through the Caribbean and leave above 2000 m, the approximate depth of the sills. Using these constraints, the zero-velocity crossover is found to be at ~1100 m, as shown in Figure 6. The reversal between 1900–2000m is an artifact introduced to bring the velocity to zero at the bottom. (The mean velocity may fall to negligibly-small values at a depth above 1900 m; these assumptions are somewhat arbitrary but are required in order to be specific.) This velocity profile suggests that there is a mean speed of ~2 cm/sec at 800 m. The transport in each layer is ~0.8 Sv. This result, based on the mean properties across the basin, gives no information about the horizontal structure of such a flow. From studying

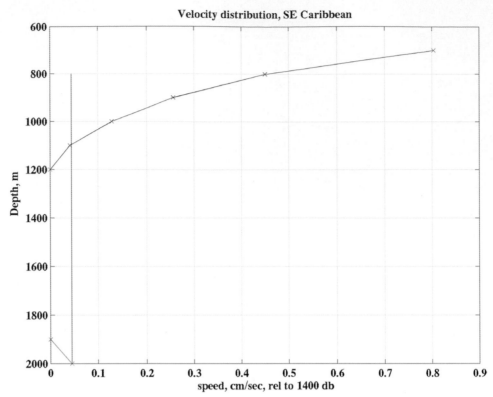

Figure 7. Mean vertical shear in the SE Caribbean basin. Absolute velocity is determined by conservation of transport between 800 and 2000m (see text).

Figure 4, however, one would conclude that the mean inflow to the Gulf, in the NW Caribbean, lies directly beneath the primary upper-layer flow. From Figure 8, the 4.9–5.9 contours suggest that the inflow comes into the Caribbean from both deep passages.

The value of ~0.8 Sv of mean deep transport into (and out of) the Gulf is a most interesting result because it can now be compared with recent mooring data. *Bunge et al.* [2002, Fig. 3] showed that the flow below ~800 m at Yucatan has great variability, with a peak-to-peak range of ~6 Sv. *Candela et al.* [2003, Fig. 2] show a 2-year mean velocity section in Yucatan. Although there are no contours below 800 m, their figure shows mean flow into the Gulf below 800 m, in a region ~45 km wide and 1200 m deep. Taking a mean velocity of ~2 cm/sec, the mean inflow is computed to be ~1 Sv. Considering the size of the error bars, the agreement between this value and the 0.8 Sv obtained from the mean vertical shear in the Caribbean is gratifying. *Rivas et al.* [2005] show a mean flow pattern in Yucatan that, in the vertical, is remarkably like that found here.

Figure 7 shows results from a similar calculation in the S-E basin. The horizontal distance is greater here, so we expect the mean velocities to be smaller. The deep flow must cross the ridge between South America and Jamaica-Hispaniola,

where the controlling depth is probably less than 1600 m. The depth at which the velocity must vanish is uncertain, so the value computed here is a lower limit. The mean transport is computed to be only ~0.4 Sv in each layer. It is barely credible that a deep mean flow of less than 0.1 cm/sec can be obtained reliably by "standard methods." Nevertheless, the sign of the horizontal slope of the deep density surfaces is well determined and the assumptions involved are simple. As in the previous case, the zero-velocity crossover is required near 1100 m.

The cartoon of Plate 2 summarizes the essential results. The primary upper-layer flow extends well below 800 m, to ~1100 m before reversing. The return flow is in a layer between ~1100 and some deep level above 2000 m. It strengthens the case considerably to find that the t, S property gradients and potential vorticity maps support these findings, as shown below.

5. THREE CAVEATS

Figures 6 and 7 should be interpreted with care. First, the slope of density surfaces across the basin is a clear, unambiguous result. The difference in total transport between the two profiles has to be accounted for by transport out of

Windward Passage, but only if we believe the profiles are reliable to perhaps 5–10%. If we take a more pessimistic view, that realistic error bars of the two profiles may overlap, then how the flow is distributed between the two deep passages is an unanswered question.

Second, the time-averaged flow here is the classic problem of a small mean over large and irregular cycles. The deep outflow during Loop Current intrusions is 5–10 Sv [*Bunge et al.,* 2002, *Candela et al.* 2003, *inter alia.*]. Furthermore, the red spectra of most geophysical processes are not kind to modern measurement methods. The time scales involved here are dominated at a minimum by the Loop Current's behavior (order of a year), probably by decadal variability in the transport of the Gulf Stream [*e.g., Sturges and Hong,* 2001], and may be affected by the multi-decadal cycles of the North Atlantic Oscillation [*e.g., Enfield et al.* 2001]. Thus one is hard pressed to know how to interpret error bars on this set of data, no matter how much care is taken in data processing, other than to assume that the error bars are large.

A third difficulty is that much of the mean flow may take place very near the boundaries, or "side walls." Despite my attempt to construct data profiles as close to the boundaries as possible, the available data are not the same as a careful set of observations "up the bottom," such as one would require in looking at a section of the Gulf Stream off the U.S. east coast. Significant fractions of a boundary flow will be missed in these profiles.

6. TRACING THE DEEP FLOW

6.1. Tracing the Upper Layer

On the basis of the previous arguments, we conclude that there is flow from the Caribbean into the Gulf at depths below the main surface inflow. Figure 8 shows potential vorticity in a layer between $\sigma_{1.5} = 34.15$ and 34.51. The layer begins within the main upper-layer flow and extends to depths near 1100 m, where the deep velocity approaches zero. Many PV contours are continuous from the western Caribbean into the Gulf, consistent with the flow pattern proposed here. The 4.9 contour extends into the central Gulf from the open Atlantic near Jungern Passage. Although the inflow from the Atlantic comes from the eastern side and the PV contours here seem fairly uniform across the basin,

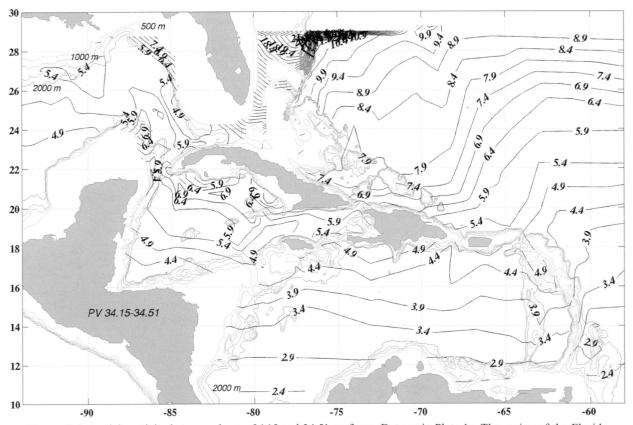

Figure 8. Potential vorticity between the $\sigma_{1.5}$ 34.15 and 34.51 surfaces. Data as in Plate 1a. The region of the Florida Straits is not resolved in these data.

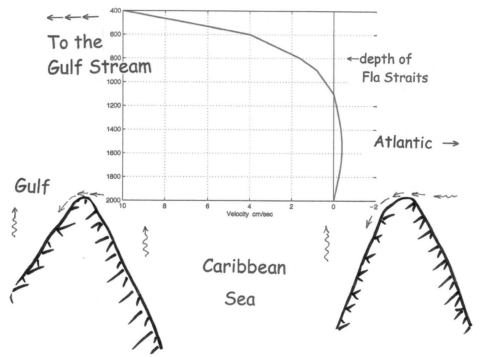

Plate 2. A schematic, or cartoon, of the flow regime described here.

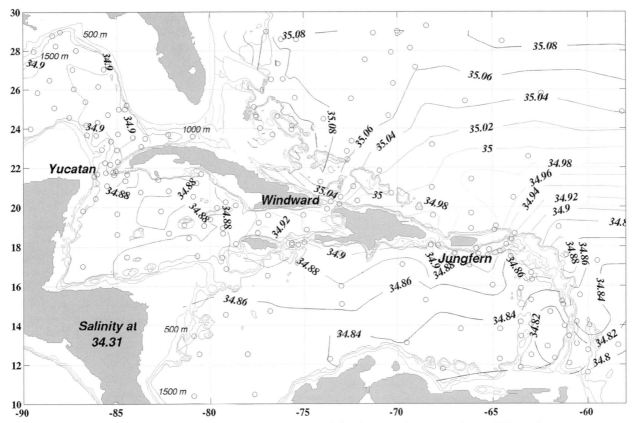

Figure 9. Salinity on the $\sigma_{1.5}$ surface 34.31, which lies between the two surfaces of Figure 8.

the shear associated with the flow into the Gulf as seen in Figure 4 clearly suggests that the inflow is concentrated on the western side of the NW Caribbean as part of the primary upper-layer pattern.

Figure 9 shows salinity within this layer, at $\sigma_{1.5}$ = 34.31, at depths of 800–950 m. Salinity contours of 34.88–34.92 psu are found in the Caribbean, so the salinity in the Gulf, 34.90, can easily originate there. Because of the need to conserve space in the main text, maps showing the depths of the sigma-theta surfaces used here and in the next section are included in Appendix.

6.2. Tracing the Lower Layer

Figure 10 is the deeper-layer analog of Figure 8, showing potential vorticity between the $\sigma_{1.5}$ 34.51 and 34.58 surfaces. One would prefer a calculation over a greater depth range, but at depths below the 34.58 surface, the signal is increasingly degraded by noise. The 1.7 contour extends from the Gulf through the Caribbean all the way back to Jungfern Passage; thus, mean flow along this contour is clearly possible. From Jungfern into the Gulf, latitude changes from 18° to 28° N, suggesting that the PV result is robust.

To provide clues about the direction of the flow, Figure 11 shows salinity on an intermediate surface, $\sigma_{1.5}$ 34.56, at depths of 1250–1300m. At these depths the flow easily crosses the Cayman Ridge between the two deep basins. (Salinity on the 34.58 surface is quite similar.) The waters of the Gulf have salinity of ~34.96 psu. This salinity can be traced through the Caribbean as far back as Jungfern Passage. However, there is no water in the open Atlantic on that density surface (near Jungfern) having salinity so fresh. The conclusion seems inescapable that these waters are formed in the Gulf by deep mixing and flow back to the open Atlantic.

6.3. The Path of Cold Renewal Water

One of the ideas I had hoped to use in this study was that the deep flow might possibly be dominated by topographic rectification, so that there would be a mean deep cyclonic flow around the edges of the basins. *DeHaan and Sturges* [2005] found that, although the entering water at Yucatan is only ~0.1°C cooler than Gulf water at ~2000 m, this temperature deficit is lost slowly by mixing in a weak mean cyclonic flow around the boundary over a distance greater than 1,000 km. Therefore I have examined the properties inside the Caribbean sills in some detail. I can find no reliable trend with distance from the sills, in temperature or salinity, in the averaged data in either of the deep basins. I suspect that the reason for this difference is in the mixing and vertical structure of the flow over the sills.

Candela et al. [2003] show that the mean inflow in Yucatan, as well as the modal structure of the variability, has remarkably weak vertical shear. Indeed, in the middle of the passage the deep flow has one sign all the way to the bottom, with return flow at the sides of the channel. By contrast, observations at the sill at Jungfern [e.g., *Sturges,* 1975, and others] show that the entering water is as much as 0.3°C colder than the surrounding Caribbean water; yet the inflow is found in a layer only ~100 m thick, with large vertical shear. Presumably this leads to greatly increased mixing and entrainment near the sills of the Caribbean. Thus the temperature deficit cannot be traced very far, as contrasted with the Gulf.

While I had expected to find evidence of a weak mean cyclonic deep flow here, these figures offer no support for that idea.

Different overflow orientations can lead to quite different flow configurations. The deep overflow into the Caribbean flows along the bottom and is rapidly mixed away, whereas the outflows from the Mediterranean and Red Seas detach from the bottom and form isolated mid-depth eddies ("Meddies" and "Reddies") that can be tracked into the interior [e.g. *Nof et al.,* 2004].

7. CONCLUSIONS

The main new idea that emerges from this study is that there is a large-scale deep overturning in the Caribbean and Gulf basins. Two sources of mean deep inflow to the Gulf are below the main upper-layer flow that becomes the Loop Current-Florida Current system. The "upper layer of deep inflow" is between the bottom of the upper-layer inflow, ~800 m, and ~1100 m, as shown in Plate 2. The potential vorticity map of Figure 8 shows that this flow is likely, and the salinity gradients are consistent with flow from east to west. The vertical shear (Figure 4) suggests that this inflow, in the mean, is directly beneath the upper-layer flow on the western side of the Caribbean.

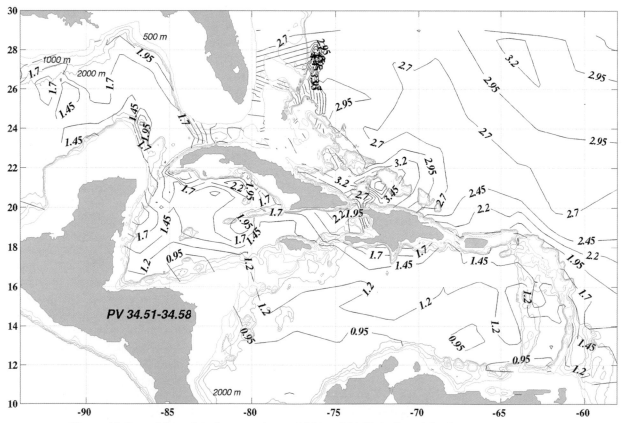

Figure 10. Potential vorticity between the $\sigma_{1.5}$ 34.51 and 34.58 surfaces. The data are as in Plate 1a.

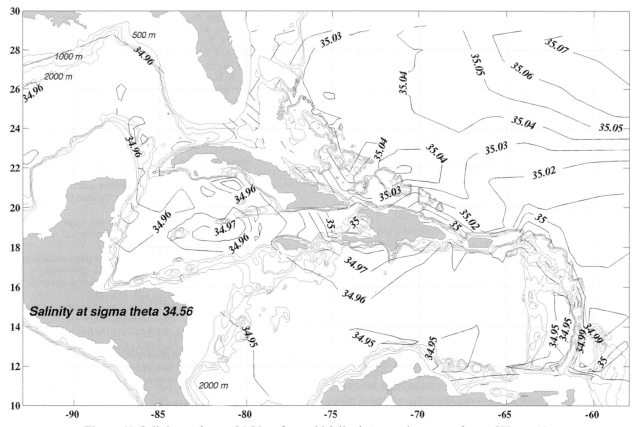

Figure 11. Salinity on the $\sigma_{1.5}$ 34.56 surface, which lies between the two surfaces of Figure 10

The second region of deep flow into the Gulf is the well-known inflow of deep water that pours over the sills at Jungfern and Windward Passage and on to the sill at Yucatan. The unambiguous slope of the deep density surfaces leads to an unmistakable vertical shear, and the argument that there can be little net mean flow into the Gulf between ~800 and 2000 m seems sound.

To achieve mass balance, there must be a return flow out of the Gulf at depths below 1100 m; at least to ~1500 m, and perhaps deeper, into the Caribbean, all the way back to Windward and Jungfern Passages and out into the Atlantic. The salinity contours near 34.96 psu support this idea, as does the potential vorticity map of Figure 10. These results arise from the mean properties in the Gulf-Caribbean system and are strongly supported by the recent current meter results in Yucatan. The idea of a deep cyclonic flow at these depths, driven by the rectification of deep topographic Rossby waves, and augmented by the waters flowing over the fortuitous geometry of the sills, is consistent with these conclusions, but the observations are not clear enough in themselves to give support to such a cyclonic flow pattern. The noise level in the data make the relative amounts of flow that leave via Windward versus Jungfern passes difficult to assess.

In their study of the mean deep circulation in the Gulf, *DeHaan and Sturges* [2005] found that the temperature deficit of the entering renewal water eroded away gradually over a distance of thousands of kilometers. No evidence is found here in the Caribbean of such a long decay scale.

These arguments are based on a conceptual model that relies on horizontal flow to the exclusion of vertical mixing. In recent years, large-scale tracer studies in the open ocean have shown that vertical mixing is greatly suppressed compared with the effects of horizontal advection [*e.g., Ledwell et al.* (1998)], although vertical mixing can be greatly enhanced near steep topography having the right horizontal length scales. At depths just above the sills, however, the vertical stratification is weak and the mean flow is small. At these depths the effects of vertical mixing as well as upwelling would therefore be expected to become dominant.

Acknowledgments. This work was begun while I was supported by a grant from the National Science foundation and was continued with support from the U.S. Minerals Management Service, for

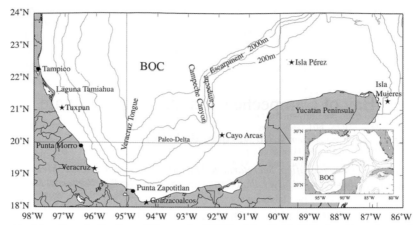

Figure 1. Geographical locations in the Bay of Campeche (BOC) of the Gulf of Mexico. Stars denote locations of special meteorological stations referred to in section 5. Bathymetry contours shown for the 200, 1000, 2000, and 3000-m isobaths. Inset shows the whole Gulf of Mexico. Cheliva Pass through the Sierra Madre mountains (not shown) is centered at about 95°W and 17°N in the Isthmus of Tehuantepec.

with the mean surface currents. We also address the probable mechanisms for the existence of a permanent cyclonic circulation in the BOC, its seasonal variation, and for the non-seasonal variability of the circulation therein. Dominant patterns of variability of the sea surface height (SSH) in the BOC are made possible via data derived from satellite altimeter sensors [*Leben et al.*, 2002].

With the exclusion of the BOC, the western Gulf of Mexico is dominated by the presence of transient energetic anticyclonic (warm core) eddies whose lifetime is about one to two years. It is well established, based on both ship surveys and remote thermal and altimeter measurements from satellites, that these transient warm core eddies are spawned by the Loop Current and translate into the western Gulf primarily along a south-westward path [e.g., *Elliott*, 1982; *Vukovich and Crissman*, 1986]. The shedding interval varies from several months to over a year with mean value of about 10 months [*Sturges and Leben*, 2000]. This is consistent with the usual presence of two warm core eddies in the western Gulf at any given time.

Cold core cyclonic eddies have been found to occur in association with the westward drifting anticyclones [*Merrell and Vázquez*, 1983; *Brooks*, 1984; *Hamilton*, 1992]. There is evidence from model studies that transient cyclonic eddies can be created by interaction of the energetic warm core eddies with the continental slope [*Smith*, 1986; *Oey*, 1995].

It is now well recognized that there also exists an underlying wind-driven anticyclonic circulation in the western Gulf north of the BOC that has an associated moderately strong western boundary current with a significant seasonal signal [*Sturges*, 1993]. But what has not been emphasized in the literature until the study of *Gutiérrez de Velasco and Winant* [1996] is that south of about 22°N the annual mean

wind stress curl is cyclonic and is asserted herein to produce a permanent cyclonic circulation in the BOC. Indeed the paper by *DiMarco et al.* [2005] confirms the existence of the anticyclonic western boundary current north of 22°N and the BOC cyclone south of that latitude. Moreover, in section 5.2 we discuss how the morphology of the Sierra Madre mountain range bordering the BOC on the west and south plays a profound role in the existence of the cyclonic wind stress curl over the BOC.

Hydrographic and satellite-derived SSH data reveal that some intrusion of transient eddies into the BOC does occur. Moreover, intrusion of warm water occasionally takes the form of a narrow ridge between the BOC cyclone and the Campeche escarpment on the eastern side of the BOC [as in VC75]. Figure 2 however shows a more typical pattern of the BOC cyclone from a February–March cruise of the *R/V Geronim*o in 1967. The dynamic height field shown in that figure is one in which horizontal scales less than about 30 km have been filtered out. The method for doing this is discussed in section 2.1.

Cyclonic eddies tend not to entrain water from their surroundings, and, as independent evidence that a permanent closed cyclonic circulation exists in the BOC, the observed oxygen minimum in the BOC is lowest in the western Gulf. The average minimum oxygen value for the BOC is 2.50 ml·L^{-1} compared with 2.70 ml·L^{-1} for the northwestern Gulf. For both regions the depth of the minimum is about 300m. The standard error of estimated difference is about 0.02 ml·L^{-1} or only 10% of the difference. Background and details of the analysis are given in [*Jochens et al.*, 2005].

The organization of the presentation is as follows. In section 2 we present the hydrographic evidence for mean

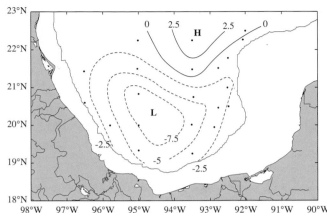

Figure 2. Contours of EOF mode 1 dynamic height (dyn cm) of the sea surface relative to 425 db for the *Geronimo* cruise 67-G-12 during 23 February–13 March 1967. See section 2.1 for discussion of EOF mode 1 dynamic height. Dots show locations of stations. The 200-m isobath is shown.

cyclonic circulation in the BOC. Section 3 then presents the evidence based on Lagrangian measurements. Discussed are the mean surface and 900-m current fields determined respectively from drifter and float measurements. Section 4 presents results of an empirical orthogonal function (EOF) analysis based on over eight years of SSHA data for the BOC provided by the University of Colorado Center for Astrodynamics Research.

Section 5 contains a discussion of the findings, the probable mechanism for the existence of the permanent cyclonic circulation and its seasonal signal in the BOC based on Ekman pumping, and related results concerning the BOC from model studies by others. Closing remarks are then given in section 6.

2. HYDROGRAPHIC EVIDENCE

2.1. Dynamic Topography of Sea Surface Relative to 425 db

The ship surveys within the BOC that were employed in the VC93 study are listed in Table 1. This excludes 5 cruises in the early 1950s that did not have reliable salinity data, but patterns of the depth of the 15°C surface indicated that a cold core eddy was present on each. The selected thirteen cruises span the period from 1958 to 1991. They include two surveys by Texas A&M University, two by the U. S. Fish and Wildlife Service, and nine surveys by the Mexican government. Table 1 includes the duration of the survey within the BOC domain and the number of stations occupied south of 22.5°N. The common deepest depth of temperature and salinity sampling for the 247 stations in the data ensemble is 425-m (or about 425 db in pressure).

Table 1. Bay of Campeche (BOC) oceanographic cruises from 1958 through 1991 that are used in the VC93 estimate of the mean surface topography field.

Vessel	Cruise Name	Duration in BOC	Total stations in BOC
Hidalgo	58-H-1	03/23/58–04/03/58	6
Hidalgo	62-H-3	03/11/62–03/24/62	17
Geronimo	67-G-12	02/23/67–03/13/67	21
Geronimo	67-G-16	08/19/67–09/19/67	19
Uribe	Cosma 70-12	10/31/70–11/13/70	22
Uribe	Cosma 71-10	05/24/71–06/09/71	23
Uribe	Cosma 71-16	08/04/71–09/03/71	29
Uribe	Cosma 71-22	10/27/71–11/10/71	23
Uribe	Cosma 72-10	04/25/72–05/18/72	20
Uribe	Cosma 73-10	05/23/73–06/04/73	9
Altair	Alt-M-87	05/08/87–05/27/87	15
J. Sierra	Circam 1	01/17/91–01/31/91	29
J. Sierra	Circam 2	07/12/91–07/24/91	14
Total			247

The sparse sampling of stations in many of the surveys make it difficult to resolve the small eddy, as illustrated in Figure 2. In order to suppress aliasing of horizontal patterns associated with unresolved scales in the data being sampled, the VC93 study employed a filtering technique introduced by Fukumori and Wunsch [1991]. In that method, the vertical distributions of those variables to be mapped horizontally are represented by the dominant mode of an empirical orthogonal function (EOF) decomposition of that data. The rationale is that the dominant EOF mode is normally associated with the largest natural horizontal scale (the first baroclinic radius of deformation), as in the case of dynamic modes [Gill, 1982]. Thus, by suppressing the higher vertical modes, one also suppresses spatial patterns of smaller horizontal scale. In the discussion to follow, we denote the surface value of the first mode of dynamic height anomaly by D1.

The vertical profiles of the first three EOF modes for the geopotential anomaly based on the ensemble of 247 stations of 425-m depth for the BOC are shown in Figure 3, along with the ensemble mean profile of the geopotential anomaly. Each modal profile is scaled by the square root of its fractional variance (EOF eigenvalue), such that the sum of the squares of the discrete profile gives a total that is proportional to the variance. We show these in this way to emphasize that the possible contribution near the sea surface by modes other than the first are not negligible in terms of amplitude, even though the contribution to the variance of

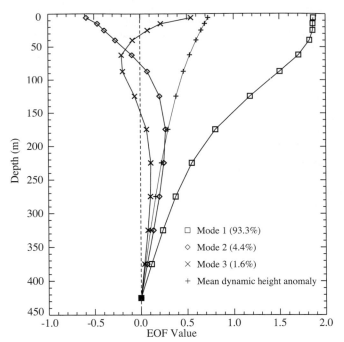

Figure 3. Vertical structure for the EOF modes 1, 2, and 3 based on 247 stations. Also shown is the mean dynamic height anomaly (dyn m). The actual signs and magnitudes of the different modes are governed by the product of the independent modal amplitudes and the corresponding modal profile.

the ensemble is small. For the BOC data the fractional variances of geopotential anomaly contributed by the first three modes are 93.3%, 4.4% and 1.6%. The radii of deformation (R_n) are about 33, 12 and 4 km, respectively.

The pattern of dynamic height for the survey illustrated in Figure 2 was contoured subjectively from the discrete array of first mode approximation of sea surface dynamic height (D1) at station positions. Another synoptic realization of the BOC eddy as depicted in terms of the contours of the first EOF mode representation is shown in Figure 4 for an October–November 1971 cruise of the R/V Uribe. In both cases the center of the cyclone is near 20°N latitude, but the second realization is somewhat smaller.

Fields of D1 were computed for all 13 cruises listed in Table 1. Of these surveys, 5 were in spring, 4 were in summer, 2 were in fall, and 2 were in winter of nine different years. Based on the combined 247 stations represented by these cruises a mean field was constructed in the following manner. The data were binned according to station location in quadrangles of 1.4° × 1.4°, centered on each 0.5° of latitude and longitude over the deep portion of the BOC domain. The average D1 of those stations in each of the overlapping bins was evaluated, and bins having less than 9 stations were ignored. The average number of stations per bin was

26, with a range from 9 to 53 (small numbers occurring near the outer limits of the bin domain). The total number of bins was 61. For each of these bins, the mean values of latitude and longitude were evaluated for plotting the location of each bin mean D1.

The resulting subjectively contoured pattern of mean D1 referenced to 425 db is shown in Figure 5. The minimum D1 within this mean cyclonic eddy is -5.1 dyn cm. Dots denote the effective centroids of the bins. As expected, the averaging over the 13 synoptic realizations of cyclones having different center locations yields a mean cyclone with larger scale than those of the individual surveys. The mean state shows a single well-defined cyclonic circulation in the whole portion of the BOC south of about 21°N.

We take as a measure of uncertainty of the mean D1, the standard deviation of D1 divided by the square root of the sample size. The standard deviation of D1 determined from the ensemble of 247 stations is about 5 dyn cm. This is amazingly close to the standard deviation of SSHA (5.5 cm for the central BOC) determined from satellite-derived altimeter data [*Nowlin et al.*, 2001, Figure 5.4–6]. If we use the mean sample size per bin of 26, then the uncertainty of the mean D1 values in Figure 5 is about 1 dyn cm. Thus the mean cyclonic circulation based on the ensemble of hydrographic data is a statistically robust estimate of the mean surface dynamic topography relative to 425 db.

2.2. Dynamic Topography Relative to 800 db and Deeper

An analysis of the mean surface dynamic height relative to 800 db and that of the 800 db surface relative to 1500 db for the Gulf of Mexico as a whole was carried out in the Deepwater Reanalysis and Synthesis study funded by the

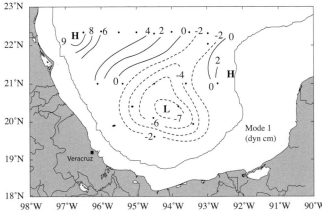

Figure 4. Contours of EOF mode 1 dynamic height (dyn cm) of the sea surface relative to 425 db for *Uribe* cruise Cosma 71-22 during 27 October–10 November 1971. Dots show locations of stations. The 200-m isobath is shown.

Minerals Management Service [*Nowlin et al.*, 2001]. We summarize important aspects of the analysis and results relevant to the BOC here. All available hydrographic data for the Gulf of Mexico having both temperature and salinity data extending at least to 800 m were assembled for QC screening. The reliability of the T/S relation, using that from such cruises as *R/V Hidalgo* 62-H-3 and *R/V Kane* 6906 as a benchmark for the Gulf, was used in the screening. The total number of accepted stations for the Gulf was 1630 of which 201 were in the BOC. Table 2 lists the cruises, their duration and the number of stations from each cruise having stations reaching at least 800 m and those reaching at least 1500 m.

The dynamic height computations for a given pressure surface in the analysis of this data was done in the standard manner, although an average value over a given domain of the Gulf may be subtracted without altering the relative values or associated geostrophic current. While the *Fukumori and Wunsch* [1991] EOF filtering method is useful in the representation of synoptic data as done in the VC93 analysis, it has little influence on fields of the mean dynamic height anomaly where binning of data from different cruises is employed. The original binning of the Gulf data was by 0.5° × 0.5° quadrangles. This leads to a field with many empty bins. So in the analysis of this modified data set for the BOC region, we used overlapping bins of 1.5° × 1.5° at spacing of 0.5° in latitude and longitude. This is similar to the procedure used in the VC93 analysis, except that in the present analysis the averages are plotted at the bin centers for simplicity.

The resulting pattern of mean dynamic height anomaly of the sea surface relative to 800 db for the BOC is shown in Figure 6, in which the area average for the BOC has been subtracted. The maximum depression of 5 dyn cm in the cyclone is comparable to that of Figure 5. Thus the main difference

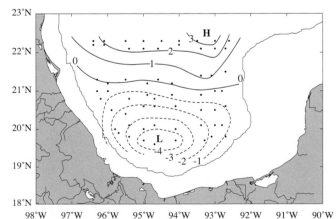

Figure 5. Contours of the mean EOF mode 1 dynamic height (dyn cm) of the sea surface relative to 425 db. Dots show centroids of stations in 1.4° × 1.4° bins from which the bin sample means were determined. The 200-m isobath is shown.

Table 2. Bay of Campeche (BOC) oceanographic cruises from 1932 through 1985 that are used in the computation of dynamic height with references at 800 db and 1500 db.

Vessel	Duration in BOC	Total stations in BOC	
		Surface re 800 db	800 db re 1500 db
Alaminos	01/21/64–01/25/64 & 03/02/64–03/03/64	8	4
Alaminos	10/07/65	1	–
Alaminos	02/13/65	1	–
Alaminos	02/28/68	1	–
Alaminos	07/26/69–07/28/69	3	–
Atlantis	04/07/35	1	–
Atlantis	11/16/68	1	–
Geronimo	02/23/67–02/25/67	8	5
Geronimo	03/05/67–03/13/67	9	3
Geronimo	08/19/67–09/12/67	21	15
Hidalgo	03/29/58–03/30/58	2	1
Hidalgo	02/25/59	2	2
Hidalgo	03/11/62–03/24/62	19	14
Justo Sierra	03/11/85–03/22/85	14	11
Kane	07/19/69	1	1
Kane	08/06/69–08/13/69	15	2
Kane	10/21/69 & 10/24/69	2	2
San Pablo	05/17/68	1	–
Taylor	04/6/32–04/16/32	11	6
Virgilio Uribe	11/03/70–11/13/70	15	
Virgilio Uribe	05/25/71–06/09/71	17	1
Virgilio Uribe	08/05/71–08/09/71	8	1
Virgilio Uribe	08/31/71–09/03/71	13	1
Virgilio Uribe	10/28/71–11/09/71	6	–
Virgilio Uribe	04/26/72	1	–
Virgilio Uribe	05/15/72–05/18/72	14	–
Virgilio Uribe	05/30/73–06/01/73	6	–
Total		201	69

in the use of the deeper reference of 800 db is to produce a distortion in the mean cyclone at the sea surface. In particular, the elongation of the major axis of the low is NW–SE in Figure 6 rather than W–E as in Figure 5. The effective sample size for the overlapping large bins varies from 8 to 40 with an average of 25. The sample standard deviation of the dynamic height anomaly is 4.5 dyn cm, again close to that

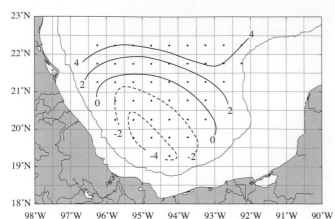

Figure 6. Contours of the mean dynamic height anomaly (dyn cm) of the sea surface relative to 800 db. The average dynamic height anomaly from all bins has been subtracted. Dots show the centers of $1.5° \times 1.5°$ bins from which the bin sample means were determined. The 200-m depth contour is shown.

for the SSHA for the BOC, so the uncertainty of the mean contours is about 1 dyn cm (as for Figure 5).

A similar analysis was carried out for the dynamic height anomaly of the 800 db pressure surface relative to 1500 db. The sample of data for this deep region is much smaller and suggests only that a weak and ill-defined circulation exists at the 800 db surface with a range of geopotential anomaly of order 2 dyn cm.

3. LAGRANGIAN CURRENT EVIDENCE

Two Lagrangian current studies, that of *DiMarco et al.* [2005] on near surface flow and that of *Weatherly et al.* [2005] on the 900-m flow are very relevant to the BOC circulation. Both papers are presented in this AGU monograph on the Gulf of Mexico circulation. Therefore since the readers have direct access to these studies, we give only those details of analysis that differ from those of *DiMarco et al.* [2005] and focus on results pertaining primarily to the BOC for each.

3.1. Mean and Seasonal Near-Surface Current Fields

In the study by *DiMarco et al.* [2005], position data from drifters drogued at depths from 6 to 200 m but mostly at 50m were available for the period 1989 through 1999. The drogued drifter data were used to produce an estimate of the ten-year mean and seasonal pattern for near surface quasi-geostrophic flow for the Gulf of Mexico. Details of the quality control are given in *DiMarco et al.* [2005].

The resulting drifter velocity estimates were determined for each 0.5° of latitude and longitude using $1.5° \times 1.5°$ overlapping bins and averaged for selected periods. Figure 7 shows the ten-year mean near surface vectors for each 0.5° latitude and longitude in the BOC, and the standard deviation ellipses are shown in Figure 8. The uncertainty ellipses, characterizing the standard error of estimate of the mean vectors, can be estimated from the ellipses in Figure 8 divided by the square root of the effective degrees of freedom, which we take as the number of 6-hr sample averages divided by the integral time scale taken as 8 days. The effective degrees of freedom for each of the overlapping $1.5° \times 1.5°$ bins are given in Figure 9. For the BOC, those vectors having less than 5 degrees of freedom are considered not statistically robust and have not been plotted in Figure 7. Vectors shown for bins centered over land (e.g., at 95.0°W, 18.5°N in Figure 7) are based on that part of the $1.5° \times 1.5°$ bin over water. The vectors for bins that include shelf regions with water depths less than 200m may be subject to considerable variance due to local variability of winds. This is probably the reason for the very large variance shown in the southern part of the BOC in Figure 8.

The ten-year mean near surface current vectors have a very clear cyclonic pattern of circulation south of 22.5°N. The vector pattern is qualitatively consistent with the geostrophic streamlines of Figure 5 but with the major axis shifted about 0.5° northward. Also the vector pattern shows a westward intensification which is more consistent with the streamlines of Figure 6. Unfortunately the eastern closure region for the northward flowing limb of the cyclonic circulation in

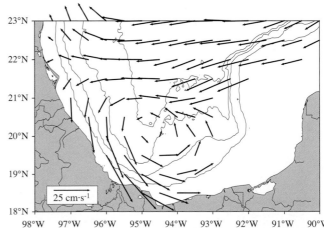

Figure 7. Average current vectors based on near-surface drogued drifter data over a ten-year period. These vectors are for overlapping bins of $1.5° \times 1.5°$ with the tail of the vectors centered on each of the overlapping bins. Only vectors having 5 or more degrees of freedom are shown (see Figure 9). The 200, 1000, 2000, and 3000-m isobaths are shown. [Adapted from *DiMarco et al.*, 2005.]

the southern BOC region of Figure 7 was not captured in a statistically dependable way.

For the BOC, circulation patterns based on separation of the drifter data into a six-month winter (October through March) and a six-month summer (April through September) were also computed and the vector plots for the BOC are shown in Figure 10. The variance ellipses for winter and summer are similar to those shown in Figure 8. The degrees of freedom for the summer case are given in Figure 11 (those for winter can be estimated from Figures 9 and 11 by difference). Clearly there is a striking difference in the summer and winter patterns. Unfortunately, the vectors in the west and south of the BOC had insufficient degrees of freedom for the summer data so it is difficult to compare the summer and winter patterns quantitatively. However comparison of the winter pattern with the overall mean pattern in Figure 7 does show a stronger western boundary current for winter (note the different velocity scales in Figures 7 and 10).

3.2. Mean Current Field at 900-m

The field of current at approximately 900 m in the Gulf of Mexico has been estimated using PALACE float data by *Weatherly et al.* [2005]. The estimates of vector flow at 900-m are based on the results of seventeen floats that were set in the northern slope region of the Gulf of Mexico and tracked for the period April 1998 to March 2002. The floats spend about 6 days of their 7 day cycle from sea surface to about 900-m at the latter depth, and each cycle produces one vector estimate provided the float is not grounded. For details regarding the lifetime of the floats, the methodology of tracking, checks for grounding, and the deep drift quantification,

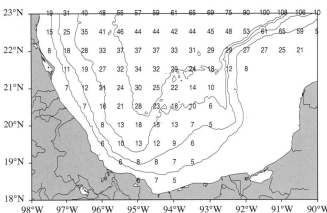

Figure 9. Values of the effective number of degrees of freedom for the near surface vectors determined from 1.5° × 1.5° overlapping bins based on an integral time scale of 8 days. [Adapted from *DiMarco et al.*, 2005.]

the reader should consult the above reference. Our use of the results of that analysis is limited strictly to the domain of the Bay of Campeche.

Figure 12 shows the estimated mean current vector field at about 900 m for the BOC region, as adapted from *Weatherly* [2004]. The vectors are centered in the 0.5° × 0.5° bins to which they apply, and only vectors based on averages over five or more (6-day) deep drift estimates are shown. Comparison of Figure 12 with the near surface vector field shown in Figure 7 shows a similarity of patterns on the western, central and southern region of the BOC, but not elsewhere. Nevertheless the 900 m field of flow provides a robust mean reference state for the mean geostrophic fields shown in Figure 6.

The gulf-wide flow pattern at 900m by *Weatherly* [2004] is also shown in *Weatherly et al.* [2005] that appears in this monograph. It shows a remarkably continuous cyclonic flow around the periphery of the gulf at the 900 m level, of which the BOC part is just a small portion.

4. SATELLITE ALTIMETER EVIDENCE

4.1 Time Sequence of SSHA for the Gulf of Mexico

An 8.5-year sequence of sea surface height anomaly (SSHA) patterns for the Gulf of Mexico beginning in January 1993 was available at the time of preparation of this paper. These patterns were derived from blending of altimeter data from the TOPEX and ERS-2 satellites by the Colorado Center for Astrodynamics Research (CCAR). The TOPEX and ERS-2 satellites have exact repeat orbits respectively of 10 and 35 days period. The details of the processing are given in *Leben et al.* [2002].

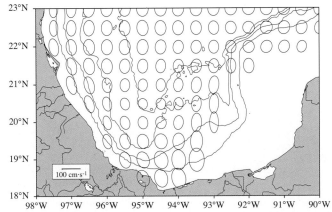

Figure 8. Standard deviation ellipses for the near-surface drogued drifter velocity data. These ellipses are for overlapping bins of 1.5°x1.5°. The 200, 1000, 2000, and 3000-m isobaths are shown. [Adapted from *DiMarco et al.*, 2005.]

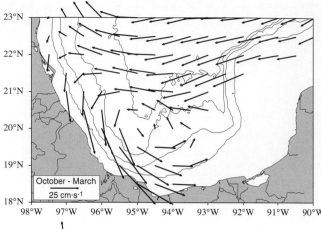

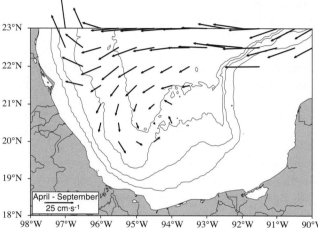

Figure 10. Seasonal current vectors from the near-surface drogued drifter data for (upper) October–March and (lower) April–September. These vectors are for overlapping bins of 1.5° × 1.5° with the tail of the vectors centered on each of the overlapping bins. Only vectors having 5 or more degrees of freedom are shown (see Figure 11). The 200, 1000, 2000, and 3000-m isobaths are shown. Note the different velocity scale from that in Figure 7. [Adapted from *DiMarco et al.*, 2005.]

Our use of the 8.5-year sequence of the SSHA data is limited to the BOC region between 90°W and 98°W, south of and including 23.0°N, and deeper than 200-m water depth. Using a 0.5° × 0.5° grid gives a total of 82 spatial nodes. The original SSHA data set gave daily samples. However, in view of the 12 day decorrelation time scale, we converted the data to be non-overlapping biweekly averages, in anticipation of carrying out EOF analysis of this space-time matrix ensemble.

Two limitations of SSHA are important in our analysis here. The first is that the SSHA gives no information about the mean sea surface height pattern. The second is that annual and semi-annual variability of the SSHA is subject to serious aliasing due to the existence of tides [*Tierney et al.*, 1998].

We calculated the annual signal and its first two harmonics using least squares cyclic reduction [*Bloomfield*, 1976; *DiMarco and Reid*, 1998] at the 82 BOC spatial nodes. The results showed a much larger semi-annual than annual signal, which is highly unrealistic for the Gulf [*Pattullo et al.*, 1955; *Whitaker*, 1971]. Therefore we removed this seasonal signal from the SSHA data, since it was likely contaminated by tidal aliasing. This is unfortunate since it does not allow an independent estimate of the observed seasonal variability in the BOC to supplement that from the drifter data. However the SSHA can be valuable in assessing the non-seasonal spatial and temporal eddy variability.

4.2 Patterns of Dominant SSHA EOF Modes

The SSHA patterns in the BOC are weak and tend to be noisy. For this reason we seek the dominant EOF spatial patterns and their temporal variability. The modified matrix of 82 spatial nodes for over 200 biweekly time average values of the modified SSHA data is readily converted, using singular value decomposition (SVD), into a matrix of spatial patterns (EOFs) by mode, a matrix of modal amplitudes by time, and a set of 81 singular values, whose squares are the fractional variances of SSHA contributed by each mode. We present below the results of applying the SVD analysis to the modified SSHA data for the BOC.

Figure 13 shows the spatial patterns of the four most dominant modes, which together represent nearly 60% of the total variance of the modified SSHA. The dominant mode, which contributes 26.3% of the variance, shows a relatively strong eddy centered at 23.0°N and 95.5°W with a weaker eddy of opposite sign centered at 20.5°N and 95.5°W. The actual signs and strength for given time depends on the sign

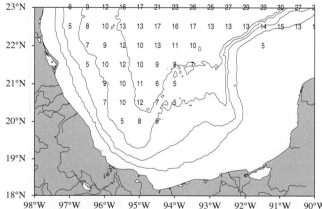

Figure 11. Values of the effective number of degrees of freedom for April–September of the near surface vectors determined from 1.5° × 1.5° overlapping bins based on an integral time scale of 8 days. [Adapted from *DiMarco et al.*, 2005.]

and magnitude of the modal amplitude, positive (negative) implying that the northern eddy is cyclonic (anticyclonic), while the southern eddy is anticyclonic (cyclonic). The mode 2 pattern also constitutes an eddy pair similar to mode 1 but shifted northward (southward) about one degree depending on whether the amplitude of mode 2 is of same (opposite) sign relative to that of mode 1. The two modes in combination can give a filtered approximation of the eddy propagation northward or southward in the northern BOC depending on the sequencing of the modal amplitudes. Mode 3 contains north and south eddies of like sign but with a ridge or trough intruding southwestward between the pair. We interpret this as a modal version of southward intrusion along the Campeche escarpment that appears in several of the cruises illustrated in VC75. But the SSHA data shows that this mode contributes only 10.8% to the total variance of the nonseasonal variability.

4.3 Temporal Variability of the Modal Amplitudes

The variations of the EOF modal amplitudes of the first four modes are shown in Plate1. As expected, we see no obvious annual or semiannual signal in these 8.5-year sequences. Note that the amplitude scale is tied to the modal patterns (Figure 13) as illustrated in the following example. For mode 1, the maximum value for the eddy centered at 23.0°N and 95.5°W at the end of August 1993 is about -0.24×-55 or $+13.2$ cm while the mode 1 eddy centered at 20.5°N and 95.5°W is about -7.7 cm.

The auto-spectra for modes 1 through 4 (not shown) confirm the visual impression from Plate1 that no significant spectral variance remains at annual or semi-annual fre-

quency. However, possible peaks occur near about 1.5, 2.5, 3.5, and 5 c·yr^{-1} among the four modes. This suggests eddy lifetimes in the range of about 2.4 to 8 months in the BOC, which also corresponds to the range of positive or negative modal amplitude durations seen in Plate 1.

Within the context of a two-mode approximation of the eddy field, we discuss several examples of propagation of transient eddies. Consistent with the patterns of modal SSHA in Figure 13, we will use the terms low and high as synonymous with cyclone and anticyclone respectively. While the lifetime of a given transient eddy mode generally does not exceed eight months, a combination of modes 1 and 2 usually shows one lagging the other in time and their combination can produce a propagating eddy field (either southward or northward) with a somewhat longer lifetime than either mode pattern by itself. We will call such a combination an episode.

Three classes of mode 1 and 2 episodes occur in terms of their time sequence of amplitudes as in Plate 1. Class A is characterized by negative cross-correlation of amplitudes for the episode lifetime, with mode 2 lagging mode 1, and has southward propagation. Class B is characterized by positive cross-correlation of amplitudes, with mode 1 lagging mode 2, and also has southward propagation. Finally class C is characterized by negative cross-correlation of amplitudes, with mode 1 lagging mode 2, and has northward propagation. Examples of the three classes of propagation episodes are identified by color-coding in Plate 1: A = green, B = blue, and C = red. For simplicity, the following descriptions illustrating the three episode classes ignore any prior or following episode in time.

An example of class A is the southward propagation of a high during the first eight months of 1997 (denoted by 1 in Plate1). During January the pattern is dominated by mode 1 and has a weak high in the northernmost region (note signs in Figures 13 and Plate 1 for mode 1). In May both modes are near full strength and imply a strong combined high with center between 21.5 and 23°N. Finally in August mode 2 dominates and shows a weak high centered at about 21.5°N. The mean time lag of mode 2 relative to mode 1 appears to be roughly two months. We regard this lag as an effective propagation time of the transient high.

An illustration of a class B southward propagating transient high is for October 1999 to June 2000 (episode 2 of Plate 1). The early pattern is dominated by mode 2 (Figure 13). In January 2000 the combination of both modes shows a strong high shifted southward, and by May mode 2 dominates with a deeply intruded but weak high within the BOC (Figure 13).

Finally, an example of a northward propagating cyclone exists for the months of May through November of 1994 (episode 3 of Plate 1).

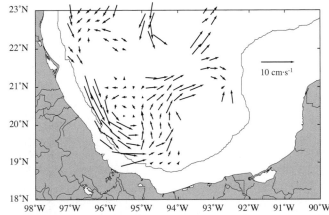

Figure 12. Averaged current vectors at about 900-m depth based on PALACE float data over a four-year period. These vectors are for overlapping 0.5° × x0.5° bins having 5 or more 6-day average samples. The 900-m isobath is shown. [Adapted from *Weatherly*, 2004.]

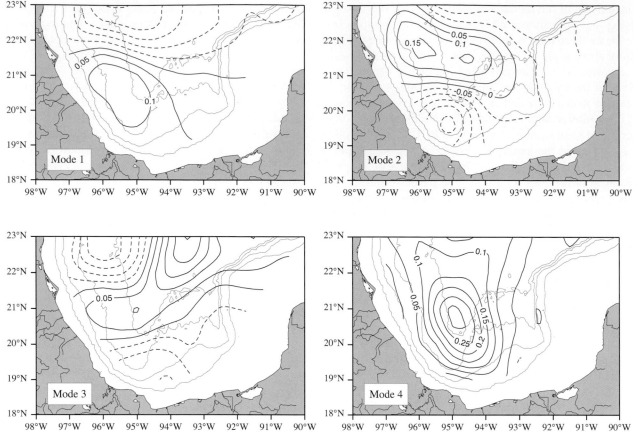

Figure 13. The first four EOF spatial patterns of SSHA based on the modified data matrix, for which the annual and semiannual temporal signals have been subtracted. The modified SSHA data matrix consists of over 220 non-overlapping biweekly values at each of 82 spatial nodes in the BOC. The fractional variances of SSHA by modes 1 to 4 are respectively 26.3%, 14.0%, 10.8%, and 8.3%.

There are clearly many more examples in Plate 1 of southward propagation of eddy energy than of northward propagation. This fact seems consistent with the relatively short lifetimes of the weak transient eddies in the BOC, compared with the very energetic Loop Current Eddies of the western Gulf.

A reviewer has suggested that an animation of the original SSHA data for the BOC region would illustrate the propagation properties more clearly. The reader does in fact have the opportunity to see actual (but noisy) animations of eddy variability using the CD-ROM supplement to *Leben* [2005] in this monograph.

5. DISCUSSION

5.1. Consistency of Hydrography and Lagrangian Currents

Figures 7 and 12 near 20°N show speeds of about 30 and 10 cm·s^{-1} for the near surface and 900m levels, so this implies the near surface speed is about 20 cm·s^{-1} relative to 900 m (or to 800 m as an approximation). A geostrophic velocity with this magnitude is feasible in the southwest part of the eddy shown in Figure 6. Thus the mean surface pattern of dynamic height relative to 800 db appears to be reasonably consistent with the mean vector pattern of flow shown in Figure 7 in a geostrophic sense.

With regard to the flow at 900 m, the study of *Weatherly et al.* [2005] shows that this is not an extension of the BOC upper layer closed cyclone but instead is probably the upper part of a deep cyclonic circulation that extends around the entire continental slope of the Gulf of Mexico (see also *Sturges et al.* [2004]).

5.2 Forcing of a Permanent BOC Cyclone

The mechanism that explains the existence of the observed long term mean cyclonic circulation within the BOC (at least in the upper 800-m) is probably that associated with wind

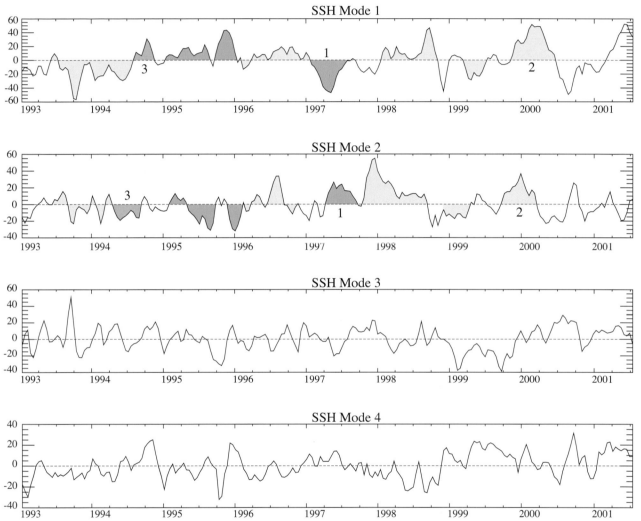

Plate 1. Time sequences of the amplitudes of the first four dominant EOF modes derived from the SSHA data. The reader is cautioned that the signs of both the amplitude and of the spatial patterns (Figure 13) for given mode must be taken into account in the identification of where cyclonic (negative SSHA) and anticyclonic (positive SSHA) eddies are located in the BOC domain. The shaded parts of the amplitudes in the upper two panels show episodes of eddy energy propagating southward into the BOC (green and blue) or northward (red). See section 4.3 of the text for discussion of episodes 1, 2, and 3.

stress curl as asserted in the VC93 study. That study used wind stress values for the Gulf of Mexico from *Elliott* [1979] that were based on the then unpublished Hellerman 106 year data base for marine winds over the world oceans by month [*Hellerman and Rosenstein*, 1983]. We revisit the wind stress curl argument here, but also make use of a three year estimate of the mean winds over the BOC from a study by *Gutiérrez de Velasco and Winant* [1996, hereinafter GW96]. In addition to the U. S. meteorological buoys in the northern Gulf of Mexico, the GW96 study made use of data from six coastal meteorological stations in the southern Gulf of Mexico in a cooperative project between Mexican and U. S. scientists for the period 1 August 1990 to 31 July 1993. The locations of all of these special Mexican stations appear as stars in Figure 1 of the present paper, and provide good marine exposure for wind estimates. The three-year mean gridded wind stress pattern over the whole Gulf is shown in Figure 14 using the *Large and Pond* [1981] relation for wind stress. As pointed out in GW96, the cyclonic turning of the wind vectors and modification of the wind speeds in the southwest region of the BOC is due to the presence of the very high coastal Sierra Madre mountain range which steers the air flow towards the Isthmus of Tehuantepec and through Chivela Pass to the Pacific Ocean. This is confirmed by *Steenburgh et al.* [1998 and references contained therein]. Moreover, a similar pattern of wind stress vectors over the BOC region occurs in the *Elliott* [1979] analysis. In contrast, the winds over the Gulf of

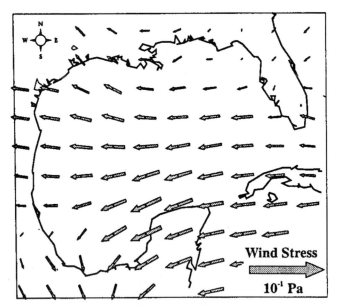

Figure 14. Averaged field of wind stress for 1 August 1990 to 31 July 1993 for the Gulf of Mexico. [Adapted from *Gutiérrez de Velasco and Winant*, 1996]. Chevila Pass through the Sierra Madre mountains is centered near the third wind stress vector from the left over the Isthmus of Tehuantepec.

Mexico by *Rhodes et al.* [1989], that were used in the study by *Sturges* [1993], are realistic north of 22°N but not in the BOC region.

Shown in the GW96 paper are patterns of three-year mean and seasonal mean wind stress curl for the whole Gulf of Mexico. These confirm the presence of cyclonic wind stress curl centered over the BOC, which is about 60% stronger (weaker) in the winter (summer) season than the annual mean. For the calculations discussed here, we used the GW96 tabulated information for the three-year mean wind stress vectors at the four coastal meteorological stations within the BOC (Tuxpan, Veracruz, Coatzacoalcos, and Cayo Arcas). These together with the locations of the four stations allowed an estimation of the area average of the wind stress curl. This was done using Stokes law, which involves numerically estimating the circulation (line integral of the tangential component) of the wind stress vector around the periphery of the rhomboid domain formed by the four BOC coastal stations and then dividing that circulation of stress by the area of the rhomboid. The resulting three-year, area average wind stress curl within the BOC is 1.3×10^{-7} Pa·m⁻¹. The maximum 3-year mean wind stress curl from GW96 (their Figure 3d) is about 0.90×10^{-7} Pa·m⁻¹ and the area average for the rhomboid region we estimate to be about 0.6×10^{-7} Pa·m⁻¹. This is only about half that which we calculated based on their Table 3 data. Moreover if we take the ratio of the maximum curl to the rhomboid average curl as 1.5 (as implied above), then our estimated maximum 3-year mean curl is about 2.0×10^{-7} Pa·m⁻¹. This is in agreement with that deduced from the *Elliott* [1979] 106-year mean wind stress curl at 20°N based on the Hellerman and Rosenstein data. Thus in summary we accept the Table 3 information from GW96 and the ratio of 1.5 to convert the rhomboid average curl to the estimated curl at 20°N. The latter corresponds approximately to the latitude of maximum curl for the BOC from GW96 as well as *Elliott* [1979].

Using 2.0×10^{-7} Pa·m⁻¹ for the curl at 20°N and an effective zonal width (W) of 400 km for the BOC gives an estimated northward volume transport by the Sverdrup relation (W curl τ/ρ)($\partial f/\partial y$)⁻¹ of 4.0×10^6 m³·s⁻¹ within the BOC at 20°N where f is the Coriolis parameter. This corresponds to a mean northward component of surface current of about 3 cm·s⁻¹ at that latitude, based on the first EOF mode for dynamic height anomaly in the upper 800 m of the BOC (this mode has an effective depth H_e of about 340 m). We assume that the wind stress curl is zero at about 22°N, where the observed re-circulating flow is westward. In order to balance the northward transport within the BOC caused by the wind stress curl, the strong western boundary current at 20°N (Figure 10) must provide an equal transport to the south. The zonal variation of the boundary current can be approximated by an exponentially decaying profile to the east with an e-folding scale L and a vertical structure as assumed for the northward flowing water. If we take L as the radius of deformation for the first baroclinic mode (about 40 km) and the maximum surface value of the western boundary current at 20° N as 30 cm·s^{-1} (Figure 7), then one gets for the southward component of the western boundary transport the required amount to balance the northward transport produced by the wind stress curl.

The question now is whether the Sverdrup mass transport due to the wind stress curl matches that of the observed state. The latter is most easily estimated from the hydrographic data. Figure 8 indicates a range of dynamic height ($\Delta\eta$) of about 6 dyn cm within the outer potentially closed contour. The cyclic transport within this eddy is estimated as $g \times \Delta\eta \times He/f$ which gives a value very close to the wind stress curl induced transport of 4.0×10^6 m^3·s^{-1}.

5.3. Seasonal Forcing

We accept the finding by both *Elliott* [1979] and GW96 that the wind stress curl over the BOC is cyclonic in all seasons and is maximum (minimum) in the 3-month winter (3-month summer). This is qualitatively consistent with the finding from the near surface drifter analysis which shows a maximum cyclonic western boundary current in the 6-month winter and possibly a minimum in the 6-month summer. We make no attempt to quantify this relationship without a more adequate drifter data base for the BOC summer. It is noteworthy however, that the maximum anticyclonic western boundary current north of the BOC occurs in summer, about one month after the wind stress curl in the northern Gulf attains its maximum anticyclonic value [*Sturges*, 1993]. Thus, we argue that the same mechanism is probably at work in the BOC but with the wind stress curl having opposite signs and about six months different in occurrence of the maximum magnitude of the wind stress curl.

5.4. Results From Model Studies

Some recent model study results have come to our attention that qualitatively support the present observational evidence that the long term mean circulation in the BOC is cyclonic. These are results from the Princeton Ocean Model (POM) and the Miami Isopycnic Coordinate Model (MICOM) reported in the study by *Sturges et al.* [2004]. Both models show mean cyclonic circulation patterns in the BOC at 400-m depth that resemble the pattern of flow in Figure 6 of this study, but are of course weaker. The MICOM model uses the Comprehensive Ocean-Atmosphere Data Set (COADS) wind climatology. We have checked the COADS wind climatol-

ogy in the BOC region and it is qualitatively similar to that of GW96, in the sense that the wind stress curl is positive. The results from these model studies with local wind forcing may or may not confirm our assertion here that the local Ekman pumping over the BOC is forcing the mean circulation. Clearly the most persuasive evidence that is needed in settling this question is observed and model evidence of similar seasonal variability of the cyclonic circulation, as in *Sturges* [1993] on the annual cycle of the western boundary current north of the BOC domain. We know that the seasonal variability of the wind stress curl is large from the studies of *Elliott* [1979] and GW96. We do not yet have sufficiently long drifter records to define the seasonal variability of the mean circulation in a statistically reliable manner for the BOC and therefore cannot give evidence for matching the observed seasonal signals of the forcing and observed response in a persuasive way. Such a comparison can be made with the models, and one would hope that their forcing in the BOC region takes into account the very significant influence of the Sierra Madre mountains.

5.5. Non-Seasonal Variability of the BOC Circulation

The eddy variability in the near-surface BOC is evidenced in several ways. One way is by the different locations, scales, and strengths of synoptic events seen in the hydrographic data. Another is through similar variations of patterns seen in the satellite derived sea surface height anomaly (SSHA) data for which independent realizations at two week intervals have been captured [*Leben*, 2005]. In fact the SVD analysis of the modified SSHA data shows the dominant eddy mode structure, and we have shown that the time evolution of these modes gives evidence of eddy energy propagation into the BOC. Finally the near-surface drifter data allows a quantification of the standard deviation of the velocity whose pattern is shown in Figure 8. The probable source of the transient eddies and their associated energy is from the more energetic anticyclonic Loop Current eddies of the northwestern Gulf which tend to break into smaller cyclones and anticyclones upon impacting the upper continental slopes [*Smith* 1986; *Oey*, 1995].

The model study by *Romanou et al.* [2004] suggests that the dominance of cyclonic circulation in the BOC is caused by accretion of cyclones generated in the western Gulf by interaction of Loop Current eddies with the continental slope. Moreover they state that the lifetime of the cyclonic eddies entering the BOC varies from 2 to 6 months. The observational evidence from the SSHA data yields lifetime estimates in the range of 2.4 to 8 months, which supports the model findings but for eddies of both positive and negative vorticity. While such transient accretion of cyclonic and anticyclonic eddies can explain the non-seasonal character of the circulation, we feel that this is simply a supplement superimposed on a permanent locally forced cyclonic circulation due to the wind stress curl over the BOC.

5.6. Evidence of Wind-Driven Shelf Circulation

The primary focus of this paper has centered on the circulation in the deep and slope waters of the BOC. The evidence presented on currents has dealt primarily with patterns of weak to moderate flow. However, the shallow waters regions of the BOC are quite different. A vivid example of high velocity was experienced by personnel on a large PEMEX Corporation oil tanker anchored in water depth of 80 m at about 19.5°N and 92°W in the BOC. Sustained currents as high as 100 cm·s^{-1} were experienced for about 3 days directed towards the south and southeast. However the location was on the wide Campeche shelf and the current was no doubt part of a wind-induced along-shelf current forced by winds out of the north with speeds of about 20 knots. Thus currents other than those associated with the cyclonic circulation over the deepwater region can be important to oil exploration and recovery operations in the BOC shelf region, just as they are on the northern slope and shelf regions of the Gulf of Mexico. Moreover, this gives a specific example to substantiate a statement made in section 3.1 concerning possible contribution to the variance of surface currents in the regions of the BOC close to the shelf regions.

6. CONCLUDING REMARKS

We have presented evidence derived from long-term mean hydrographic data and surface drifter data to support the claim of a permanent cyclonic circulation in the upper 800 m of the Bay of Campeche. The estimated mean fields of dynamic height and of associated currents have sufficiently small uncertainty to assure statistically valid patterns. Moreover these patterns are reasonably consistent in a geostrophic sense. The near surface drifter vectors show a westward intensification in the BOC, as does the surface dynamic height relative to 800 m.

The physically logical forcing mechanism for maintaining the permanent cyclonic circulation and its observed western intensification is Ekman pumping associated with wind stress curl over the BOC. The mean wind stress curl over the BOC from the GW96 study is consistent with that of *Elliott* [1979]. These indicate a cyclonic cyclic volume transport of about 4×10^6 m^3·s^{-1}, which is the same as that indicated by the hydrographic data relative to 800 db. The cyclonic curl associated with the winds in this region is strongly governed by the steering of the airflow by the Sierra Madre mountains southward along the western part of the BOC and towards Chivela Pass where it exits to the Gulf of Tehuantepec.

The near surface drifter data analysis by *DiMarco et al.* [2005] when stratified into a 6-month winter and a 6-month summer does show evidence of a strengthening of the western cyclonic boundary current in the BOC during the 6-month winter. This is the time when the cyclonic wind stress curl over the BOC is at its maximum. The 6-month summer drifter pattern shows an ill-defined cyclonic pattern during the period of weak cyclonic wind stress curl. Unfortunately the summer pattern is inconclusive because insufficient drifters were seen in the BOC western boundary region.

We have addressed the non-seasonal variability of the eddy structure in the BOC both spatially and temporally via EOF analysis of the satellite-derived SSHA data. The dominant spatial patterns show cyclonic and anticyclonic features of scales somewhat smaller than the permanent cyclonic eddy scale. The total variance contributed by these transient eddies is less than 30 cm^2, which is far less than that in the northwest region of the Gulf of Mexico. The temporal variability of the modal amplitudes indicates life times in the range 2.4 to 8 months, and the BOC is essentially an energy sink for those transient eddies that enter.

The presence of an isolated closed wind-driven cyclone within the upper 800m of the BOC is in striking contrast with the deeper region of the BOC that appears to be part of a Gulf wide mean cyclonic flow along the abyssal boundary as evidenced by the remarkable findings of *Weatherly et al.* [2005]. In fact one wonders if there might be some influence of one on the other. Added temporal variability in the upper BOC caused by remotely forced topographic waves existing in the deeper BOC region is a possibility.

Acknowledgments. We wish to thank Mathew K. Howard for providing assistance in the assembly, quality control, and analysis of section 2.2, and Robert Leben for suggesting looking at the temporal variability of the SSHA signals in the western Gulf of Mexico. We acknowledge the use of some new results reported in prior reports of our colleagues, particularly Georges Weatherly, that have an important bearing on this study of the Bay of Campeche circulation. Also we thank the reviewers whose suggestions have greatly helped us clarify the presentation and perhaps made the paper more persuasive. The first author wishes to acknowledge his close associates in the Mexican Navy who encouraged his early oceanographic studies of the western Gulf of Mexico and the Bay of Campeche in particular. Finally, we wish to express our thanks to the Minerals Management Service of the U. S. Department of the Interior for supporting the publication of this special AGU monograph on the Gulf of Mexico circulation.

REFERENCES

Bloomfield, P. (1976), *Fourier Analysis of Time Series: an Introduction*, pp. 22–41, John Wiley, New York.

Brooks, D. A. (1984), Current and hydrographic variability in the northwestern Gulf of Mexico, *J. Geophys. Res.*, *89*, 8022–8032.

DiMarco, S. F., and R. O. Reid (1998), Characterization of the principal tidal current constituents on the Texas–Louisiana shelf, *J. Geophys. Res.*, *103*, 3093–3109.

DiMarco, S. F., W. D. Nowlin, Jr., and R. O. Reid (2005), A statistical description of the near-surface velocity field from drifters in the Gulf of Mexico, this volume.

Elliott, B. A. (1979), Anticyclonic rings and the energetics of the circulation of the Gulf of Mexico, Ph.D. dissertation, 188 pp, Department of Oceanography, Texas A&M University, College Station, Texas.

Elliott, B. A. (1982), Anticyclonic rings in the Gulf of Mexico, *J. Phys. Oceanogr.*, *12*, 1292–1309.

Fukumori, I., and C. Wunsch (1991), Efficient representation of North Atlantic hydrographic and chemical distributions, *Progress in Oceanography*, *27*, 111–195.

Gill, A. E. (1982), *Atmosphere-Ocean Dynamics*, Academic Press, 662 pp.

Gutiérrez de Velasco, G., and C. D. Winant (1996), Seasonal patterns of wind stress and wind stress curl over the Gulf of Mexico, *J. Geophys. Res.*, *101*, 18,127–18,140.

Hamilton, P. (1992), Lower continental slope cyclonic eddies in the central Gulf of Mexico, *J. Geophys. Res.*, *97*, 2185–2200.

Hellerman, S., and M. Rosenstein (1983), Normal monthly wind stress over the world ocean with error estimates, *J. Phys. Oceanogr.*, *13*, 1093–1104.

Jochens, A. E., L. C. Bender, S. F. DiMarco, J. W. Morse, M. C. Kennicutt II, M. K. Howard, and W. D. Nowlin, Jr. (2005), Understanding the Processes that Maintain the Oxygen Levels in the Deep Gulf of Mexico: Synthesis Report, U.S. Dept. of the Interior, Minerals Management Service, Gulf of Mexico OCS Region, New Orleans, Louisiana.

Large, W. G., and S. Pond (1981), Open ocean momentum flux measurements in moderate to strong winds, *J. Phys. Oceanogr.*, *11*, 324–336.

Leben, R. R., G. H. Born, and B. R. Engebreth (2002), Operational altimeter data processing for mesoscale monitoring, *Marine Geodesy*, *25*, 3–18.

Leben, R. R., Altimeter-derived Loop Current metrics (2005), this volume.

Merrell, W. J., Jr., and A. M. Vázquez (1983), Observations of changing mesoscale circulation patterns in the western Gulf of Mexico, *J. Geophys. Res.*, *88*, 7721–7723.

Nowlin, W. D., Jr., A. E. Jochens, S. F. DiMarco, R. O. Reid, and M. K. Howard (2001), Deepwater Physical Oceanography Reanalysis of Historical Data: Synthesis Report, 528 pp, OCS Study MMS 2001-064, U. S. Dept. of the Interior, Minerals Management Service, Gulf of Mexico OCS Region, New Orleans, Louisiana.

Oey, L.-Y. (1995), Eddy- and wind-forced shelf circulation, *J. Geophys. Res.*, *100*, 8621–8637.

Pattullo, J., W. Munk, R. Revelle, and E. Strong (1955), The seasonal oscillation in sea level, *J. Mar. Res.*, *14*(1), 88–155.

Rhodes, R. C., J. D. Thompson, and A. J. Wallcraft (1989), Buoy-calibrated winds over the Gulf of Mexico, *J. Atmos. Oceanic Technol.*, *6*(4), 608–623.

Romanou, A., E. P. Chassignet, and W. Sturges (2004), Gulf of Mexico circulation within a high-resolution numerical simulation of the North Atlantic Ocean, *J. Geophys. Res.*, *109*, C01003, doi: 10.1029/2003JC001770.

Smith, D. C., IV (1986), A numerical study of Loop Current eddy interaction with topography in the western Gulf of Mexico, *J. Phys. Oceanogr.*, *16*, 1260–1272.

Steenburgh, W. J., D. M. Schultz, and B. A. Colle (1998), The structure and evolution of gap outflow over the Gulf of Tehuantepec, Mexico, *Mon. Weather Rev.*, *126*, 2673–2691.

Sturges, W. (1993), The annual cycle of the western boundary current in the Gulf of Mexico, *J. Geophys. Res.*, *98*, 18053–18068.

Sturges, W., and R. Leben (2000), Frequency of ring separation from the Loop Current in the Gulf of Mexico: a revised estimate, *J. Phys. Oceanogr.*, *30*, 1814–1819.

Sturges, W., E. Chassignet, and T. Ezer (2004), Strong mid-depth currents and a deep cyclonic gyre in the Gulf of Mexico: Final report, 89 pp., OCS Study MMS 2004-040, U. S. Dept. of the Interior, Minerals Management Service, Gulf of Mexico OCS Region, New Orleans, Louisiana.

Tierney, C. C., M. E. Parke, and G. H. Born (1998), An investigation of ocean tides derived from along-track altimetry, *J. Geophys. Res.*, *103*, 10273–10287.

Vázquez de la Cerda, A. M. (1975), Currents and waters of the upper 1200 meters of the southwestern Gulf of Mexico, M.S. Thesis, 106 pp., Department of Oceanography, Texas A&M University, College Station, Texas.

Vázquez de la Cerda, A. M. (1993), Bay of Campeche Cyclone, Ph.D. Dissertation, 91 pp., Department of Oceanography, Texas A&M University, College Station, Texas.

Vukovich, F. M., and B. W. Crissman (1986), Aspects of warm core rings in the Gulf of Mexico, *J. Geophys. Res.*, *91*, 2645–2660.

Weatherly, G. (2004), Intermediate Depth Circulation in the Gulf of Mexico: PALACE Float Results for the Gulf of Mexico between April 1998 and March 2002, OCS Study MMS 2004-013, 51 pp., U. S. Dept. of the Interior, Minerals Management Service, Gulf of Mexico OCS Region, New Orleans, Louisiana.

Weatherly, G. L., N. Wienders, and A. Romanou (2005), Intermediate-depth circulation in the Gulf of Mexico estimated from direct measurements, this volume.

Whitaker, R. E. (1971), Seasonal variation of steric and recorded sea level of the Gulf of Mexico, Technical Report 71-14T, 80 pp., Department of Oceanography, Texas A&M University, College Station, Texas.

S. F. DiMarco, A. E. Jochens, and R. O. Reid, Department of Oceanography, Texas A&M University, 3146 TAMU, College Station, TX 77843-3146 USA

A. M. Vázquez de la Cerda, Instituto Oceanografico, Secretaria de Marina, El Salado, Veracruz C.P. 95263 MEXICO (albertomvc@yahoo.com)

Wind and Eddy-Related Shelf/Slope Circulation Processes and Coastal Upwelling in the Northwestern Gulf of Mexico

Nan D. Walker

Department of Oceanography and Coastal Sciences, Coastal Studies Institute, Louisiana State University, Baton Rouge, Louisiana

Wind- and eddy-driven circulation on the Louisiana-Texas shelf and slope are investigated during 1993–1994 by integrating measurements from satellite sensors (sea surface temperature, sea surface height) and in situ measurements from satellite-tracked drifting buoys and midwater shelf current meters. Four distinct circulation regimes were identified during autumn/winter including a down-coast jet on the west Louisiana and Texas inner shelf regions, cross-shelf entrainment within the "Texas jet," seaward entrainment by slope eddies, and weak cyclonic circulation on the outer Louisiana shelf. The inner shelf coastal jet (20–98 cm s^{-1}), associated with strong northeasterly winds, developed between the 10 m and 30 m isobaths seaward of the main sea-surface temperature front. Maximum current speeds were measured along the Texas coast between 29° N and 26° N. Although down-coast flow occurred 80% of the time, current reversals of 1–3 days resulted from northwesterly winds during frontal passages. A high velocity cross-shelf jet (13–59 cm s^{-1}) was identified at 28° N, where an anticyclone/cyclone eddy pair entrained surface drifters onto the slope. During summer, prolonged current reversals with surface velocities of 20–50 cm s^{-1} occurred on the Louisiana-Texas shelf as a result of strong southerly wind stress over the south Texas shelf and moderate southerly/southwesterly wind stress over the Louisiana shelf. A ~20-km wide zone of relatively cool waters (23°C–26°C) was detected along the coast between 23° and 28° N, subsequent to the current reversal. In situ measurements over three consecutive summers revealed that this coastal upwelling lasts four to six weeks in summer along the south Texas and north Mexican coast.

1. INTRODUCTION

The large-scale circulation in the Gulf of Mexico (GOM) is dominated by the Loop Current [*Cochrane*, 1972], the large warm-core anticyclonic rings that separate from it [*Elliot*, 1982; *Forristall et al.*, 1992], and associated smaller cold-core cyclonic eddies [*Hamilton*, 1992]. The anticyclonic rings usually separate from the Loop Current after it has surged north beyond 27° N latitude [*Forristall et al.*, 1992]. The ring shedding cycle is variable; however separations have occurred most often at 6 and 11 months [*Sturges and Leben*, 2000]. Anticyclones typically have diameters of 200–400 km and maximum surface currents of 100–200 cm/s [*Huh and Schaudt*, 1990; *Cooper* et al., 1990; *Forristall et al.*, 1992]. They drift westward at about 5 cm s^{-1} [*Wiseman and Sturges*, 1999] where they eventually collide against the continental shelf slope, lose energy, and often move northward towards the Louisiana-Texas shelf [*Lewis and Kirwan*, 1985; *Vidal et al.*, 1992]. These features are large reservoirs of heat, extending to a depth of at least 800 m,

Circulation in the Gulf of Mexico: Observations and Models
Geophysical Monograph Series 161
10.1029/161GM21

which can intensify tropical and extra-tropical low pressure systems that move over the northern GOM [*Schumann et al.*, 1995]. Cold-core cyclonic eddies have diameters of 50–150 km and are often formed from interaction of anticyclonic rings with the continental shelf slope [*Lewis et al.*, 1989; *Hamilton*, 1992], producing cyclonic-anticyclonic eddy pairs that influence surface current variability along the outer continental shelf edge [*Merrell and Morrison*, 1981; *Brooks and Legeckis*, 1982; *Oey*, 1995].

Circulation on the Louisiana-Texas (LATEX) continental shelf (Figure 1) is controlled mainly by the regional wind field and by the slope eddy field. On the inner LATEX shelf (< 30m water depth), surface circulation is primarily wind driven. Strong coherence has been found between alongshore wind stress and alongshore currents for coastal locations downstream of Atchafalaya Bay and along the Texas coast [*Smith*, 1978; *Crout et al.*, 1984; *Lewis and Reid*, 1985; *Cochrane and Kelly*, 1986; *Murray et al*, 1998]. Easterly wind stress results in flow to the west on the inner shelf of Louisiana and southward flow along the Texas coast [*Cochrane and Kelly*, 1986; *Murray et al*, 1998; *Cho et al.*, 1998]. The term "down-coast" refers to flow from Louisiana towards Mexico along the coast. "Up-coast" refers to flow

from Texas towards the Mississippi delta (Figure 1). The Atchafalaya and Mississippi Rivers flow onto the Louisiana shelf with a combined annual average discharge of 18,400 $m^3 s^{-1}$ and sediment input of 210 million tons [*Milliman and Meade*, 1983] (Figure 1). These river discharges exhibit an annual cycle with strongest flow onto the shelf in spring and weakest flow in late summer and autumn [*Walker*, 1996]. This buoyancy flux produces pressure gradients that are balanced by the Coriolis force [*Munchow and Garvine*, 1993], resulting in an enhancement of down-coast flow along the Louisiana coast.

The generalized large-scale circulation on the LATEX shelf was first described by *Cochrane and Kelly* (1986) as a semi-permanent cyclonic gyre that occupies the shelf from September through May. A wind-driven coastal jet forms along the inshore limb of this gyre [*Cochrane and Kelly*, 1986; *Cho et al.*, 1998]. During summer, a major current reversal is experienced over the LATEX shelf, with up-coast flow from south Texas to the Mississippi delta [*Cochrane and Kelly*, 1986; *Murray et al.*, 1998; *Cho et al.*, 1988], but no evidence of a coastal jet.

The main goal of this paper is to investigate wind- and eddy-driven circulation over the LATEX shelf and slope

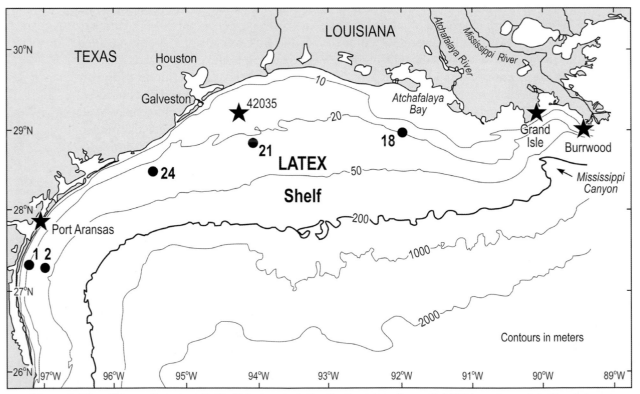

Figure 1. Map of the northwestern Gulf of Mexico, depicting the Louisiana-Texas (LATEX) continental shelf and slope, the location of the coastal and shelf meteorological stations (black stars) and the LATEX-A current meter moorings (black dots).

during winter and summer of 1993–1994. One of the main catalysts for this study was a movie loop of surface-drifter tracks [*Johnson and Niiler*, 1994], which revealed complex circulation patterns in the northwestern GOM that required high quality satellite measurements of sea surface temperatures (SSTs) and sea surface height (SSH) for interpretation. Thus, in this study we have integrated SST, SSH, and drifter data from the Surface Current and Lagrangian-drift Program (SCULP) [*Ohlmann et al.*, 2001] to better assess surface shelf and slope circulation processes during the autumn and winter of 1993–1994 when the drifter data were most abundant. We have also used selected current measurements obtained during the Texas-Louisiana Physical Oceanography study (LATEX-A) [*DiMarco et al.*, 1997] to quantify aspects of variability of up-coast and down-coast flow on the LATEX inner shelf during the autumn and winter and the summer of 1993–1994. The Eulerian fixed frame of reference provided by the moored meters facilitated the study of wind-related changes in circulation. Satellite sea surface temperature (SST) data were used to investigate seasonal shelf temperature changes, to identify and characterize water masses, and to study how shelf and slope currents relate to surface thermal fronts. Since SST data are readily available, the relationships we discuss may prove useful in assessing or predicting surface currents in the absence of in situ measurements. The SSH data were used primarily to detect cyclonic eddies, features which do not always have unique SST signatures.

2. METHODOLOGY

2.1. Satellite Measurements

Satellite sea-surface temperature (SST) measurements, obtained by the NOAA AVHRR (Advanced Very High Resolution Radiometer), provided a frequent source of high resolution SST data (1.1 km^2 pixels) covering the study area. These data were received in real time via antennae and processed at the Earth Scan Laboratory, Coastal Studies Institute, Louisiana State University. SSTs were computed using the MCSST technique described in *McClain et al.* (1985).

Sea surface height anomaly data were obtained from Dr. Bob Leben at the Colorado Center for Astrodynamics Research (CCAR), University of Colorado, Boulder. TOPEX and ERS-1 satellite data were used to create 10-day anomaly maps which were referenced to the OSUMMS95, a mean sea surface calculated at the Ohio State University by Dick Rapp and Yuchan Yi [*Leben and Born*, 1993]. The contoured sea surface height (SSH) anomalies were superimposed on clear-sky satellite SST imagery to provide additional information on the locations of cyclones and anticyclones. The SSH data were unavailable from 13 December 1993 to 4 May 1994, due to sensor malfunction.

2.2. SCULP-1 Surface Drifter Data

From October 1993 through October 1994, approximately 300 satellite-tracked ARGOS drifting buoys were released on the Louisiana shelf during the SCULP I drifter program [*Johnson and Niiler*, 1994; *Ohlmann et al.*, 2001; *Ohlmann and Niiler*, 2005]. (See SCULP movie in Johnson directory on the CD-ROM included with this volume.) Daily positions (1115 UTC) of these surface-drogued drifters (to 1 m) were available to LATEX researchers as ASCII files. These positions were superimposed on the satellite SST data to aid in the interpretation of circulation processes in relationship to shelf SST fronts and the offshore eddy field. Details of the computation of daily positions can be found in *Ohlmann et al.,*(2001).

2.3. LATEX-A Current Meter and Temperature Data

The LATEX-A current meter and temperature measurements were extracted from the CD-ROM NODC-92 produced by Texas A&M [*DiMarco et al.*, 1997]. The moorings near the 20m isobath were chosen since they were closest to the SCULP drifter tracks during 1993–1994 and they also yielded better data return than those closer to the coast. Inner shelf circulation was investigated using moorings 2, 24, 21, and 18 in water depths ranging from 22 to 37 m (Figure 1, Table 1). Each mooring had two current meters; one at midwater column and one near the bottom. The mid-column meters were used in this study, with depths ranging

Table 1. Station numbers, water depths, vertical depth of instruments, and positions of LATEX A current meter moorings.

Mooring number	Water depth (m)	Top CM depth (m)	Latitude (°N)	Longitude (°W)
18	22	11	28.963	91.983
21	24	14	28.837	94.080
24	28	10	28.474	95.437
2	37	12	27.284	96.980

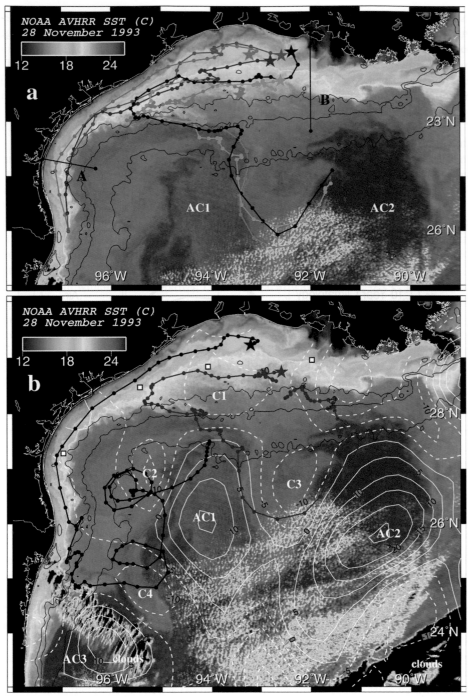

Plate 1. NOAA AVHRR SST image of 11/28/93 superimposed with: (a) the tracks of six SCULP drifters released in 10/93 (20440, 20481, 20396, 20412, 20480, 20469) on the LATEX shelf and slope to demonstrate circulation within the coastal current, and seaward entrainment by the Texas jet and slope eddies. Drifter tracks are color coded, start locations are depicted with stars, and daily positions are shown with small dots. Profile lines A and B are illustrated with black lines. The two slope anticyclones are labeled AC1 and AC2. The 50m, 100m and 1000m isobaths are shown with solid black lines; (b) sea surface height (SSH) anomaly data, depicted using solid white lines for positive SSH and dashed white lines for negative SSH. LATEX moorings 2, 21, 24, and 18 are shown with white squares. The centers of three anticyclones (AC1, AC2, AC3) and four cyclones (C1, C2, C3, C4) are annotated. Drifter tracks 20440 and 20480 are shown for the 10/93–1/94 period in relationship to the anticylones and cyclones.

Table 2. Summary of percentage of drifters that were observed in each of the four LATEX shelf and slope circulation regimes. Data are compiled by month of deployment from 10/93–1/94. Percentages do not always add to 100, since single drifters could move in and out of circulation regimes.

Circulation regime	10/93	11/93	12/93	1/94	10/93–1/94
1. Coastal current (Louisiana to Mexico)	36%	41%	52%	73%	51%
2. Texas jet (28°N)	44%	46%	19%	0%	27%
3. Slope eddy entrainment (includes some Texas jet drifters)	28%	34%	19%	4%	21%
4. Louisiana outer shelf cyclonic gyre	11%	0%	4%	0%	4%
Total drifters in counts	36	41	27	26	130

The tracks of three SCULP drifters deployed in October 1993 are shown in Plate 1a. These drifters were among the five fastest drifters, deployed in October and November 1993, to reach the Texas/Mexico border within the LATEX coastal current. Drifters released on the Louisiana shelf in waters depths of 15–25 m moved westward initially in close proximity or towards the inner shelf SST front, that marked the outer edge of the cool coastal band (Plate 1a). Drifters deployed in deeper water also moved westward towards the Texas shelf, reaching the SST front near the 20 m isobath at a later time. Upon encountering the main SST front of the cool coastal band along the Texas inner shelf, 25–30 km from the coast, the drifters turned to the southwest, exhibiting motion parallel to the main SST front. Many of the drifters continued to move down-coast within the coastal jet, formed at the convergence of river and estuarine out-flow plumes with the wind-driven shelf waters. The down-coast movement of two drifters, deployed on 22

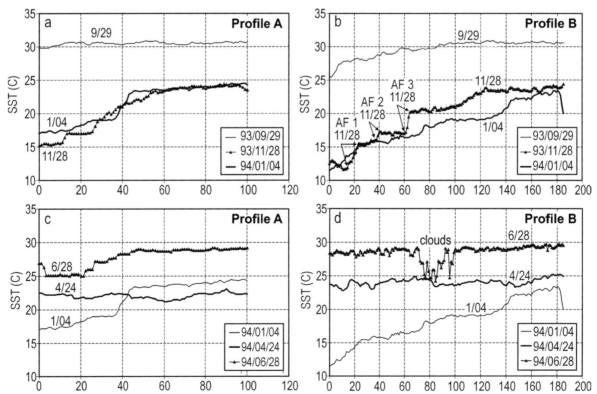

Figure 3. (a) SSTs extracted from profile line A (Location, Plate 1) on 9/29/93, 11/28/93, and 1/4/94; (b) SSTs extracted from profile Line B (Location, Plate 1) for same days as 'a' above. Atchafalaya plume fronts are shown as AF1, AF2, and AF3; (c) SSTs extracted from Line A (Location Plate 2) on 1/4/94, 4/24/94, and 6/28/94; and (d) SSTs extracted from Line B (Location Plate 2) on same days as 'c' above. All distances are measured from the coast in km.

ber of drifters from the December 1993 and January 1994 that were entrained seaward within the Texas jet (Table 2). In this same time frame, AC3 moved northward and C2 appeared to strengthen and/or move closer to the coast as a new region of offshore entrainment of coastal current drifters was observed at 26.5° N in January 1994. During December 1993 and January 1994, 15% and 23% of drifters were entrained seaward in this secondary jet that developed close to Brownsville.

One additional circulation pattern was exhibited by the drifter data. A slow cyclonic circulation was observed between 92° and 94° W, in water depths of 40–200 m on the Louisiana shelf [*Walker*, 2001]. It was only exhibited by 11% of drifters deployed in October 1993 and 4% of those deployed in December 1993 (Table 2).

3.5. Up-Coast Currents and Coastal Upwelling in Summer

Between 4 January 1994 and 24 April 1994, satellite SST measurements revealed that widespread surface warming had occurred on the LATEX shelf. Along the inner shelf, SSTs increased 4°C to 12°C and the cool coastal band from winter had disappeared (Plate 2a; Figures 3c,d). Little SST frontal structure was detected on the shelf, as SSTs ranged from 21°C–25°C. Warmest shelf waters (25° C) were detected on the Louisiana shelf. By 28 June, additional surface warming of 5° C–8 °C was detected, and much of the Louisiana shelf exhibited SSTs as high as 28°C–30°C (Plate 2b; 3c,d). The climatological data for Grand Isle (Figure 2) revealed that even warmer coastal temperatures would be expected in July and August.

The progressive vector diagram of winds (Figure 6) revealed that the predominant northeasterly winds of autumn 1993/winter 1994 gradually shifted to southeasterly during April and May. In June and July 1994, southerly and southwesterly winds prevailed on the Louisiana shelf, while strong southeasterly winds prevailed on the south Texas shelf. By August, easterly winds had returned to the Louisiana coast, and Port Aransas winds had regained more of an easterly component. This major change in wind direction during summer resulted in a 180° reversal in shelf currents during June across the study area as revealed by the current displacements at the four mooring locations (Figure 9).

The time-series of subtidal along-shelf currents at moorings 21 and 2 showed prolonged up-coast flow during three separate events in June, July, and August (Figure 10). Over the three month period, up-coast flow prevailed 61% of the time at mooring 21 and 54% of the time at mooring 2. The wind record from NDBC 42035 (Figure 10) revealed that upwelling favorable southerly winds prevailed during much of May and early June. However, up-coast currents were not

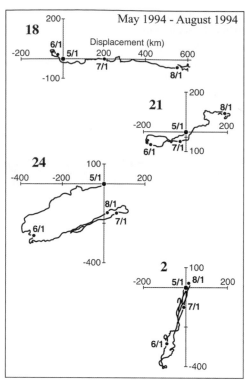

Figure 9. Progressive vector diagram of currents from 5/1/94–8/31/94 at LATEX moorings 18, 21, 24, and 2. The vertical axes show north-south displacement with north positive. The horizontal axes show east-west displacement with east positive. Distances are in km.

well established until 10 June at mooring 2 (in the south) and 16 June at mooring 21 (Figure 10). Thus, a lag of several weeks was observed between initiation of wind forcing and sustained reversal of the coastal current across the Texas shelf. A closer temporal association was observed between up-coast flow and westerly winds at 42035, especially during the second and third up-coast flow periods, 21–29 July and 20–27 August. During the three major up-coast current episodes, shown with arrows in Figure 10, currents were notably stronger at the south Texas station. The strongest current speeds (40 cm s⁻¹) were measured at mooring 2 during the first up-coast flow period (10 June–6 July) (Figure 10).

Tracks of three SCULP drifters deployed in June 1994 are shown in Plate 2b, revealing eastward flow along the central and eastern Louisiana shelf. Little informative drifter data was available on the Texas shelf during the summer. Drifter 21723 (blue line) was deployed on 4 June and was observed to move eastward for several days, then westward, and finally eastward towards the Mississippi delta. Its initial eastward motion corresponded with weak up-coast flow at mooring 21, located about 50 km west of it. It exhibited a mean speed of 16 cm s⁻¹ between 4 June and 13 July on the

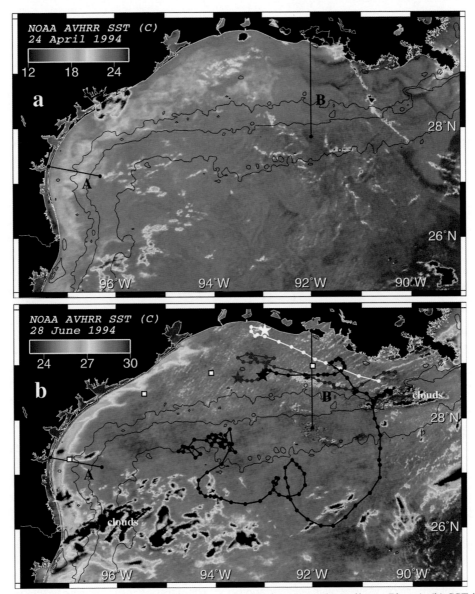

Plate 2. (a) NOAA AVHRR SST image of 4/24/94 enhanced with the same color pallet as Plate 1; (b) SST image of 6/28/94 enhanced for the summer SST range. Upwelled water along the coast of Texas is depicted with blue and green tones. Profile lines A and B, from which SSTs were extracted, are shown in both panels (see data in Figure 3). Mooring locations are indicated with boxes in 'b'. Three selected drifters released in June 1994 are shown, revealing up-coast flow on the Louisiana coast and, in one case, seaward entrainment by slope eddies. See text for descriptions of speeds and start/end times.

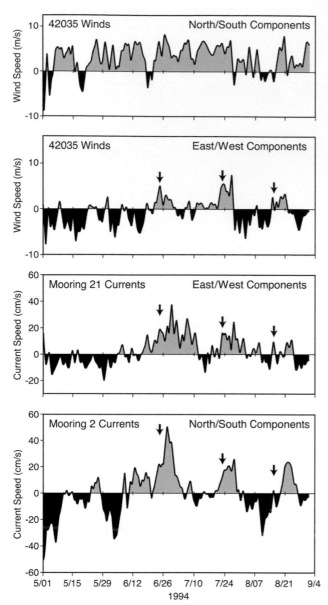

Figure 10. Winds and currents from 5/94 through 8/94 including: (upper 2 panels) 40-hr low-pass filtered north/south (north positive) and east/west wind components (east positive) from 42035; (middle panel) 40-hr low pass filtered east/west along-shelf components of currents at mooring 21 (east positive); and (lower panel) 40-hr low pass filtered north/south along-shelf currents at mooring 2 (north positive). Darker shading was applied to the down-coast currents. Up-coast events described in the text are identified with arrows.

Louisiana shelf. It exhibited speeds exceeding 40 cm s^{-1} on 18 June and 28 June during the strongest period of up-coast flow as measured at both moorings 21 and 2 (Figure 10). Drifter 21728 (white line) was deployed on 15 June. It traveled down-coast initially, but reversed course on 17 June and

continued its eastward motion until 29 June. Its mean speed was 24.2 cm s^{-1} and its maximum daily speed was 46.4 cm s^{-1} on 29 June. Drifter 21681 (black line) was deployed on 30 June and traveled eastward on the Louisiana shelf with a mean speed of 12.8 cm s^{-1} and a maximum speed of 38.5 cm s^{-1} on 7 July. Its path from June 30 to 3 November is shown in Plate 2b, revealing off-shelf entrainment by slope eddies and subsequent net motion to the west on the LATEX slope and rise. The drifter data demonstrated relatively strong eastward currents (> 40 cm s^{-1}) on the inner and middle regions of the Louisiana shelf during the strongest period of up-coast flow as identified in the mooring data (Figure 10).

The SST image of 28 June 1994 revealed an extensive band of cool waters close to the coast from Galveston Bay (29.2° N) to Pt. Jerez, Mexico (23° N) (Plate 2b), and persistence of this cool water through July (Plate 3). Time-series temperature data from LATEX mooring 1 (Figure 11) revealed distinct cold-water events on the south Texas inner shelf, beginning in May 1994 and lasting through July 1994. Coastal upwelling is the most likely explanation for these cool water events. Hydrographic measurements, obtained on the LATEX shelf towards the end of the 1994 upwelling season (26 July to 7 August), revealed the upward tilting of the isopleths of temperature, chlorophyll a and nutrients (Jochens et al., 1998), typical of coastal upwelling regions (Leming and Mooers, 1981), even though the cruise occurred toward the end of a cool water event.

A sequence of three SST images from 28 June through 26 July are shown in Plate 3 to illustrate the spatial structure of the coastal upwelling along the Texas and Mexican coastlines and their temporal changes during summer 1994. The lowest coastal temperatures were detected on 28 June when a distinct band of cool water (24°C–26°C) was observed from 23° N to 29.2° N (Plate 3a). This image corresponded in time with the coldest water temperatures (22° C) at mooring 1 (Figure 11) and the strongest up-coast currents along the Texas and Louisiana shelf (Figure 10). In this image, the upwelling zone along the south Texas coast appeared as a cool near-shore band extending at least 20 km offshore and northward as far as Galveston Bay. North of Matagorda Bay, filaments of cool water, up to 50 km long, extended towards the east and northeast on the inner shelf (Plate 3a).

The image acquired on 8 July revealed a large reduction in the coastal upwelling, as cool coastal waters were not observed north of Port Aransas (Plate 3b). The current measurements showed little northward flow at mooring 2 at this time, although eastward flow was measured at mooring 21 (Figure 10). Although active upwelling had abated along the Texas coast, filaments of cool water were apparent north of Port Aransas between the 20 and 50 m isobaths. Along the Mexican coast near 24.5° N, an elongated filament of cool

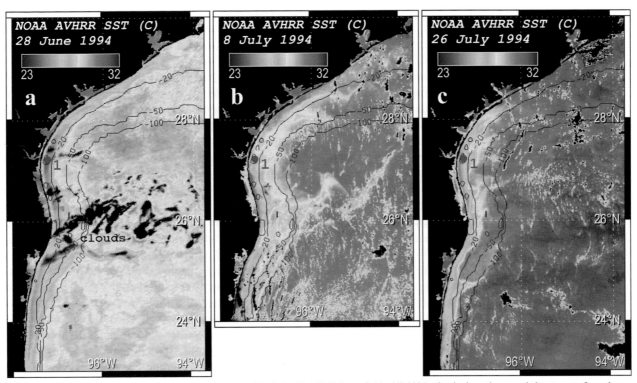

Plate 3. Sequence of 3 satellite images: (a) 6/28/94; (b) 7/8/94; and (c) 7/26/94; depicting the spatial extent of cool upwelled waters along the Texas coast (blue and green tones). LATEX mooring 1, within the core of the summer upwelling region, is shown with a red dot. The 20m, 50m, and 100m isobaths are shown.

close to the shelf edge, during November and early December 1993. Surface velocities associated with the slope eddy pairs averaged 45 cm s⁻¹ with a range of 13–75 cm s⁻¹ (Table 3).

A second major pathway for the seaward entrainment of inner shelf water was detected along the Mexican coast near 25° N, in December 1993 and January 1994. Another eddy pair with a prominent anticyclone centered at 24° N entrained 60% of coastal current drifters (released in October and November 1993) over the Mexican slope. Although initiation mechanisms for the seaward advection of water across this shelf were not investigated, wind stress or convergence of shelf flows are probable causes. *Brooks and Legeckis* [1982] obtained hydrographic measurements through an eddy pair in a similar location in April 1980, demonstrating that the cyclone extended to 600m depth and the anticyclone to 800m or more. Temperature imagery enabled the detection of a cool shelf-water mass that extended from the coast between the two counter-rotating features. The shipboard measurements revealed relatively low salinities within its center. A detailed evaluation of the movement of eddies along the western shelf slope edge in December 1993–January 1994 demonstrated that a 20,000 km² parcel of cool shelf water detached from the shelf in early January and moved northwards into the northwest GOM, as anticyclones AC1 and AC2 moved southeast away from the LATEX shelf [*Walker,* 2001].

4.3. Summer Up-Coast Flow and Coastal Upwelling

Southerly winds are most prevalent across the LATEX shelf from May through August, causing reversal in shelf circulation from down-coast to up-coast flow. Resultant currents are northward on the Texas shelf and eastward on the Louisiana shelf. *Murray et al.* [1998] demonstrated that the development of the up-coast flow regime across the LATEX shelf is not a simple wind-driven process, but is also controlled by along-shelf slopes in water level. Subsequent analyses of *Jarosz and Murray,* [this volume] confirm the importance of along-shore sea surface slopes as an important driving force for the eastward flow reversal on the Louisiana shelf in summer. Although time lags were detected between wind reversals and current reversals, up-coast flow became established from south Texas to the Mississippi River delta by mid-June. Current reversals close to 180 degrees were experienced at all moorings with the transition from down-coast to up-coast. During summer 94, three main up-coast flow events were experienced across the LATEX inner shelf, the longest of which lasted one month, from 10 June to 10 July (Figure 10). Up-coast flow along the Louisiana coast lagged that on the Texas coast by several days. From June through August 1994, up-coast currents occurred 50–60% of the time at both stations.

The near isothermal shelf temperature structure of spring 1994 was replaced by a very different surface temperature distribution in summer. Whereas most of the LATEX shelf was covered with waters of 30°C–32°C in summer, a coastally extensive 20–30 km wide band of cool surface water (24°C–27°C) was detected from 23° N to 29° N along the northern Mexican and southern Texas coast. The development of this cool coastal band of upwelled water was preceded by reversal in the Texas coastal current from down-coast to up-coast flow in June 1994. Based on satellite observations, temperature measurements at LATEX mooring 1, and LATEX-A hydrographic measurements [*Jochens et al.*, 1998] we conclude that the cool water upwells from deeper regions along the south Texas and Mexican shelves during summer. The southerly wind stress, up-coast currents, and Ekman processes result in northward and northeastward movement of the upwelled waters along and across the shelf. Although upwelling was detected in late May 1994, the major upwelling event of summer 1994 occurred uninterrupted from 24 June to 6 July 1994. The 28 June 1994 satellite image captured the upwelling event at its peak, revealing cool waters extending from at least 23° N to 29° N (Galveston Bay). Although the 1994 summer upwelling season was abnormally intense (coastal temperatures < 23° C) and influenced more of the Texas inner shelf than is normal, temperature records from 1992 and 1993 confirm that coastal upwelling (coastal temperatures < 25° C) can be expected for about four to six weeks during summer on the south Texas shelf. Surface current velocities ranging from 20–50 cm s⁻¹ were observed along the Texas and Louisiana shelf regions during this up-coast flow regime.

Acknowledgements. Adele Babin and Shreekanth Balasubramanian are gratefully acknowledged for their able assistance with data processing. This study was made possible by the simultaneous data collection efforts of several groups. In particular, Drs. Robert Leben, Pearn Niiler, Walter Johnson, and the LATEX-A research team are thanked for sharing their datasets. This study was supported primarily by the Minerals Management Service study MMS 98-0040, contract no. 14-35-0001-30660-19942, and by the Minerals Management Service-LSU Coastal Marine Institute Cooperative Agreement 309511. The anonymous reviewers are also thanked for their comments.

REFERENCES

Barron, C. N., and A. C. Vastano (1994), Satellite observations of surface circulation in the northwestern Gulf of Mexico during March and April 1989, *Cont. Shelf Res.*, *14*, 607–628.

Brooks, D. A., and R. V. Legeckis (1982), A ship and satellite view of hydrographic features in the western Gulf of Mexico, *J. Geophys. Res.*, *87*, 4195–4206.

Cho, K. R., O. Reid, and W. D. Nowlin (1998), Objectively mapped stream function fields on the Texas-Louisiana shelf based on 32 months of moored current meter data., *J. Geophys. Res.*, *103*, 377–10, 390.

Cochrane, J. D. (1972), Separation of an anticyclone and subsequent developments in the Loop Current (1969), in *Contributions on the Physical Oceanography of the Gulf of Mexico, Texas A&M Univ. Oceanographic Study, vol. 2*, edited by L. R. A. Capurra and J. L. Reid, pp. 91–106, Houston, Texas.

Cochrane, J. D. and F. J. Kelly (1986), Low frequency circulation on the Texas Louisiana continental shelf, *J. Geophys. Res.*, *91*, 10,645–10,659.

Cooper, C., G. Z. Forristall, and T. M. Joyce (1990), Velocity and hydrographic structure of two Gulf of Mexico warm-core rings. *J. Geophys. Res.*, *95*, 1663–1679.

Crout, R. L., W. J. Wiseman, Jr., and W. S. Chuang (1984), Variability of wind-driven currents, west Louisiana inner continental shelf, *Contrib. Mar. Sci.*, *27*, 1–11.

DiMarco, S. F., A. E. Jochens, and M. K. Howard (1997), *LATEX Shelf Data Report, Current meter moorings, April 1992 to December 1994, Texas-Louisiana Shelf Circulation and Transport Processes Study*, 57pp, U. S. Minerals Management Service, New Orleans, LA.

Elliot, B. A. (1982), Anticyclonic rings in the Gulf of Mexico, *J. Phys. Oceanog.*, *12*, 1293–1309.

Forristall, G. Z., K. J. Schaudt, and C. K. Cooper (1992), Evolution and kinematics of a Loop Current eddy in the Gulf of Mexico during 1985, *J. Geophys. Res.*, *97*, 2173–2184.

Hamilton, P. (1992), Lower continental slope cyclonic eddies in the central Gulf of Mexico, *J. Geophys. Res.*, *97*, 2185–2200.

Hsu, S. A. (1988), *Coastal Meteorology*, Academic Press, Inc., San Diego, California.

Huh, O. K. and K. J. Schaudt (1990), Satellite imagery tracks currents in Gulf of Mexico, *Oil Gas J.*, *88*, 70–76.

Huh, O. K., W. J. Wiseman, Jr., and L. J. Rouse, Jr. (1978), Winter cycle of sea surface thermal patterns, Northeastern Gulf of Mexico, *J. Geophys. Res.*, *83*, 523–529.

Jarosz, E. and S. Murray (2005), Velocity and transport characteristics of the Louisiana-Texas coastal current, *this volume*.

Jochens, A. E., D. A. Wiesenburg, L. E. Sahl, C. N. Lyons, and D. A. DeFreitas (1998), *LATEX Shelf Data Report, Hydrography April 1992 through November 1994, Texas Texas-Louisiana Shelf Circulation and Transport Processes Study*, U. S. Minerals Management Service, New Orleans, LA.

Johnson, W. R. and P. P. Niiler (1994), SCULP drifter study in the Northwest Gulf of Mexico, AGU Fall Meeting, San Francisco. O52C-8.

Leben, R. R. and G. H. Born (1993), Tracking Loop Current eddies with satellite altimetry, *Adv. Space Res.*, *13*, 325–333.

Leming, T. D. and C. N. K. Mooers (1981), Cold water intrusions and upwelling near Cape Canaveral, Florida, in, *Coastal Upwelling*, edited by F. A. Richards, pp. 63–71, AGU, Washington, D.C.

Lewis, J. K. and A. D. Kirwan (1985), Some observations of ring topography and ring-ring interactions in the Gulf of Mexico, *J. Geophys. Res.*, *90*, 9017–9028.

Lewis, J. K. and R. O. Reid (1985), Local wind forcing of a coastal sea at subinertial frequencies, *J. Geophys. Res.*, *90*, 934–944.

Lewis, J. K., A. D. Kirwan, and G. Z. Forristall (1989), Evolution of a warm-core ring in the Gulf of Mexico, Lagrangian observations, *J. Geophys. Res.*, *94*, 8163–8178.

McClain, E. P., W. G. Pichel, and C. C. Walton (1985), Comparative performance of AVHRR-based multi-channel sea surface temperatures, *J. Geophys. Res.*, *90*, 11,587–11,601.

Merrell, W. J., Jr. and J. Morrison (1981), On the circulation of the western Gulf of Mexico with observations from April 1978, *J. Geophys. Res.*, *86*, 4181–4185.

Milliman, J. D., and R. H. Meade (1983), World-wide delivery of river sediment to the ocean, *J. Geol.*, *91*, 1–21.

Muller-Karger, F. E., J. J. Walsh, R. H. Evans, and M. B. Meyers (1991), On the seasonal phytoplankton concentration and sea surface temperatures cycles of the Gulf of Mexico as determined by satellites, *J. Geophys. Res.*, *96*, 12,645–12,665.

Munchow, A. and R. W. Garvine (1993), Buoyancy and wind forcing of a coastal current. *Journal of Marine Research*, *51*, 293–322.

Murray, S. P., E. Jarosz, and E. T. Weeks, III (1998), Physical oceanographic observations of the coastal plume, in *An observational study of the Mississippi-Atchafalaya coastal plume*, edited by S. P. Murray, pp. 5–105, U. S. Minerals Mgmt. Service, New Orleans, LA.

N.O.A.A. (2003 and 2004), Daily Weather Maps, U. S. Dept. of Commerce, National Oceanic and Atmospheric Administration, Washington, DC.

Nowlin, W. D. and C. A. Parker (1974), Effects of a cold-air outbreak on shelf waters of the Gulf of Mexico, *J. Geophys. Res.*, *4*, 467–486.

Oey, L.-Y (1995), Eddy and wind-forced shelf circulation, *J. Geophys. Res.*, *100*, 8621–8637.

Ohlmann, J. C., P. P. Niiler, C. A. Fox, and R. R. Leben (2001), Eddy energy and shelf interactions in the Gulf of Mexico, *J. Geophys. Res.*, *106*, 2605–2620.

Ohlmann, J. C. and P. P. Niiler (2005), Circulation over the continental shelf in the northern Gulf of Mexico. *Progress in Oceanography*, *64*, 45–81.

Rabalais, N. N., R. E. Turner, Q. Dortch, D. Justic, V. J. Bierman, and W. J. Wiseman, Jr. (2002), Nutrient-enhanced productivity in the northern Gulf of Mexico: past, present and future, *Hydrobiologia*, *475/476*: 39–63,.

Sahl, L. E., W. J. Merrell, and D. C. Biggs (1993), The influence of advection on the spatial variability of nutrient concentrations on the Texas-Louisiana continental shelf, *Cont. Shelf Res.*, *13*, 233–251.

Schumann, S., G. A. Johnson, J. Moser, N. D. Walker, and S. A. Hsu (1995), An overview of a strong winter low in the Gulf of Mexico, March 12–13, 1993, *National Weather Digest*, *20*, 11–25.

Smith, N. P. (1978), Low-frequency reversals of nearshore currents in the north-western Gulf of Mexico, *Contrib. Mar. Sci*, *21*, 103–115.

Sturges, W., and R. Leben (2000), Frequency of ring separations from the Loop Current in the Gulf of Mexico, a revised estimate, *J. Phys. Oceanog.*, *30*, 1814–1819.

Sweitzer, N. B. Jr. (1898), Origin of the Gulf Stream and circulation of waters in the Gulf of Mexico, with special reference to the effect on jetty construction, *Trans. Am. Soc. Eng.*, *40*, 86–98.

Vidal, V. M., F. V. Vidal, and J. M. Perez-Molero (1992), Collision of a Loop Current Anticyclonic ring against the continental shelf slope of the western Gulf of Mexico, *J. Geophys. Res.*, *97*, 2155–2172.

Walker, N. D. (1996), Satellite assessment of Mississippi River plume variability, Causes and predictability, *Remote Sens. Environ.*, *58*, 21–35.

Walker, N. D. (2001), Wind and eddy-related circulation on the Louisiana/Texas shelf and slope determined from satellite and in-situ measurements, October 1993–August 1994, *OCS Study MMS 2001–025*, U. S. Dept. of Interior, Minerals Mgmt. Service, Gulf of Mexico OCS, New Orleans, La., 56 pp.

Walker, N. and A. Hammack (2000), Impacts of winter storms on circulation and sediment transport, Atchafalaya-Vermilion Bay region, Louisiana, USA, *J. Coastal Res.*, *16*, 996–1010.

Wiseman, W. J., Jr. and W. Sturges (1999), Physical Oceanography of the Gulf of Mexico, processes that regulate its biology, in *The Gulf of Mexico Large Marine Ecosystem*, edited by Kumpf, et al., pp. 77–92, Blackwell Science, Inc., Malden, MA.

Nan D. Walker, Department of Oceanography and Coastal Sciences, Coastal Studies Institute, Louisiana State University, Baton Rouge, Louisiana 70803 (nwalker@lsu.edu)

Intermediate-Depth Circulation in the Gulf of Mexico Estimated from Direct Measurements

Georges L. Weatherly

Department of Oceanography, Florida State University, Tallahassee, Florida USA

Nicolas Wienders

Department of Oceanography, Florida State University, Tallahassee, Florida USA

Anastasia Romanou

Department of Applied Physics and Applied Mathematics, Columbia, University, New York, New York USA

Data from 17 PALACE floats set in the Gulf of Mexico sampling the intermediate-depth ($\approx$ 900 dbar) flow from April 1998 to February 2002 indicate a mean cyclonic circulation along the northern and western edges of the Gulf of Mexico. This flow intensified into a $\approx$ 0.10 m/s current in the western Bay of Campeche and was deflected around a topographic feature, called here the Campeche Bay Bump, in the southern Bay of Campeche. Floats launched in the eastern Gulf of Mexico tended to stay there, and those launched in the western Gulf tended to stay in the western Gulf, suggesting restricted connection at depth between the eastern and western Gulf of Mexico. Not surprisingly, the measured flow was stronger when the measured flow was under the Loop Current and warm-core rings, but the direction of the intermediated depth currents bore no apparent relation to the surface flow inferred from satellite altimeter maps. However, comparing the floats' surface drifts to their intermediate depth drifts, the floats at depth ended to track the surface flow in the Loop Current, and both indicate a cyclonic gyre in the Bay of Campeche.

1. INTRODUCTION

This is a report of inferred intermediate depth circulation in the Gulf of Mexico obtained from 17 PALACE (Profiling Autonomous Lagrangian Circulation Explorer) floats [*Davis et al.*, 2001]. The floats were acquired and set by John Blaha of the Naval Oceanographic Office, Stennis Space Center,

Circulation in the Gulf of Mexico: Observations and Models
Geophysical Monograph Series 161
10.1029/161GM22

MS, as part of a National Oceanographic Partners Program study of the Gulf of Mexico [*Blaha et al.*, 2000]. The authors of this report were invited to examine the PALACE data after the floats were set.

PALACE floats are a relatively new instrument for inferring intermediate depth flow. They were first used for this purpose in the Drake Passage [*Davis et al. 1992*]. More recently they have been used in studies in the tropical and south Pacific [*Davis*, 1998], in the northern Atlantic, e.g., *Lavender et al.* [2000] and western tropical Atlantic [*Kwon and Riser, 2005*], in the southwestern Indian Ocean, e.g.,

Chapman et al. [2004], in the Southern Oceans [*Gille, 2003*], in the Bering Sea [*Johnson et al. 2004*], and in the Okhost Sea [*Oshinin et al. 2004*]. The floats are not tracked as they drift at depth as the SOFAR (SOund Fixing And Ranging) and RAFOS (SOFAR spelled backwards) floats are: see, for example, *Owens* [1991] and *Richardson and Garzoli* [2003]. Instead they are tracked at the surface by satellite shortly before they descend to the drift depth and, after drifting at depth for a fixed interval, which ranged in the studies above from about four to twenty-five days, they are tracked again after they surface. It is these surface fixes which are used to estimate the intermediated-depth drift.

In the above PALACE studies of *Davis et al.* [1992], *Davis* [1998], *Lavender et al.* [2000], and *Gille,* [2003], the floats were at the surface for about 21 hours. This relatively long time was chosen so that enough surface fixes could be obtained so that the surface drift would be objectively extrapolated to the times the float surfaced and submerged. Then the start and stop location of each intermediate-depth drift was estimated assuming a geostrophic shear and knowing the time it takes the float to sink and to rise. The technique is quite involved [*Davis et al,.1992*] and requires the surface drift time be longer than the inertial period. With a surface fix error of ± 2 km and submerged time of 25 days *Davis* [1998] estimates an error of ± 0.001 m/s in the intermediate depth drift.

In some cases, such as this study, where the study region is semi-enclosed such as the Bering Sea [*Johnson et al. 2004*], and the Okhost Sea [*Oshinin et al. 2004*] or in regions of topography such as the western Labrador Sea [*Fischer and Schott, 2002*] and the western, tropical Atlantic [*Kwon and Riser, 2005*], a shorter surface time of about 12 hours is chosen to reduce the risk that the floats drift into water shallower than their programmed drift depths while at the surface. *Johnson et al.*[2004] *Kwon and Riser* [2005],and *Oshinin et al.* [2004] assumed that the last surface fixed before submerging is a reasonable estimate of the beginning location of the intermediate depth drift, and the first surface fix after resurfacing is a reasonable estimate of the end location of the intermediate level drift. They estimate that this introduces an error of about 10% in their intermediate depth drifts. *Fischer and Schott* [2002] did carefully, hand-edited polynomial fits to the surface fixes and extrapolation to estimate the float positions when they surfaced and their positions when they sank. With a surface fix error of ±0.5 km and submerged time of 5 or 10 days they estimate an uncertainty of ~ 0.01 m/s in the intermediate depth drifts.

The floats examined here drifted at ≈ 900 dbar for 6.23 days and surfaced for 11.5 h once every week. They were launched in the northern Gulf of Mexico (the launch coordinates and dates are given in *Weatherly* 2004), but drifted throughout the Gulf of Mexico (Plate 1). The flow in much of the open Gulf of Mexico, here taken to be in water deeper that the ≈ 900 m of the floats' drift depth, is dominated by the Loop Current and Loop Current rings in the eastern Gulf and Loop Current rings elsewhere (e.g., *Schmitz* [this issue]), Loop Current rings here meaning cyclonic rings as well as the more familiar anticyclonic ones. Since these features extend down to about the drift depth of the floats (e.g., *ibid.*), the floats should at times respond to these features if below them. Though the floats drifted throughout the Gulf of Mexico, the majority of the intermediate-depth drifts were from the lower continental margin (Plate 1). The lower continental margin is a region of energetic topographic Rossby waves [*Hamilton and Lugo-Fernandez,* 2002, *Oey and Lee,* 2003]. Therefore the floats drifted in regions rich with transient, strong currents with time scales ≈ a month.

The data which these floats yielded over a four-year period beginning in April 1998 (Figure 1) make possible long-term mean circulation estimates in water deeper than ≈ 900 m. The floats drift depth is near the transition depth for an upper layer and lower layer flow regimes. We now review some of what is thought to occur in each regime. Drifter observations obtained over a ten-year interval indicate an anticyclonic basin-wide, excluding the Bay of Campeche (the bay in the southwestern Gulf), surface-flow pattern [*DiMarco et al.* this issue] although in the northern Gulf along the continental margin this flow is weak and ill defined. These

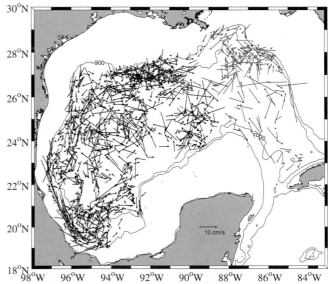

Figure 1. The 900-dbar velocities estimated from the intermediate-depth drifts. The vectors mid-points are centered at the mid-point of the intermediated-depth drifts. The black (grey) vectors are from floats launched west (east) of 89°W. The 900-m and 2000-m isobaths are shown.

data also indicate that in the winter this gyre may split into an eastern one and a western one at about 88°W. An average of a nine-year numerical simulation, shown in *Lee and Mellor* [2003], shows a basin-wide anticyclonic flow pattern near the surface. At about 1000 m depth this and two other numerical studies [personal communication by *Welsh* [2002] and Groupment Mercator] indicate a weaker basin-wide cyclonic flow pattern in the Gulf of Mexico. Consistent with a 1000-m depth cyclonic flow pattern are the mean current records of duration ≥ 1 year from the northern and eastern Gulf [*Hamilton,* this issue]. *DeHaan* [2003], using historic hydrographic and current meter data presented in *Hamilton* [1990], also inferred a deep cyclonic flow. In the Bay of Campeche *Vazquez de la Cerda* [1993] and *Vazquez de la Cerda et al.* [this issue] describe a quasi-permanent upper layer cyclonic gyre in the Bay of Campeche which they inferred was wind driven. The drifter data reported in *DiMarco et al.* [this issue] also indicate a cyclonic gyre in this Bay. *Vazquez de la Cerda et al.* [this issue] infer using results presented in this study that this cyclonic flow extends down to ≈ 1000-m depth. No western intensification of the surface flow is apparent in the Bay of Campeche [*Vazquez de la Cerda,* 1993, *Vazquez de la Cerda et al,.*this issue, *DiMarco et al.,* this issue].

2. DATA AND DEEP DRIFT ESTIMATES

The floats were set to drift at 800 dbar so the temperature profiles they made as they surfaced each week complemented those made by 800-m XBTs [personal communication, *Sturges*, 2002]. Their drift depths varied from 830 dbar to 1040 dbar, with the mean depth near 900 dbar [*Weatherly,* 2004]. Their drift depths increased with time at a rate about 23 dbar/yr ± 10 m/yr. Some of the floats drifted over three year, so that towards the end some were drifting at depths about 100 dbar deeper than when they started. In this study we refer to their drift depth collectively as 900 dbar. The geostrophic shear in the depth range in which the floats drifted is weak [*DeHaan*, 2003].

About 30% of the Gulf of Mexico is shallower than 900 m, and even with a relatively short surface time (11.5 hours every 7 days), it was not unusual for a float while at the surface to drift into water shallower than the drift depth (Plate 1, the trajectories shoreward of the 900 m isobath). All float trajectories which had a depth <830 dbar, the minimum of the initial depths to which the floats descended, were assumed to be grounded, and were excluded. As many of the trajectories were along the continental margin, the depth versus time plots of the floats were also examined for suspiciously shallow values, i.e., values > 830 dbar but notably shallower than their neighbors. If such values occurred adjacent to the 900

m isobath in Plate 1 they were considered grounded and also excluded. We obtained 1315 intermediated depth drifts.

The challenge with PALACE floats is to estimate the location at the beginning and end of each intermediate depth drift from surface fixes. The method we used to estimate the start and stop locations is somewhat between that used in Johnson et al..[*2004*], Oshinin et al. [*2004*], and Kwon and Riser, [*2005*], who assume the former is the last surface fix before submerging and the latter is the first fix after surfacing, and that used by *Fischer and Schott* [2002], who assume the former is the surface location at the time of sinking and the latter is the surface location at the time of first surfacing. As noted earlier *Fischer and Schott* [2002] used hand-edited polynomial fits of the surface fixes to extrapolate these locations. Here we estimated diving and surfacing positions by making linear extrapolations based on the first and last fix each time the float surfaced. To estimate the time at which the floats surfaced and sank we used a rise time of 2.5 h and a drift time at the surface of 11.5 h; to estimate when they reached drift depth we used a sinking time of 4.5 h. These times were provided by the manufacturer of the floats, Webb Research, Falmouth MA USA.

Three to six satellite fixes were usually made at the surface, and we plotted all surface drifts based on these positions. These surface drifts look very much like those seen in Plate 1 which are based on the first and last fixes. Because there was some curvature in the drifts we re-estimated the sinking and rising positions on the basis of linear extrapolations on the two nearest fixes (the first two for the surfacing position and the last two for the dive position). The resulting intermediate-depth drifts are essentially the same as those presented here, and the conclusions presented here about the intermediate depth circulation do not change.

In the estimates of intermediate-depth drifts presented here we have attempted to remove the sub-surface drift which resulted when floats descended to their drift depth from the surface as well as that which occurred when they rose from drift depth to the surface. To do so we assumed that there was a linear decrease from the surface current (estimated from the surface fixes) to the current at the drift depth. We first estimated the location where the float reached drift depth and its location where it began to surface by assuming the current at the intermediate depth was zero. The intermediate depth drift was then estimated. Using that estimate, the drift during sinking and rising were then re-estimated. The procedure was repeated three more times. The deep-drift estimates essentially did not change after the second iteration, and the first estimate was generally within 10% of the final estimate. We also estimated the deep drifts assuming there was zero drift while sinking and rising and got essentially the same flow results.

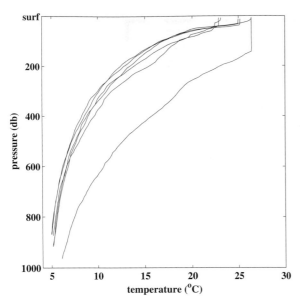

Figure 4. All the temperature profiles from a 1° latitude by 1° longitude bin centered at 26.5°N, 89.0° W. The right-most profile was made in a warm-core Loop Current ring.

Ocean using a comparable integral time scale. Generally, if $n \geq 5$, the flow direction appears resolved in the western Gulf of Mexico and about the Campeche Bay Bump, but it is only marginally resolved elsewhere (Figure 3).

To estimate when the above-noted flow features were sampled during the ≈ four-year duration study period, the intermediate-depth velocities were plotted for sequential six-month periods. These figures are shown in *Weatherly*

[2004] and in this issue's CD; they are summarized in the table. The southerly flow along the western edge of the Gulf of Mexico, the intensified southerly flow along the western boundary of Bay of Campeche, and the cyclonic flow about the Campeche Bay Bump were sampled during much of the study period.

3.2 Variability

To see whether some of the velocities were related to inter-mediate-depth extensions of the Loop Current and Loop Current rings, we superposed the current vectors onto satellite altimeter maps. We tried two variants: monthly satellite maps with all currents within two weeks of the map and biweekly satellite maps with all currents within a week of the map. At times the intermediate-depth flow under the Loop Current and its rings appeared to extend to intermediate depth; at other times it did not (some examples of each are found in *Weatherly* [2004]). We could not tell ahead of time what to expect. For example, we could find no pattern if we considered only the Loop Current, newly formed Loop Current rings, or older rings.

There is some uncertainty with the float data and the satellite altimeter maps. To see if we could get more consistent results we restricted ourselves to velocity measurements made when both the temperature data from the float and the altimeter map indicated that the float had been under the Loop Current or an anticyclonic, warm-core Loop Current ring. As an example, Figure 4 shows all temperature profiles made from the 1° latitude by 1° longitude box centered at

Table. A tabulation, using the velocities presented in Figures A1–A8 in this issue's CD (also shown in [*Weatherly*, 2004]), of when different regions along the edge of the Gulf of Mexico had their intermediate depth flow measured. Under (a) NE is the region in the northeast Gulf of Mexico east of 89°W, N is the region in the northern Gulf of Mexico between 89°W and 95°, NW is the western Gulf of Mexico north of 23°N, WCB is the western Gulf of Mexico in Bay of Campeche, and CBB is the Campeche Bay Bump. Under (b) is the cyclonic, gyre in the western Bay of Campeche centered near 20.5°N, 95.5°W.

Period	Cyclonic flow along slope					Campeche Bay Gyre
	(a) NE	(a) N	(a) NW	(a) WCB	(a) CBB	(b)
04/98 ≤time <10/98	x	x	x			
10/98 ≤time <04/99	x	x	x	x	x	x
04/99 ≤time <10/99		x	x	x	x	x
10/99 ≤time <04/00		x	x	x	x	x
04/00 ≤time <10/00		x	x	x	x	
10/00 ≤time <04/01	x	x	x		x	x
04/01 ≤time <10/01	x	x	x	x		x
10/01 ≤time <04/02		x	x		x	

22.5°N, 79.0°W. The profile farthest to the right is conspicuously warm indicating that this profile was made either in the Loop Current or a Loop-Current ring. Comparing the location of this profile (22.2°N, 77.9°W) with a satellite altimeter map for the date of this profile (25 November 1999) found at http://ww-ccar.colorado.edu/~realtime/gom/gom_nrt.html indicated that this profile was made in a Loop Current warm-core ring.

We found 103 cases in which a float's temperature profile and the associated altimeter map for the appropriate day indicated that the float was in the Loop Current or in a warm-core Loop Current ring. For these cases the intermediate-depth flow was stronger (Figure 5). However, we could discern no consistent pattern between the flow direction at intermediate depth and that indicated at the surface in the altimeter maps (not shown).

4. SUMMARY AND DISCUSSION

During the ≈ four-year period during which the floats drifted, the average intermediate (900 dbar) depth flow was generally cyclonic along the continental margins in the northern and western portions of the Gulf of Mexico. There was too little data from the eastern and southern Gulf of Mexico to say whether this pattern extends to these regions. This flow strengthened to ≈ 0.1 m/s in the western Bay of Campeche and was diverted around the Campeche Bay Bump in the southern Bay of Campeche.

The above results are not overly sensitive to the manner in which we estimated the intermediate depth drifts from the surface fixes. Taking last and first surface fixes as reasonable estimates of the start and end intermediate depth drift locations (as do *Johnson et al.* [2004], *Kwon and Riser,* [2005], and *Oshinin et al.* [2004]) gave the same flow pattern as in Figure 2 except that the current magnitudes were increased by about 10%. The latter is consistent with the 10% error reported with this assumption (*Johnson et al.* 2005). We tried two linear extrapolation methods to estimate the surface locations of surfacing and sinking and got essentially the same intermediate depth drift results for each. We did do polynomial fits to the to the surface drift fixes for two floats to estimate surfacing and sinking locations for these floats. We found, similarly to what *Fischer and Schott* [2004] found, that extensive editing was required to yield reasonable results. Most of the editing was automated (e.g., to exclude fixes separated in time by less than an hour) but other were done by hand (e.g., when the first surface fix was obtained five or so hours after surfacing). Because the intermediate depth drift for these two floats were nearly identical to those estimated earlier, and because the majority of the surface drifts were nearly linear we do not think the results we present would change appreciably if we had estimated sinking and surfacing locations by polynomial fits to the surface fixes.

In our estimated drifts we attempted to correct for possible drift while sinking from the surface to drift depth and while rising from drift depth to the surface. However, neglecting such drift (i.e., assuming the float went straight down and up) we got essentially the same flow pattern as in Figure 2 and essentially the same velocity histogram as in Figure 5 (solid curve) except the average value increased by 0.015 m/s.

We found that the conclusions of southerly flow along the western edge of the Gulf of Mexico and the diversion of the flow about the Campeche Bay Bump in Figure 2 were statistically significant. While the westerly flow along the northern edge of the Gulf of Mexico was marginally significant it was consistent with long-term moored current measurements from this region [*Hamilton,* this issue]. We attempted to account for some of the uncertainty in the intermediate-depth drift estimates resulting from an ± 1 km uncertainy in the surface fixes (this is intermediate to the ± 2 km uncertainty of *Davis,* [1998] and the ± 0.5 km uncertainty of *Fischer and Schott* [2002]). We found that the estimates are meaningless for inferred intermediate-depth velocities < 0.01 m/s, and for velocities ≈ 0.01 m/s, ≈ 0.05 m/s, ≈ 0.10 m/s the flow direction is resolved to ≈ ± 45°, ± 13°, ± 7°, respectively. We excluded from the averages shown in Figure 2 all those with inferred intermediate depth currents ≤ 0.01 m/s and got again the same flow pattern shown in Figure 2. Thus the conclusion of statistically significant southerly flow along the western edge of the Gulf of Mexico and diversion of this flow about the Campeche Bay Bump

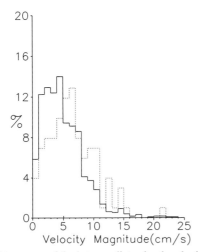

Figure 5. Histograms of the intermediate-depth velocity magnitude of all the measurements (solid line) and only those measurements estimated to be made under the Loop Current or a warm-core Loop Current ring (dashed line).

Lavender, K., R. E. Davis and B. Owens, 2000: Mid-depth recirculation observed in the interior Labrador and Irminger Seas by direct velocity measurements, *Nature*, **407**, (7 Sept,), 66–69.

Lee, H.-C. and G. L. Mellor, 2003: Numerical simulation of the Gulf Stream System: the Loop Current and the deep circulation, *J. Geophys. Res.*, **108** (C2) 10.1029/2000JC000736,

Oey, L.-Y. and H.-C. Lee, 2003: Deep eddy energy and topographic Rossby waves in the Gulf of Mexico, *J. Phys. Oceanogr.*, **32**, 3499–3527.

Oey, L.-C., T. Ezer, and H.-C. Lee, 2005: Loop Current, rings, and related circulation in the Gulf of Mexico: a review of numerical models and future challenges, this issue.

Ohshima, K. I., D. Simizu, M. Itoh, G. Mizuta, Y. Fukamachi, S. C. Riser and M. Wakatsuchi, 2004: Sverdrup Balance and the Cyclonic Gyre in the Sea of Okhotsk. *J. Phys. Oceanogr.*, **33**, 3499–3527, 513–525.

Owens, W. B., 1991: A statistical description of the mean circulation and eddy variability in the northwestern Atlantic using SOFAR floats, *Prog. Oceanogr.*, **28**, 257–303.

Richardson, P. L. and S. L. Garzoli, 2003: Characteristics of intermediate water flow in the Benguela Current as measured by RAFOS floats, *Deep-Sea Res. II,* **50**, 87–118.

Schmitz, W. J., Jr, 2005. Cylcones and westward propagation of in the shedding of anticyclonic rings from the Loop Current, this issue.

Vazquez de la Cerda, A. M, 1993: Bay of Campeche Cyclone, Ph.D. Dissertation, Dept. of Oceanography, Texas A.& M. Univ., College St., Texas, 91 pp.

Vazquez de la Cerda, A. M. , R. O. Reid, S. F. DiMarco, and A. E. Jochens, 2005; Bay of Campeche: an update, this issue.

Weatherly, G., N. Wienders and R. Harkema, 2003. Temperature inversions in the open Gulf of Mexico, *J. Geophys. Res.,* **108**, (C6) 10.1029/2002JC001680.

Weatherly, G. L., 2004 Intermediate depth circulation in the Gulf of Mexico: PALACE Float Results between April 19988 and March 2002, U.S. Dept. of the Interior, Minerals Management Service, New Orleans, LA, 41 pp, available at http://www.gomr.mms.gov/homepa/regulate/environ/studies/2004/2004–13.pdf.

Welsh, S. E., 1996: A numerical modeling study of the Gulf of Mexico under present and past environmental conditions, Ph.D. Dissertation, Louisiana St. Univ., Baton Rouge, Louisiana, 206 pp.

Welsh, S. E. and M. Inoue, 2000: Loop current rings and deep circulation in the Gulf of Mexico, *J. Geophys. Res.*, **105**, 1,951–16,959.

Georges L. Weatherly, Department of Oceanography, Florida State University, Tallahassee, FL 232306-4320 USA

West Florida Shelf Circulation on Synoptic, Seasonal, and Interannual Time Scales

Robert H. Weisberg[1,] Ruoying He[2], Yonggang Liu[1], and Jyotika I. Virmani[1]

The continental shelf ocean circulation is driven by a combination of local and deep ocean forcing, where the local forcing is by surface momentum and buoyancy (heat and fresh water) fluxes and by land-derived fresh water fluxes, and the deep ocean forcing is by momentum and buoyancy fluxes transmitted across the shelf break. While all continental shelves share the same response physics, their geometries, river distributions, and boundary currents make their individual behaviors unique. Here we consider the responses of the West Florida Continental Shelf (WFS) to external forcing on time scales ranging from the passage of synoptic scale weather systems to interannual anomalies. Our approach is to combine measurements made over several years with applied numerical model experiments, and we describe several features of the time and space varying circulation that may have relevance to the ecological workings of the WFS. These include the role of the bottom Ekman layer in transporting cold, nutrient-rich waters of deep-ocean origin to the nearshore; seasonal reversals of the inner shelf currents; seasonal reversals of the outer-shelf currents that can occur independent of the Gulf of Mexico Loop Current to provide a WFS pathway for Mississippi River water in spring and summer; the formation of a spring cold tongue and the associated "Green River" phenomenon; and ventilations by deep-ocean water that may occur interannually.

1. INTRODUCTION

Continental shelves are the relatively shallow coastal regions that separate the land from the abyssal ocean. All with differing geometry, they share in common a point of bottom slope change, called the shelf break, seaward of which the water depth rapidly increases across the shelf slope toward the abyss. The coastal ocean circulation on all continental shelves is externally forced by a combination of momentum and buoyancy, input either locally (by surface winds, surface heat and fresh water fluxes, and by river or spring-fed fresh water fluxes) or from the deep ocean (by momentum and buoyancy fluxes transmitted across the shelf break). The responses to such external forcing factors are governed by the same physical principles on all continental shelves, but differences in geometry and in the estuarine and boundary current interactions render each continental shelf environment unique.

Much of the West Florida Continental Shelf (WFS), depicted in Figure 1, has a relatively simple geometry, with a gently sloping inner shelf that is dynamically separated from the shelf slope by virtue of the large shelf width. Geometrical complexities arise in the south, where the Florida Keys provide both a barrier to the flow and a convergence of isobaths west of the Dry Tortugas, and in the north, where the isobaths again converge in the DeSoto Canyon before making a right angle bend into the Mississippi Bight. The DeSoto Canyon is also the region of narrowest shelf width.

Fresh water enters the WFS from various sources. From northwest to southeast there occurs a seasonal succession

[1]College of Marine Science, University of South Florida, St. Petersburg, Florida

[2]Woods Hole Oceanographic Institution, Woods Hole, Massachusetts

Circulation in the Gulf of Mexico: Observations and Models
Geophysical Monograph Series 161
10.1029/161GM23

the vicinity of the shelf break such that flows imposed on the shelf tend to be barotropic and weak. Similarly, flows locally forced on the shelf tend not to penetrate onto the shelf slope. These studies conclude (on the basis of linear theory) that deep ocean forcing should be relatively ineffective at generating shelf currents.

A case study confirming these predicted behaviors is provided for the WFS by *He and Weisberg* [2003a]. Figure 2 shows the geostrophic currents (relative to the bottom) calculated from two hydrographic sections sampled about a month apart and observed at a time when the LC impacted the shelf slope offshore of Tampa Bay. The LC is constrained to the shelf slope with its baroclinic shear diminishing to zero within a Rossby radius of deformation (estimated to be about 25 km), and on the shelf the flow is weak. Moored acoustic Doppler current profilers (ADCP) confirmed these behaviors and provided absolute current determinations for separating the (smaller) barotropic and (larger) baroclinic portions of the flow. Simplified (constant density) numerical model simulations (Figure 3) for the combined effects of deep ocean (LC) and local (wind) forcing further demon-

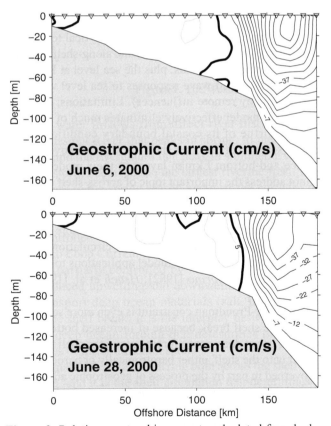

Figure 2. Relative geostrophic currents calculated from hydrographic sections sampled offshore from Sarasota Florida on June 6th and June 28th 2000. The contour interval is 5 cm s⁻¹, and solid lines denote southeastward flow (from *He and Weisberg*, 2003a).

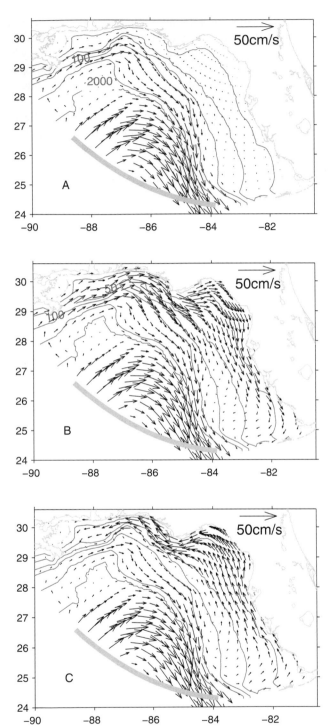

Figure 3. Depth-averaged current maps calculated from idealized LC and wind-forced model experiments: a. LC only, b. LC plus 0.1 N m⁻² upwelling-favorable winds, c. LC plus 0.1 N m⁻² downwelling-favorable winds. The thick line along the open boundary is where the LC is controlled. The depth contours are the 20 m, 50 m, 100 m, 200 m, 1000 m, and 2000 m isobaths (from *He and Weisberg*, 2003a).

strates the decoupling of these two effects, consistent with other ADCP observations made across the inner shelf.

Despite the above constraints there are instances when deep ocean effects are quite apparent on the WFS. We may account for these instances through the effects of the frictional Ekman layers (even in the above case landward advection did occur in the bottom Ekman layer; we will return to this later), the region of relatively narrow shelf width within and east of the DeSoto Canyon, and the region of convergent isobaths at the Dry Tortugas. Narrow shelf width is important because features with sufficiently large relative vorticity (large Rossby number) can overcome the small Rossby number, Taylor-Proudman constraint over the relatively short distance. Examples of eddies in the vicinity of the shelf break abound on the U.S. southeast coast, where Gulf Stream spin-off eddies are regularly observed; e.g., see the review by *Boicourt et al.*, [1998], and *Paluszkiewicz et al.* [1983] provides an example for the WFS. At the DeSoto Canyon the shelf is sufficiently narrow such that eddy effects can permeate the entire canyon head [*Huh et al.*, 1981].

The Dry Tortugas region is particularly interesting for two reasons. First, it is the most upstream location on the WFS with respect to the propagation direction for continental shelf waves and second, deep- and shallow-water isobaths converge there because of the partial closure of the WFS by the Florida Keys. Consistent with continental shelf-wave dynamics, a pressure perturbation will propagate along isobath with the coastline on the right. Thus a deep ocean pressure perturbation impacting shallow isobaths at the Dry Tortugas can potentially affect the entire WFS (as included in the *Mitchum and Clarke*, [1986b] analysis). *Csanady* [1978] provides the basis for how this may impact the shelf in the presence of frictional damping, and *Hetland et al.* [1999] expands on this idea for the WFS. With the LC regularly penetrating into the Gulf of Mexico before shedding a large anticyclonic eddy and collapsing back to the south [*e.g., Sturges and Leben,* 2000], such LC impacts on the shelf slope near the Dry Tortugas occasionally occur, and when they do there is now ample evidence for their direct influence from the shelf break to the beach [*Weisberg and He*, 2003 and *He and Weisberg*, 2003b]. We will return to this in section 5, along with the finding that local forcing in the DeSoto Canyon region can combine with deep ocean forcing at the Dry Tortugas to cause highly anomalous conditions over the entire WFS.

3. SYNOPTIC VARIABILITY

A property of all continental shelves is a correlation between the local winds (the along-shelf component in particular) and sea level, and as a corollary (by geostrophy) there is a correlation between the winds and the along-shelf currents. The wind

fluctuations on the WFS tend to be largest with the passage of synoptic-scale weather fronts in winter months. Frontal passage [e.g., *Fernandez-Partegas and Mooers*, 1975] generally entails southerly followed by northerly winds, as weather systems propagate across the region from west to east with a stochastic interval of some days to weeks. Reports on the WFS responses to synoptic scale weather are given by *Niiler* [1976], *Marmorino* [1982, 1983], *Mitchum and Sturges* [1982], *Cragg et al.* [1983], and *Weatherly and Thistle* [1999]. With the exception of sea level these studies are of limited duration and water column sampling. More recent observations using ADCPs over longer duration and with larger vertical resolution show spatial structures and response behaviors to be discussed in this section.

A starting point is the examination of the WFS responses to upwelling and downwelling favorable winds since frontal passages have phases of upwelling (northerly) and downwelling (southerly) directed wind stress. *Li and Weisberg* [1999a,b] respectively address the kinematics and dynamics of WFS upwelling and downwelling responses in the context of a primitive equation numerical model [using the Princeton Ocean Model (POM) of *Blumberg and Mellor*, 1987] with constant density, forced by spatially uniform, impulsively started winds. The focus is primarily on the inner shelf, initially defined, as in *Mitchum and Clarke* [1986a], as the region of overlapping surface and bottom Ekman layers. A primitive equation model allows us to examine the structure of the inner shelf and discuss its behavior relative to the entire shelf. While subtle, the topographic variations of the WFS are important, as shown through the various cross sections sampled at regions of different shelf width, and these behaviors are also sensitive to the form of vertical turbulence parameterization. For flow-dependent parameterizations, such as the model default Mellor-Yamada level 2.5 closure scheme [*Mellor and Yamada*, 1982], the flow fields are realistic, whereas with constant-eddy coefficients the flow fields are unrealistic.

In these impulsively started wind experiments the coastal ocean spins up over a pendulum day, the velocity field, even with constant density, is fully three-dimensional, and the surface and bottom Ekman layers, the geostrophic interior, and their interactions (through the vertical velocity component) are distinguishable. Offshore of Tampa Bay the inner shelf extends out to about the 50 m isobath as a region of across shelf sea level slope initially set up by a surface Ekman layer divergence. Owing to shelf width this inner shelf region is separated from the shelf break and slope regions where the flow is much more variable due to eddy-like motions (caused by the interplay between planetary and relative vorticity). The midshelf region between the inner shelf and the shelf break is influenced by the partial closure of the Florida Keys, and this results in a return flow and a reversal in the across-shelf pressure gradient. These dynamically distinctive regions near-surface currents at the 50m isobath were directed perpendicular to the coastline and to the right of the wind stress. With both time and depth the currents rotated cyclonically to the left in response to the generation of a geostrophic current in balance with the across-shelf pressure gradient. Additional cyclonic turning near the bottom resulted in onshore flow within the bottom Ekman layer induced by the geostrophic flow, thereby completing the Ekman-geostrophic route to coastal ocean spin-up.

This upwelling case study also considered model simulations under both constant density and stratified conditions to aid in explaining the position of maximum upwelling seen in a concomitant satellite image. By inhibiting vertical mixing and concentrating the bottom Ekman layer flow closer to the bottom, stratification shifts the region of cold water outcrop farther to the south and closer to the beach than the results obtained under constant density conditions. Confirmation of such stratified inner shelf behaviors are provided by *Weisberg et al.* [2004].

Momentum budget analyses are also informative since they help in interpreting the across-shelf distribution of the along-shelf flow and in identifying response asymmetries that may arise by stratification. *Weisberg et al.* [2001] describe a succession of fronts and their associated coastal ocean responses during April 1998, along with numerical model simulations of these using spatially uniform, but temporally varying, winds. Term-by-term comparisons of the along-shelf momentum balances, both in across-shelf sectional views at specific times and in time and depth sequences at a specific location, show that the primary balances in these simulations are between the pressure gradient, Coriolis, and vertical friction terms. The friction terms arise primarily from the surface and bottom Ekman layers so upon vertical integration the momentum balance is geostrophic as indicated earlier. The vertical distribution of the currents, with a middepth minimum, is consistent with the linear summation of the (surface and bottom) Ekman layers with the geostrophic interior, and the relative near-bottom maximum is consistent with the oscillatory behavior of these flows. An unexpected finding is an asymmetry between responses to upwelling and downwelling favorable winds. Under stratified conditions the upwelling responses were found to be of larger magnitude and to extend farther offshore than the downwelling responses, both in the data and in the model simulations. Analyses of the along-shelf component of vorticity shows this to be a consequence of thermal wind effects in relation to the bottom Ekman layer. Buoyancy torque adds constructively (destructively) with planetary vorticity tilting under upwelling (downwelling) requiring that the bottom Ekman layer be more (less) developed under upwelling (downwelling). Since the evolution of

the across-shelf surface slope depends on the surface and bottom Ekman layer divergences operating in tandem, one feeding the other, if the bottom Ekman layer can extend farther offshore, so can the entire coastal ocean response. Thus in the stratified case the offshore scale of the inner shelf is no longer accounted for by the region of overlapping Ekman layers; rather it is the region of interacting Ekman layers through divergence. By stratification allowing for increased bottom Ekman-layer response under upwelling the entire coastal ocean spin-up can extend farther offshore and with larger magnitude.

In combination, the observed velocity profiles and the modeled momentum balances provide information on the vertical scales of the surface and bottom boundary layers under varying stratification conditions. Figure 5, for instance, suggests a surface Ekman layer of only about 10–15 m. Term-by-term momentum analyses in *Weisberg et al.* [2001, 2002] suggest that the surface and bottom Ekman layers merge out to about the 50 m isobath under constant density conditions (consistent with the flow-dependent eddy viscosity calculated in *Li and Weisberg* [1999b]), whereas these layers are less than 10 m deep under stratified conditions. Even without overlap, however, the surface and bottom Ekman layers may interact through divergence such that the scale of the inner shelf under synoptic scale oscillatory winds is similar with and without stratification. As noted above, however, stratification changes the scale for upwelling relative to downwelling winds.

Additional definition of the inner shelf follows a posteriori from a vertically integrated vorticity analysis. The inner shelf corresponds to the region in which the primary balance is between the bottom pressure torque (the across-isobath flow that stretches planetary vorticity filaments) and the bottom stress torque, whereas the outer shelf is the region in which the primary balance is between the bottom pressure torque and the rate of change of relative vorticity indicated by eddy-like motions, and these two regions are dynamically separate over the wide portions of the WFS.

Expanding data sets and improved model capabilities are leading to more realistic simulations and more extensive model and data inter-comparisons. *He and Weisberg* [2002] and *He and Weisberg* [2003b] consider spring and fall season transitions, respectively, to be discussed in section 4, but they also provide quantitative comparisons between sea level sampled along the coast from Pensacola to Naples and velocity data sampled at several locations across the inner shelf from the 10 m to the 50 m isobaths. Good agreements are generally achieved with the sea level and currents by using realistic time- and space-varying forcing fields, and these agreements justify the use of the model simulations for diagnosing the temperature budgets and how these partition between surface heating and the advection and mixing by the ocean circulation. At synoptic weather scales the fully three-dimensional ocean advection effects are important, and data analyses, including in situ measurements of the surface heat flux [*Virmani and Weisberg*, 2003] support the model findings. In addition to the vertical distribution of turbulence affecting the Ekman depth, buoyancy effects can also alter the mixed layer depth. Hence in fall, when the heat flux changes sign from warming to cooling, convective overturning can deepen the mixed layer beyond that of the Ekman depth by direct wind mixing alone. Related observations on the ocean circulation and surface heat flux influences on the North Carolina inner shelf temperature budgets are given by *Austin and Lentz* [1999] and *Austin* [1999].

The three-dimensionality of the WFS velocity and temperature fields necessitates the use of three-dimensional models for WFS water property predictions. For instance, the cold water observed to upwell between Tampa Bay and Charlotte Harbor in the *Weisberg et al.* [2004] case study originated at the shelf break some 300–400 km away to the northwest. To see how this may happen, consider as an example the modeled upwelling response to a synoptic-scale weather event sampled at the narrowest part of the WFS offshore of DeSoto Canyon. The Figure 6 panels show snapshots of temperature, and across-shelf and along-shelf velocity components sampled on four different days during a March 1998 synoptic scale event response. Beginning with an upwelling transition from a previous downwelling event, the March 10th panels show onshore flow and an accelerating eastward coastal jet. By March 12th the coastal jet peaks, strong onshore flow continues at the shelf break, and deep waters of 18^0–19^0 C temperature are upwelled onto the shelf. Upwelling continues, and by March 14th the entire DeSoto Canyon shelf is in the range of 18^0–19^0 C and waters of 17^0–18^0 C are broaching the shelf break. Downwelling then commences, peaking on March 19th. Independent of any deep ocean forcing (the LC was omitted from the Figure 6 model simulation) this sequence shows how deep-water properties may be advected onto the WFS on the basis of local forcing only. The DeSoto Canyon region, as the narrowest part of the WFS, is a prime region for upwelling across the shelf break since the "inner shelf" response there has a scale that extends out beyond the shelf break (Figure 6). Coupling this geometry argument with the intense southeast-directed coastal jet response to upwelling and the small aspect ratio between the vertical and along-shelf velocity components it may be concluded that the entire region from DeSoto Canyon to the Florida Big Bend is one in which cold, nutrient-rich waters of deep ocean origin may broach the shelf break. Once on the shelf these waters can be further advected toward the beach within the bottom Ekman layer.

With numerical models shown to be effective at hindcasting locally forced, synoptic scale variability on the WFS it is natural to question the principal limitations to model prediction. For instance, do these reside with the models themselves, through parameterizations and numerics, or are they a consequence of the initial and boundary (surface and deep ocean forcing) conditions. The simplest place to begin testing this is on the broad part of the shelf and in shallow water. *He et al.* [2004] address this question by performing hindcast analyses using different forcing functions. The starting point is the NCEP reanalysis wind and heat flux fields (with a heat flux correction based on observed sea surface temperature). Because of relatively coarse spatial resolution these wind fields do not always agree with winds observed at the coast. In an attempt to remedy this a second wind field was constructed by merging the NCEP reanalysis with all of the nearshore and coastal wind observations (including all NDBC buoy and CMAN and NOS NWLON stations) using optimal interpolation (OI). Figure 7 shows two sets of experimental results, each comparing modeled with observed velocity sampled over the water column at the 10 m isobath offshore of Sarasota. The first is for the model forced with the NCEP reanalysis fields, and the second is for the model forced with the OI merged fields. Both show quantitative agreements for the synoptic scale variations over the analyzed two-month interval, March through April 2001. For the NCEP winds there is good visual correlation except toward the end of the record, with vector correlation coefficients greater than 0.7 and vector regression coefficients of around 0.5 (the modeled velocity underestimates the observed velocity). Significant improvements are gained using the OI merged winds. The vector correlations increase to 0.9, the regression coefficients exceed 0.7, and visual agreement occurs toward the end of the record. Across- and along-shelf momentum analyses, performed independently on the model simulations and the observations, further demonstrate that the correlations have a correct dynamical basis. From this we may conclude that presently available public domain numerical models, such as the POM used here, can provide reasonably accurate shelf circulation simulations if forced with accurate enough winds and heat fluxes. Initial conditions are also important. Initialization in the above cases was with a horizontally uniform stratification (estimated by hydrographic survey). Baroclinicity subsequently developed in balance with the wind and surface heat flux forcing. Sensitivity studies show that if the heat flux is initially directed out of the ocean (as it is in late winter) the baroclinic adjustment over the shallow shelf occurs rapidly through convective mixing so that the initial conditions are rapidly lost in the late fall through early spring seasons. The need for improv-

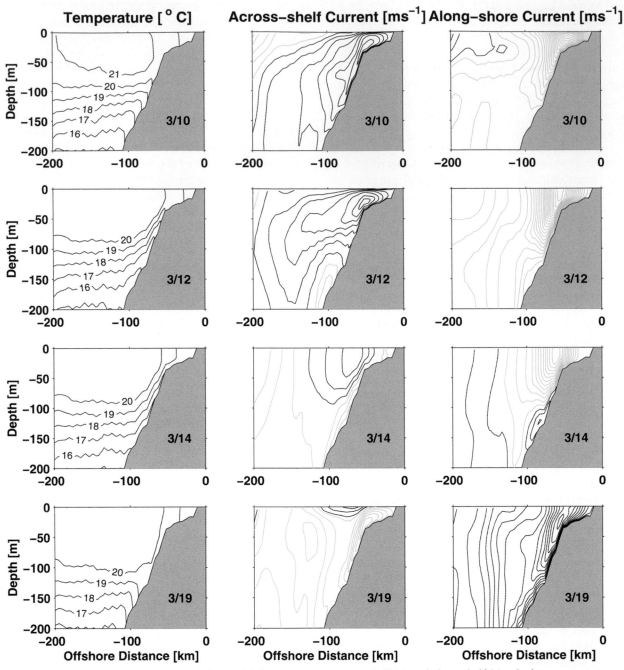

Figure 6. Across-shelf transects of the modeled temperature, across-shelf (u) and along-shelf (v) velocity components sampled at DeSoto Canyon on March 10th, 12th, 14th, and 19th for the case of local forcing only. The contour interval on the velocity components is 0.02 ms[-1]; black lines denote onshore and westward flows; and gray lines denote offshore and eastward flows. [from *Weisberg and He*, 2003].

ing winds and heat fluxes over the coastal ocean for the purpose of accurately simulating coastal ocean circulation responses provides strong justification for the coastal ocean observing systems that are emergent nationwide.

Given the predominantly wind-forced nature of the inner shelf at synoptic weather scales, how well can the vertically integrated along-shelf current at a given point on the inner shelf be estimated on the basis of readily observable winds and coastal

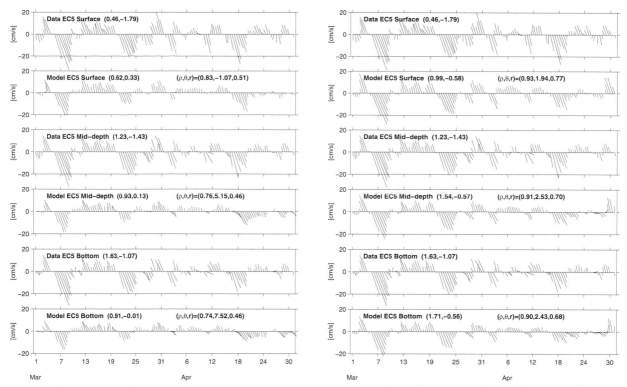

Figure 7. Observed and modeled currents at the 10 m isobath (mooring EC5) sampled near-surface, at mid-water column, and near-bottom. The left hand panels are for the case with NCEP reanalysis winds and the right hand panels are for the case with OI merged (NCEP and observed) winds. Quantitative comparison metrics are: the two-month mean east and north velocity components (left hand couplet) for each time series, and the vector correlation coefficient, angular deviation (in degrees measured counterclockwise), and vector regression coefficient (right hand triplet) for each pair of modeled and observed time series. [from *He et al.*, 2004].

sea level alone. After diagnosing the along-and across-shelf momentum balances at several locations across the shelf and concluding that the bulk of the synoptic scale fluctuations are statistically accountable by the terms in the momentum equations, *Liu and Weisberg* [2005a] test two nowcast techniques. The first considers the across-shelf balance in which the lead terms are the vertically integrated Coriolis acceleration by the along-shelf current (plus to a lesser degree the across-shelf acceleration), the vertically integrated across-shelf pressure gradient, and the vertically integrated across-shelf stress (surface minus bottom). The second considers the along-shelf balance in which the lead terms are the along-shelf components of stress and the along-shelf pressure gradient, also vertically integrated. Similar results are obtained for both, but with the along-shelf components more readily available, the latter analysis is presented here. The nowcast equation (1)

$$\overline{v}(t) = \overline{v}_0 \exp(-\frac{rt}{H}) + \int_0^t \frac{\tau_s^y}{\rho_0 H} \exp[-\frac{r(t-t')}{H}]dt'$$
$$- \int_0^t \frac{1}{\rho_0}\left\langle\frac{\partial p}{\partial y}\right\rangle\exp[-\frac{r(t-t')}{H}]dt'$$

(1)

follows that of *Lentz and Winant* [1986] and *Hickey et al.* [2003] in which bottom friction (resistance coefficient divided by water depth) sets the convolution integral time scale and hence the influences from prior winds and pressure gradients. Five panels are given in Figure 8 for nowcast models that consider: *a*. along-shelf wind alone, *b*. along-shelf pressure gradient alone (from bottom pressure measurements), *c*. the sum of *a* and *b*, *d*. along-shelf pressure gradient alone (from coastal sea level measurements), and *e*. the sum of *a* and *d*. The correlation and regression coefficients provide measures of model goodness of fit to the observations. With wind stress alone the correlation is high, but the model overestimates the along-shelf current relative to the observations. With pressure gradient alone there is an inverse correlation. With both wind stress and pressure gradient the correlation is high and the regression coefficient is improved. The interpretation is that an along-shelf pressure gradient sets up in opposition to the along shelf wind stress. This pressure gradient effect decelerates the along-shelf currents relative to those that would exist under wind stress alone. Using the coastal sea level instead of bottom pressure

Clarke, A.J. and K. Brink (1985), The response of stratified, frictional flow of shelf and slope waters to fluctuating large-scale, low frequency wind forcing, *J. Phys. Oceanogr.*, 15, 439–453.

Clarke, A.J. and S. Van Gorder (1986), A method for estimating wind-driven frictional, time-dependent, stratified shelf and slope water flow, *J. Phys. Oceanogr.*, 16, 1013–1028.

Cooper, C. (1987), A numerical modeling study of low-frequency circulation on the West Florida shelf, *Coastal Engineering*, 11, 29–56.

Cragg, J. G.T. Mitchum, and W. Sturges (1983), Wind-induced sea surface slopes on the West Florida shelf, *J. Phys. Oceanogr.*, 13, 2201–2212.

Csanady, G.T. (1978), The arrested topographic wave, *J. Phys. Oceanogr.*, 8, 47–62.

Dowgiallo, M.J., ed. (1994), Coastal oceanographic effects of the summer 1993 Mississippi River flooding, Special NOAA report, March, 1994, 77pp.

Ekman, V.W. (1905), On the influence of the Earth's rotation on ocean-currents, *Arkiv. Matem., Astr. Fysik.*, 2, 1–53.

Fan, S., L-Y Oey, and P. Hamilton (2004), Assimilation of drifter and satellite data in a model of the northwestern Gulf of Mexico, *Cont. Shelf Res.*, 24, 1001–1013.

Fernandez-Partegas, J. and C.N.K. Mooers (1975), A synoptic study of winter cold fronts in Florida, *Mon. Wea. Rev.*, 103, 742–744.

Gilbes, F., C. Tomas, J.J. Walsh, and F.E. Muller-Karger (1996), An episodic chlorophyll plume on the West Florida shelf, *Cont. Shelf Res.*, 16, 1201–1224.

Gill, A.E. (1982), *Atmosphere-Ocean Dynamics*, p408. Academic Press.

Gill, A.E. and E.H. Schumann (1974), The generation of long shelf waves by wind, *J. Phys. Oceanogr.*, 4, 83–90.

He, R and R.H. Weisberg (2002a), West Florida shelf circulation and temperature budget for the 1999 spring transition, *Cont. Shelf Res.*, 22, 719–748.

He, R and R.H. Weisberg (2002b), Tides on the west Florida shelf, *J. Phys. Oceanogr.* 32, 3455–3473.

He, R and R.H. Weisberg (2003a), A Loop Current intrusion case study on the West Florida Shelf, *J. Phys. Oceanogr.* 33, 465–477.

He, R and R.H. Weisberg (2003b), West Florida shelf circulation and temperature budget for the 1998 fall transition, *Cont. Shelf Res.* 23, 777–800.

He, R., R.H. Weisberg, H. Zhang, F. Muller-Karger, and R.W. Helber (2003), A cloud-free, satellite-derived, sea surface temperature analysis for the West Florida Shelf, *Geophys. Res. Letts.*, 30, doi:10.1029/2003GL017673.

He, R., Y. Liu, and R.H. Weisberg (2004), Coastal ocean wind fields gauged against the performance of a coastal ocean circulation model, *Geophys. Res. Letts.*, 31, L14303, doi:10.1029/2003GL019261.

Hellerman, S. and M. Rosenstein (1983), Normal monthly wind stress over the world ocean with error estimates, *J. Phys. Oceanogr.*, 13, 1093–1104.

Hetland, R.D., Y. Hsueh, R.R. Leben, and P.P. Niiler (1999), A Loop Current-induced jet along the edge of the West Florida shelf, *Geophys. Res. Lett.* 26, 2239–2242.

Hickey, B.M., E.L. Dobbins, and S.E. Allen (2003), Local and remote forcing of currents and temperature in the central Southern California Bight, *J. Geophys. Res.*, 108, C3, 3081, doi:1029/2000JC000313.

Hill, A.E. (1998), Buoyant effects in coastal and shelf seas, in The Sea, 10, 21–62, K.H. Brink and A.R. Robinson eds., Wiley, N.Y.

Hsueh, Y., G.O. Marmarino, and L.L. Vansant (1982), Numerical model studies of the winter storm response of the West Florida Shelf, *J. Phys. Oceanogr.*, 12, 1037–1050.

Huh, O.K., W.J. Wiseman, and L.J. Rouse (1981), Intrusion of Loop Current waters onto the west Florida continental shelf, *J. Geophys. Res.*, 86, 4186–4192.

Janowitz, G.S. and L.J. Pietrafesa (1980), A model and observations of time-dependent upwelling over the midshelf and slope, *J. Phys. Oceanogr.*, 10, 1574–1583.

Kelly, K.A. and D.C. Chapman (1988), The response of stratified shelf and slope waters to steady offshore forcing, *J. Phys. Oceanogr.*, 18, 906–925.

Lentz, S. (1994), Current dynamics over the northern California inner shelf, *J. Phys. Oceanogr.*, 24, 2461–2478.

Lentz, S. and C.D. Winant (1986), Subinertial currents on the South California shelf, *J. Phys. Oceanogr.*, 16, 1737–1750.

Li, Z. and R.H. Weisberg (1999a), West Florida Shelf response to upwelling favorable wind forcing, Part 1: Kinematics, *J. Geophys. Res.*, 104, 13,507–13,527.

Li, Z. and R.H. Weisberg (1999b), West Florida Shelf response to upwelling favorable wind forcing, Part 2: Dynamics, *J. Geophys. Res.*, 104, 23427–23442.

Liu, Y. and R.H. Weisberg (2005a), Momentum balance diagnoses for the West Florida shelf, *Cont. Shelf Res.*, in press.

Liu, Y. and R.H. Weisberg (2005b), Patterns of ocean current variability on the West Florida Shelf using the Self-Organizing Map, *J. Geophys. Res.*, 110, C06033, doi:10.1029/2004JC002786.

Lopez, M. and A.J. Clarke (1989), The wind-driven shelf and slope water flow in terms of a local and a remote response, *J. Phys. Oceanogr.*, 19, 1091–1101.

Marmorino, G.O. (1983a), Wind-forced sea level variability along the West Florida shelf, *J. Phys. Oceanogr.*, 12, 389–404.

Marmorino, G.O. (1983b), Variability of current, temperature, and bottom pressure across the west Florida continental shelf, winter 1981–1982, *J. Geophys. Res.*, 88, 4439–4457.

Mellor, G.L. and T. Yamada (1982), Development of a turbulence closure model for geophysical fluid problems, *Rev. Geophys. Space Phys.*, 20, 851–875.

Meyers, S.D., E.M. Siegel, and R.H. Weisberg (2001), Observations of currents on the west Florida shelf break, *Geophys. Res. Lett.*, 28, 2037–2040.

Mitchum, G.T. and A.J. Clarke (1986a), The frictional nearshore response to forcing by synoptic scale winds, *J. Phys. Oceanogr.*, 16, 934–946.

Mitchum, G.T. and A.J. Clarke (1986b), Evaluation of frictional wind-forced long waves theory on the West Florida shelf, *J. Phys. Oceanogr.*, 16, 1029–1037.

Mitchum, G.T. and W. Sturges (1982), Wind-driven currents on the West Florida Shelf, *J. Phys. Oceanogr.*, 12, 1310–1317.

Morey, S.L., P.J. Martin, J.J. O'Brien, A.A. Walcraft, and J. Zavala-Hidalgo, Export pathways for river discharged fresh water in th northern Gulf of Mexico (2003), *J. Geophys. Res.*, 108, C10, 3303. doi:10.1029/2002JC001674.

Muller-Karger, F.E. (2000), The spring 1998 northeastern Gulf of Mexico (NEGOM) cold water event: Remote sensing evidence for upwelling and for eastward advection of Mississippi water, *Gulf of Mexico Science*, 18, 55–67.

Niiler, P.P. (1976), Observations of low-frequency currents on the West Florida continental shelf, *Mem. Soc. R. Sci.* Liege X, 6, 331–358.

Nowlin,W.D., A.E. Jochens, M.K. Howard, S.F. DiMarco (1998), Nearshore bottom properties over the northeast shelves of the Gulf of Mexico as observed during early May 1998, *TAMU Oceanography Technical Report No. 98-3-T*, Dept. of Oceanography, Texas A&M university, College station TX, 77843, Dec. 1998, 60pp.

Nowlin, W.D., A.E. Jochens, M.K. Howard, S.F. DiMarco, and W.W. Schroeder (2000), Hydrographic properties and inferred circulation over the northeast shelves of the Gulf of Mexico during spring to midsummer of 1998, *Gulf of Mexico Science*, 18, 40–54.

Ortner, P.B., T.N. Lee, P.J. Milne, R.G. Zika, M. E. Clarke, G.P. Podesta, P.K. Swart, P.T. Tester, L.P. Atkinson, and W.R. Johnson (1995), Mississippi River flood waters that reached the Gulf Stream, *J. Geophys. Res.*, 100, 13595–13601.

Paluszkiewicz, K., L.P. Atkinson, E.S. Posmentier, and C.R. McClain (1983), Observations of a Loop Current frontal eddy intrusion onto the west Florida shelf, *J. Geophys. Res.*, 88, 9639–9651.

Sturges, W., and R. Leben (2000), Frequency of ring separation from the Loop Current in the Gulf of Mexico: A revised estimate, *J. Phys. Oceanogr.*, 20, 1814–1819.

Talbert, W.H., and G.G. Salzman (1964), Surface circulation of the eastern Gulf of Mexico as determined by drift-bottle studies, *J. Geophys. Res.*, 69, 223–230.

Toner, M., A.D. Kirwan, A.C. Poje, L.H. Kantha, F.E. Muller-Karger, and J. Ckpt (2003), Chlorophyll dispersal by eddy-eddy interactions in the Gulf of Mexico, *J. Geophys. Res.*, 108, C4, doi:10.1029/2002JC001499.

Virmani, J.I. and R.H. Weisberg (2003), Features of the Observed Annual Ocean-Atmosphere Flux Variability on the West Florida Shelf, *J. Climate*, 16, 734–745.

Walsh, J.J., R.H. Weisberg, D.A. Dieterle, R. He, B.P. Darrow, J.K. Jolliff, G.A. Vargo, G. Kirkpatrick, K. Fanning, T.T. Sutton, A. Jochens, D.C. Biggs, B. Nababan, C. Hu and F. Muller-Karger (2003), The phytoplankton response to intrusion of slope water on the west Florida shelf: models and observations, *J. Geophys. Res.*, 108, C6, 15, doi:10.1029/2002JC001406.

Wang, D.P., L-Y Oey, T. Ezer, and P. Hamilton (2003), Near-surface currents in DeSoto Canyon (1997–99): Comparison of current meters, satellite observations, and model simulation, *J. Phys. Oceanogr.*, 33, 313–326.

Weatherly, G.L. and D. Thistle (1997), On the wintertime currents in the Florida Big Bend region, *Cont. Shelf Res.*, 17, 1297–1319.

Weisberg, R.H., B.D. Black, and H. Yang (1996), Seasonal modulation of the West Florida continental shelf circulation, *Geophys. Res. Lett.*, 23, 2247–2250.

Weisberg, R.H., B. Black, Z. Li (2000), An upwelling case study on Florida's west coast, *J. Geophys. Res.*, 105, 11459–11469.

Weisberg, R.H., Z. Li, and F.E. Muller-Karger (2001), West Florida shelf response to local wind forcing: April 1998, *J. Geophys. Res.*, 106 (C12) 31,239–31,262.

Weisberg, R.H. and R. He (2003), Local and deep ocean forcing contributions to anomalous water properties on the West Florida Shelf, *J. Geophys. Res.*, 108, C6, 15, doi:10.1029/2002JC001407.

Weisberg, R.H., R. He, G. Kirkpatrick, F. Muller-Karger, and J.J. Walsh (2004), Coastal Ocean Circulation Influences on Remotely Sensed Optical Properties: A West Florida Shelf Case Study, Oceanography (special issue on Coastal Ocean Optics and Dynamics), 17, 68–75.

Williams, J., W.F. Grey, E.B. Murphy, and J.J. Crane (1977), Memoirs of the Hourglass cruises, Rep. IV, Mar. Res. Lab, Fla. Dep. of Nat. Res., St. Petersburg FL.

Yang, H. and R.H. Weisberg (1999), West Florida continental shelf circulation response to climatological wind forcing, *J. Geophys. Res.*, 104, 5301–5320.

Zheng, L and R.H. Weisberg (2004), Tide, buoyancy, and wind-driven circulation of the Charlotte Harbor estuary: A model study, *J. Geophys. Res.*, 109, C06011, doi:10.1029/2003JC001996.

Robert H. Weisberg, College of Marine Science, University of South Florida, 140 7th Ave. South, St. Petersburg, FL 33701 (weisberg@marine.usf.edu)